JOHN GRAY PARK LIBRARY
Kent School
Kent, Connecticut 06757

Historical Atlas of the

Arctic

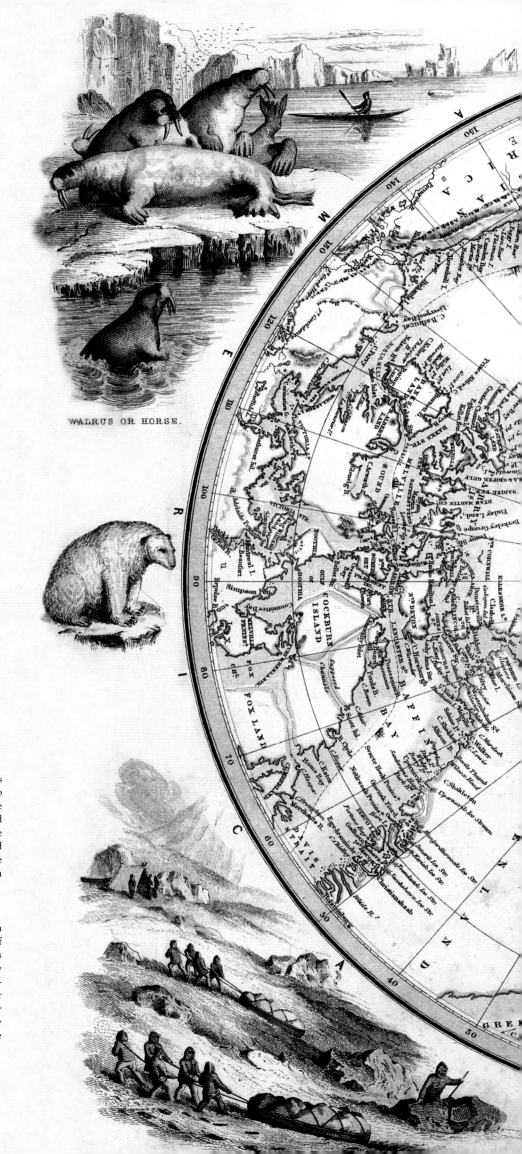

MAP 1 (*previous page*).
Gerard Mercator's concept of the north polar regions is shown in this early map. Originally drawn in 1569 (see MAP 17, page 11), the map was very influential and widely copied—probably because no one else had any idea what lay at the Pole and Mercator's idea seemed as good as any. The polar concept shown here was derived from a Carmelite monk named Nicholas of Lynn, who wrote an account of supposed voyages in his *Inventio Fortunata*, now unfortunately lost. But the map has been updated to show the discoveries of Willem Barentsz in 1594–96.

MAP 2 (*this page*).
One of a host of colorful illustrated maps produced when the euphoria for polar exploration was at its height in Britain and literally dozens of expeditions were sent out to search for the lost John Franklin. This map used new geographic information brought back to England by Robert M'Clure, who had recently traversed the Northwest Passage—though not with his ship (see page 99). The map is entitled *The Arctic Regions showing the North-West Passage of Cap. R. McClure and other Arctic Voyagers* and was published by British mapmaking company Fullarton & Co. in 1856. The top part of the sheet showed details of the Northwest Passage (see MAP 171, pages 106–7).

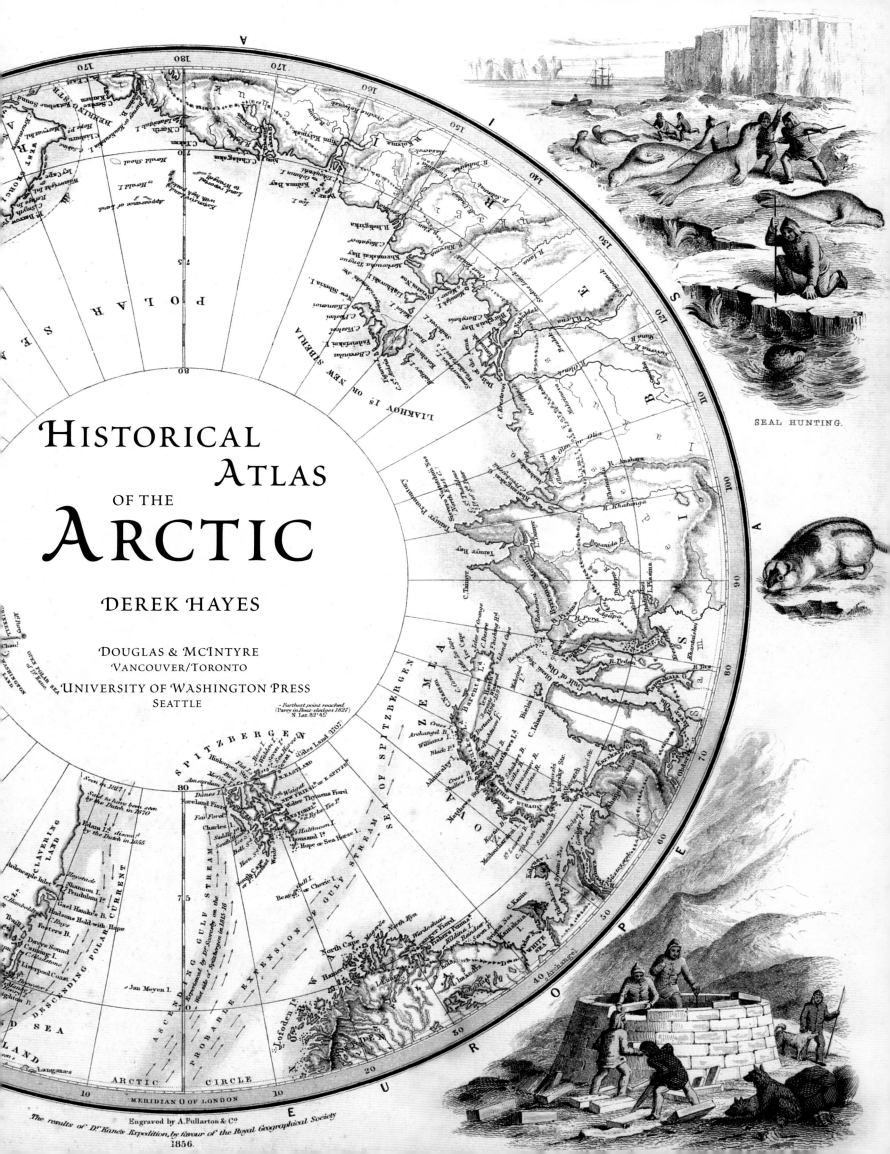

Historical Atlas of the Arctic

DEREK HAYES

Douglas & McIntyre
Vancouver/Toronto

University of Washington Press
Seattle

Copyright © 2003 by Derek Hayes

03 04 05 06 07 5 4 3 2 1

All rights reserved. No part of this book may be reproduced, stored in a retrieval system or transmitted, in any form or by any means, without the prior written consent of the publisher or a licence from the Canadian Copyright Licensing Agency (Access Copyright). For a copyright licence, visit www.accesscopyright.ca or call toll free to 1-800-893-5777.

Douglas & McIntyre Ltd.
2323 Quebec Street, Suite 201
Vancouver, British Columbia v5t 4s7
www.douglas-mcintyre.com

National Library of Canada Cataloguing in Publication Data

Hayes, Derek, 1947–
   Historical atlas of the Arctic / Derek Hayes.

   Includes bibliographical references and index.
   ISBN 1-55365-004-2

   1. Arctic—Discovery and exploration—Maps.
   2. Arctic—Historical geography—Maps.
   I. Title.
   G3051.S12H39 2003   911'.98   C2003-910548-2

Published in the United States of America by
The University of Washington Press,
P.O. Box 50096, Seattle, Washington 98145-5096
ISBN 0-295-98358-2

Design and layout by Derek Hayes
Copyediting by Naomi Pauls
Jacket design by George Vaitkunas
Printed and bound in Hong Kong, P.R.C.,
by C &C Offset
Printed on acid-free paper

We gratefully acknowledge the financial support of the Canada Council for the Arts, the British Columbia Arts Council, and the Government of Canada through the Book Publishing Industry Development Program (BPIDP) for our publishing activities.

To contact the author:
www.derekhayes.ca
derek@derekhayes.ca

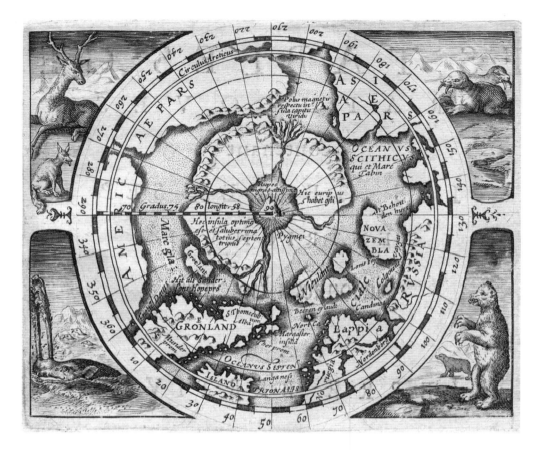

MAP 3.
A beautifully engraved rendering of the north polar regions by Dutch mapmaker—and professor of mathematics—Petrus Bertius from 1618. The geography follows that of Gerard Mercator, whose concept of the North Pole, derived from medieval reports, was widely copied; four rivers flow outwards from a polar mountain. The map incorporates an outline of the western coast of Novaya Zemlya, mapped by Willem Barentsz in 1594–96 (see page 18).

MAP 4 (*right*).
This beautiful polar map by Frederick de Wit was published in 1715—using updated printing plates that had been passed down from Henricus Hondius (see MAP 35, page 25, for a 1642 edition). The whaling scenes around the edges are untouched, but the islands of Novaya Zemlya are shown almost as a peninsula of the Asian continent, following the sentiment of the times; the geography has been "fudged" due to uncertainties in this area. In fact, Hondius's 1642 edition was more accurate that this later map. Otherwise, the geography shown on both maps is very similar, reflecting the fact that little new exploration of the Arctic took place between the voyages of Luke Foxe and Thomas James in 1630–31 and that of Christopher Middleton in 1743.

Acknowledgments

A book such as this could not have been created without the assistance and co-operation of the many institutions and individuals who hold the maps that have been reproduced.

In particular, I should like to thank Cameron Treleaven of Calgary for allowing me such generous access to his superb collection of Arcticana; and Dick Fitch, of Santa Fe, who lent me a large number of maps from *Petermann's Geographische Mitteilungen*, the German serial that carried so many excellent maps in the second half of the nineteenth century. Special thanks also to Louis Cardinal at the National Archives of Canada (now Library and Archives of Canada), who has, apart from other assistance, spent many hours trying to track down the whereabouts of the original maps of Otto Sverdrup (see page 156).

Others I should like to thank include Andrew Cook at the British Library; Tammy Hannibal, formerly at the Hudson's Bay Company Archives, Winnipeg; Edward Redmond, Geography and Map Division, Library of Congress; Carol Urness, formerly of the James Ford Bell Library, University of Minnesota; Pat Morris, Newberry Library, Chicago; Jovanka Ristic, American Geographical Society Collection, University of Wisconsin-Milwaukee; Catherine Hoffmann, Bibliothèque nationale de France, Paris; Alec Parley, Beach Maps, Toronto; Joel Sartorius, Rare Book Division, Free Library of Philadelphia; James Delgado, Vancouver Maritime Museum; Glenbow Museum, Calgary; Hatfield House, U.K.; United Kingdom Hydrographic Office, Taunton, U.K.; Special Collections, University of British Columbia; Yale University Library, New Haven, Connecticut; United States National Archives, Washington, D.C.; National Maritime Museum, Greenwich, U.K.; Public Record Office, London; and the Vancouver Public Library, Vancouver.

Finally, thanks to Naomi Pauls, who made many corrections and suggestions for improvement that have much improved the original text, and Peter Cocking, who made typographical suggestions. Special thanks to my publisher, Scott McIntyre, who originally suggested that the book be written.

Reference Maps

The reader will be assisted in having at hand a good map of the Arctic, but there are a number of maps in the present volume that have very considerable detail and are thus also useful for reference purposes. They include the following:

MAP 136, page 81, and MAP 217, page 139. There are large maps of all or most of the Arctic on pages 148–49 (MAP 229) and pages 162–63 (MAP 248). Detailed maps of the Canadian Arctic are MAP 139, pages 82–83; MAP 164, pages 100–101; and MAP 171, pages 106–7. A good general map of the Russian Arctic is MAP 220, pages 142–43. Peary's 1906 map of northern Greenland and Ellesmere Island (MAP 237, pages 154–55) is a useful reference also. Finally, a good modern map of the Arctic Ocean is MAP 288, pages 186–87.

Contents

Acknowledgments / 4

Introduction / 6
To the Edge of the Arctic / 7
A Passage to Cathay / 8
Lost Colonies and Found Whales / 21
To the Other Ocean / 26
Russian Exploration of the North / 34
The Great Northern Expedition / 38
That Illusive Passage / 44
Cook Probes the Ice / 50
The Arctic by Land / 52

To Sail over the Top of the World / 55
For Glory Not Gold / 58
For Franklin—and Glory / 84
Discoveries in the Russian Arctic / 112
The Path to the Pole / 124
The Northeast Passage Achieved / 140
The Exploration of Greenland / 145
Nansen's Drift / 148
Peary's Push for the Pole / 152
New Land / 156
The Northwest Passage at Last / 158
The Pole Achieved? / 160

Claiming the Arctic / 165
New Land in the Russian Arctic / 168
The Arctic by Air / 170
The Last New Land / 183
The Pole by Many Means / 184
Science, Sonar, and Satellites / 186

Map Catalog / 194
Bibliography / 201
Index / 204

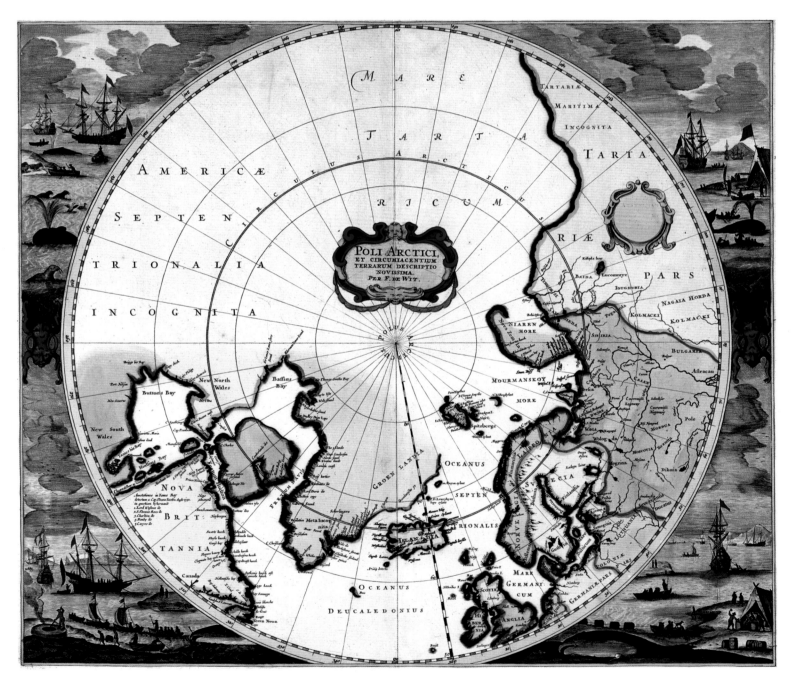

Introduction

Writing the history of the Arctic is a daunting undertaking, for there have been many more expeditions to the Arctic than can be reasonably covered in a book of this scope and size. I have thus necessarily tried to select those that are historically significant and those that produced interesting maps.

Even defining the Arctic can be a problem, for the area north of the Arctic Circle is not the region with a uniformly cold climate. The lowest temperature ever recorded in the northern hemisphere is –78°C, at Oimekon, Siberia, which is not even within the Arctic Circle. I have included the area north of the Arctic Circle but also those colder parts below it, in particular the shores of Hudson Bay.

Nomenclature has radically changed in the Arctic regions in the last fifty years, with many European names being changed back to native names. I have generally tried to used the modern name, with the old name in parentheses, at least for the first reference. But some names I felt were too ingrained to change. Few would recognize Kalaallit Nunaat, the modern official name for Greenland. Similarly Svalbard is still usually referred to by its old name of Spitsbergen, now technically applying only to the westernmost large island of the group. And even that is anglicized. Apart from the odd exception I have tried to follow the nomenclature used in Clive Holland's 1994 classic Arctic encyclopedia.

This book is really about the *exploration* of the Arctic, and thus tends to concentrate on expeditions that added knowledge of some kind. I have used the term "discovery" quite freely, recognizing that there are plenty of areas where the lands "discovered" were already inhabited by the indigenous peoples, although, uniquely in polar exploration, there were vast tracts that were in fact discovered in every sense of the term, the land being unknown to anybody before. I often refer to expeditions by the names of their leaders, for this is the normal label attaching to them, but this of course fails to recognize the often considerable numbers of other persons without whom the expedition would have got nowhere. One cannot fail to feel for the teams of enlisted men man-hauling incredible loads over near-impassable country in near-impossible conditions while their officers got all the credit just for walking alongside them.

Five hundred years ago, the Arctic was a complete unknown to Western man, a huge blank on the map supporting nothing but theories as to what was there. Total open water, all land, land with rivers and magnetic mountains: all were conceptions of the Arctic. Not to mention the supposedly easy shortcuts, east or west, to the fabled riches of the East. Indeed it was the lure of the shortcut to these riches that enticed men, unsuspecting, into the icy wastes in the first place. But then there must be motivation for exploration; otherwise all our explorers would have stayed cozily at home. But they ventured north, often losing their lives because of their unpreparedness. It would take centuries, but an increase in knowledge and the advance of technology would mean that slowly, the reality and resources of the true Arctic would be revealed.

Map 5.
For four hundred years before the nineteenth century, theories abounded as to the nature of the North Pole. One particularly tenacious fallacy was that of the "Open Polar Sea"; the twenty-four hours of daylight in summer, it was held, would not permit an entire ocean to freeze. When attempts were made to sail to the Pole, the ice encountered was dismissed as merely an annular-like barrier which had to be overcome. And then there were the hoaxes, like the one illustrated here. One John Shelden purports to have discovered the Pole in 1869 and found it to be cone-shaped, surrounded by a very small circular Polar Sea. His published "map" shows a full-rigged ship sailing on this sea, together with rowers in a box-like boat. How the ship was supposed to have traversed the solid ice surrounding it was deliberately overlooked. These theories died a natural death, but concepts such as Shelden's live on to amuse us.

To the Edge of the Arctic

Parts of the Arctic have been home to humans for perhaps 10,000 years, but vast areas have never been inhabited. In the Eurasian Arctic many cultural groups probably arrived in the northern region pushed by competitive populations farther south. The major groups were the Lapps, Nenets, Evenki, Yakut, and Chukchi. In the North American Arctic, humans almost certainly migrated east from Asia, arriving in Alaska 4,000 years ago and slowly venturing farther east still. The Aleut inhabited the islands of Alaska, and others pushed farther east in waves of migration to AD 1000. Today they are collectively the Inuit—"the people"; until recently they were known as Eskimos. The most remote were the Polar Inuit of northwest Greenland, the "Arctic Highlanders" discovered by John Ross in 1818. Disconnected from other groups, they numbered only 150 persons and had lost the use of kayaks and bow and arrow. If they drew maps, none have survived. They would likely have been ephemeral, drawn in the snow with a spear. But, as many later European explorers were to find, the Inuit had a superbly developed sense of the geography around them, and could and did draw useful maps when requested.

The first European explorer to venture towards the Arctic is usually considered to be Pytheas of Massalia, a Greek city standing where Marseilles, France, does today. Virtually nothing certain is known about Pytheas's voyage, for the book he wrote about his travels was lost. During his voyage, which reportedly took place around 320 BC, he is said to have reached a land he called Thule, only one day's sail from the frozen sea, somewhere north of Britain.

Of equal doubt is the voyage of Saint Brendan, about AD 570. According to surviving accounts, all of which are again secondhand, Brendan ventured west, out into the North Atlantic, in nothing more than a skin-covered boat. There are tales of coagulated seas, which could indicate pack ice, and icebergs, but where these were encountered is anybody's guess. And accounts of Mass held on the back of a whale tend to the realm of fantasy.

We do know, however, that the Norse explored northwards and westwards starting in the ninth century, finding and settling in Iceland around AD 870. The east coast of Greenland, 320 km across the Denmark Strait, was reached in AD 980, and the west coast was explored two years later by Eiríkr Thorvaldsson, Erik the Red. Settlement followed, possible because of a probable warmer climate than today. In AD 985 or 986 Bjarni Herjolfsson was blown westwards in a storm to North America, a journey retraced by Leifr Eiríksson (or Leif Ericsson) about 1001. Baffin Island is thought to have been his first landfall before he sailed south.

Due to a cooling climate, Norse settlements in Greenland declined during the fourteenth century and became extinct in the fifteenth.

The Norse had no cartographic tradition, but some maps seem to have been made from written sources; some of these were copied in the early seventeenth century on the orders of the Danish king, who sent out expeditions at that time to find out if any of his former subjects had survived (see page 21). Two of these maps are shown here. They give a good idea of the Norse geographical concept of the northern regions.

MAP 7 (above, top).
The Sigurdur Stefánsson map, copied in 1670 from an earlier one, now lost, drawn about 1590 (not 1570, as in the cartouche). It seems to have been one of the maps prepared before the king of Denmark sent an expedition to Greenland to search for his lost subjects. It shows the idea that Greenland is merely a peninsula in a continuous coastline reaching from northern Europe to North America.

MAP 8 (above).
A map drawn about 1640 by Jón Gudmonson, an Icelander. He was working from earlier maps, now lost. Greenland extends far to the east. Spitsbergen (Svalbard) at this time was usually thought to be "East" Greenland, an idea continued from the Norse concept. In truth it was hard to distinguish land from the pack ice that often clogged the Greenland Sea between Greenland and Spitsbergen.

MAP 6 (right).
This map was drawn by Arctic explorer Fridtjof Nansen in 1911 to show Norse ideas of the geography to the north.

A Passage to Cathay

Sailors did not brave the Arctic chill for nothing. Arctic exploration until the nineteenth century is largely the history of attempts to find a route to the riches of the East, either by going northeast or by going northwest. For the English, and then the Dutch, who were just throwing off the mantle of Spanish domination, the northern routes were the obvious ones to try, given that the Spanish and the Portuguese had already tied up the southern routes round the tips of Africa and South America.

A Northeast Passage had been shown on maps for a long time. The map of Macrobius, a fifth-century Roman philosopher, presumed the world consisted of two principal islands, with the Eurasian-African landmass balanced by a southern continent (Map 9). As an island, Eurasia was surrounded, even on its northern shores, by water. The map was reproduced in printed form as early as 1483.

Later some maps showed a possible Northeast Passage requiring a transit via the west coast of Greenland, because that island was shown attached to the Eurasian landmass.

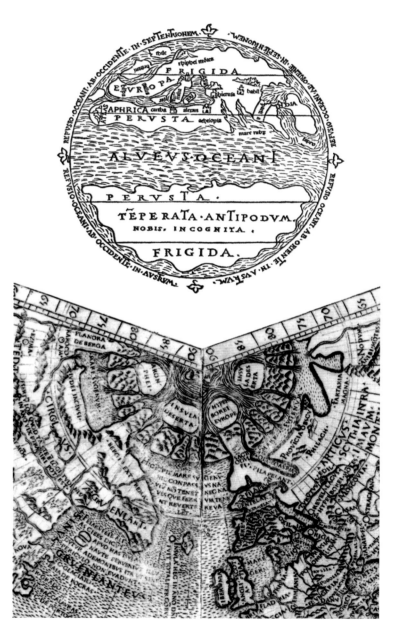

With the European discovery of North America, maps for a while showed that continent as part of Asia, and when it was shown separately it got in the way of a direct sail to Cathay, necessitating the invention of a Northwest Passage either through or around the continent. It was a voyage said to have been made by Sebastian Cabot in 1508–09 that may have been responsible for what appears to be the first depiction of a Northwest Passage on a map, that of the so-called Ambassadors' globe, known to have been created about 1530 because

Map 9 (above, left).
The world map of Macrobius, originally fifth century, printed about 1483.

Map 10 (left).
Part of the world map of Johann Ruysch, published in Rome in 1507 or 1508. This shows a medieval-style Arctic with outflowing rivers and sufficient northward-pointing peninsulas to kill any thought of a passage across the top of the world. America is shown as an extension of Asia.

Map 11 (above).
The first world map of Gerard Mercator, published in 1538, depicted the Arctic, *Septentrio Polus arcticus*, as a enormous extension of Asia, "Scythia," a geography which would have encouraged a belief in a Northwest Passage but not in a Northeast.

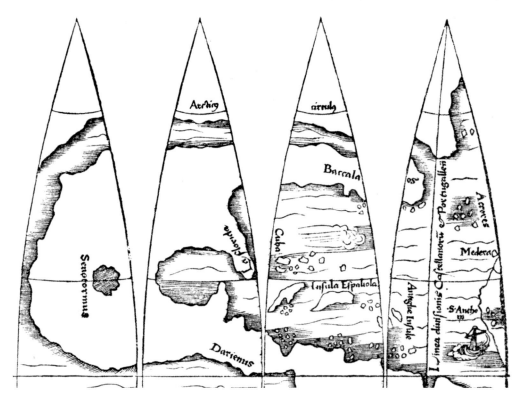

MAP 12 (*above*).
The first map to depict a true Northwest Passage consists of these globe gores of unknown origin but, it seems, based on the voyage—or claims to a voyage—of Sebastian Cabot in 1508–09. The map was drawn about 1530. This is the geography that Martin Frobisher would have had in mind when he sailed into Frobisher Bay and claimed that Asia was on his right and America on his left.

just his claims to such, the map stands as the first depiction of a true Northwest Passage—an image that one can be fairly certain Martin Frobisher and other early passage-seekers would have had in their minds.

About AD 980 a Norse chief from northern Norway, Ottar, is reported to have sailed round the Kola Peninsula (Kol'skiy Poluostrov) and into the White Sea (Beloye More), but it is not clear whether his voyage was known to the English in the sixteenth century, despite the story having been recorded by their own King Alfred of Wessex.

Sebastian Cabot was in the 1550s the most informed in England about world geography, as a result of long service in Spain. In 1553 he was Pilot Major of England, and it was he who recommended that the English search to the northeast for a passage to China. He was the first governor of a trading company called the "Merchants Adventurers of England for the discovery of lands, territories, isles, dominions and seignories unknown," which in 1553 sponsored a voyage under the command of Hugh Willoughby.

Three ships sailed in May of that year: Willoughby on *Bona Esperanza*, Richard Chancellor on *Edward Bonaventure*, and Cornelius Durfoorth on *Bona Confidentia*. Willoughby and Durfoorth lost contact with Chancellor off the coast of Norway. They continued east until, on 14 August, they sighted the west coast of Novaya Zemlya. Thinking it was the mainland they turned north intending to round it, but Willoughby soon realized that he would have to winter and try again next season. They returned to the coast of the Kola Peninsula and settled down to wait. But they were totally unprepared for the extreme rigors of an Arctic winter, and the entire company—about sixty-six men—perished. Their deaths, like others that were to follow, may have been caused by carbon monoxide poisoning caused by fires in a tightly sealed habitation rather than the cold per se.

Richard Chancellor fared somewhat better, finding his way to the White Sea and the mouth of the Dvina River, where the Russian city of Arkhangel'sk now stands. From here he struck out overland to Moscow, where he established trade relations with the tsar, Ivan the Terrible. This led to the formation of the Muscovy Company, giving the merchants of London some return for their investment.

it is shown in a Hans Holbein painting—of ambassadors—known to have been painted in 1533. The globe no longer exists, but a set of gores for it does (MAP 12).

Cabot is said to have sailed to 67° 30′ N, somewhere in Davis Strait, before being forced to return by a mutinous crew. His Northwest Passage may well have originated with Davis Strait, or Hudson Strait, Frobisher Bay, or Cumberland Sound, for he also reportedly sailed westwards into a broad strait which opened between 61° and 64° N. The problem is that no contemporary account of Cabot's voyage exists, and there is some doubt as to whether it occurred. Nevertheless, whether originating from Cabot's travels or

A number of other voyages followed, to trade with the Russians and explore farther along the coast: Chancellor in 1555, during

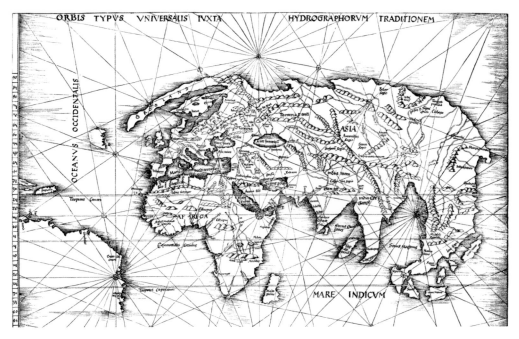

MAP 13 (*above*).
The map of the world drawn by Martin Waldseemüller in 1513, called the *Carta Marina*. John Cabot's discovery of Newfoundland is shown with an east coast only, in the middle of the North Atlantic. Greenland—*Gronland*—is depicted as attached to the Eurasian landmass, meaning that a Northeast Passage would only be possible by first sailing west. But beyond this first hurdle, a passage to Cathay along the northern shore of Asia looks tantalizingly possible.

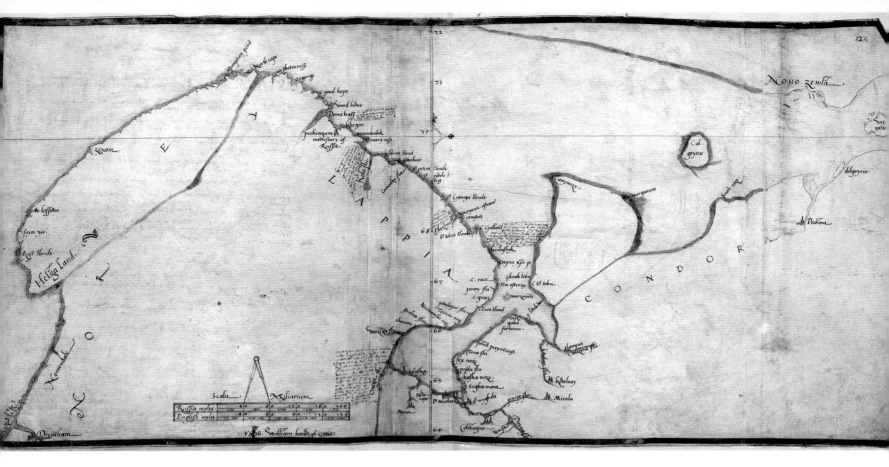

MAP 14 (*above*).
A Muscovy Company map of northern Scandinavia drawn by Stephen Borough's brother William, who accompanied his brother on his 1556–57 voyage. It is today in an atlas otherwise containing maps of English counties belonging to Lord Burghley, one of the financial backers of the Muscovy Company. Novaya Zemlya is extended far to the west.

MAP 15 (*below*).
This burned fragment is a map drawn by Hugh Smyth from his journal of the voyage of Arthur Pet and Charles Jackman in 1580. It shows Pet and Jackman's ships in the Kara Sea, the first English ships to reach this far east. Vaygach Island (Ostrov Vaygach) is shown, together with the straits to the north (Proliv Karskiye Vorota) and south (Proliv Yugorskiy Shar). The latter was discovered by Pet and Jackman. English standards are depicted flying on Vaygach and Novaya Zemlya. At the top is a recognizable portion of another similar map of the region.

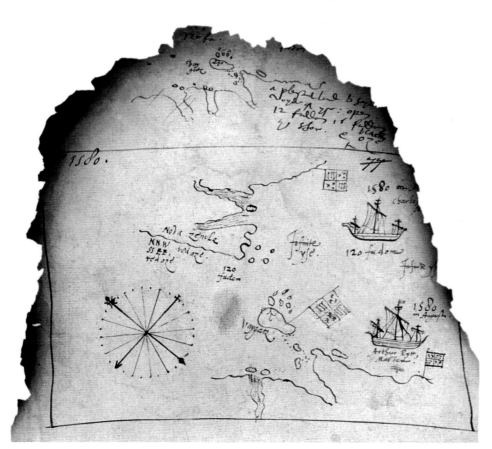

which he located Willoughby's two ships; John Buckland in 1556 (one of whose ships was wrecked, drowning Chancellor, who had wintered in Russia). Stephen Borough, who had been with Chancellor on the first voyage, in 1556 attempted to sail into the Kara Sea but was stopped by ice at Vaygach Island (Ostrov Vaygach). His map of the region is shown as MAP 14. The following year Anthony Jenkinson reached the mouth of the Dvina and traveled to Moscow. Here he stayed, superintending Muscovy Company business, until 1558, when he began an attempt to reach Cathay overland. He reached as far as Bukhara (now in Uzbekistan), in 1559, before giving up and returning. His map is shown as MAP 16.

From 1557 onwards, the Muscovy Company continued to send trading ships to the White Sea, and, except for one other expedition, the English seemed content with their discovery of Russian trade over finding a passage to Cathay.

The last English expedition to search for a Northeast Passage in the sixteenth century was that of Arthur Pet and Charles Jackman, dispatched by the Muscovy Company in May 1580. They discovered Yugorskiy Shar, between Ostrov Vaygach and the mainland, passing through it into the Kara Sea. Apart perhaps from Russian voyages (see page 34), they were the first to enter the Kara Sea. It did them no good, for the ice prevented much farther eastward progress. Pet managed to return to England late in 1580 but Jackman was

Map 16.

Anthony Jenkinson's map of his travels, drawn in 1562, was published by Abraham Ortelius in his 1570 atlas *Theatrum Orbis Terrarum*. The White Sea is at the top, with the Dvina River (*Duina fl.*) flowing into it. *Mare Septentrionale* is the Barents Sea, and to the east of Novaya Zemlya and Vaygach Island is the Kara Sea. *Biarmia* is the Latin form of the Norse name for the region.

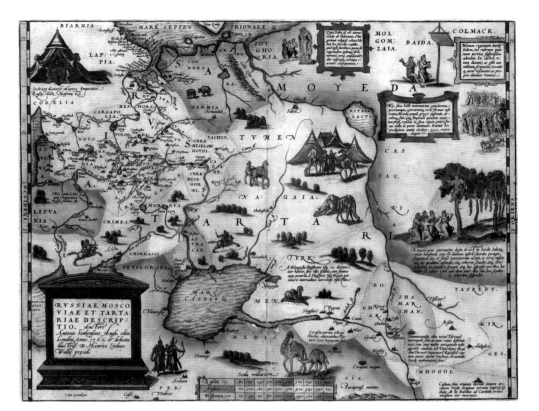

not so lucky. Caught in the ice and forced to overwinter, his expedition was lost at sea. A map by Hugh Smyth, who was with Pet, survives, and is the first English map to show Ostrov Vaygach and the straits north and south of it. Both within the Muscovy Company and outside it there were other merchants who thought that the northwest held a better chance of a route to Cathay. In 1566 Humfray Gilbert wrote a book entitled *A Discourse of a Discoverie for a New Passage to Cataia*; although not published until 1576, it was enormously influential. Gilbert argued that it would be easy for a ship to sail through the Northwest Passage to Cathay. He was not without detractors, however; his cousin Richard Grenville correctly pointed out that the passage might be impractical even if it existed, "consideringe the seas and ayre under the Artike circle are so congeled that they are navigable only 3 monethes in the yeare." But this was a technicality. There was a certainty abroad in Elizabethan England that there must be a passage at the northern end of the Americas to somehow "balance" the one that the Spaniard Ferdinand Magellan had discovered at the southern end in 1519. The concept of planetary balance, which was to last until the eighteenth century, was responsible, among other things, for the idea that there must also be a large southern continent to balance the rotation of the earth.

Into this environment came Martin Frobisher, an opportunist seemingly obsessed with the pursuit of adventure, fame, and wealth. At best he could be termed a privateer, at worst a pirate. But at the time, even pirates were potentially heroes, provided they were properly connected. Frobisher seized on the ideas of Gilbert as his path to glory. He would find the passage for England, "knowing this to be the onlye thing of the Worlde that was left undone whereby a notable mind mighte be made famous and fortunate."

The principal financial promoter of a route to Cathay via the northwest was Michael Lok, since 1571 the London agent for the Muscovy Company and by all accounts a well-connected merchant of means. He first met Martin Frobisher in 1574 when the latter

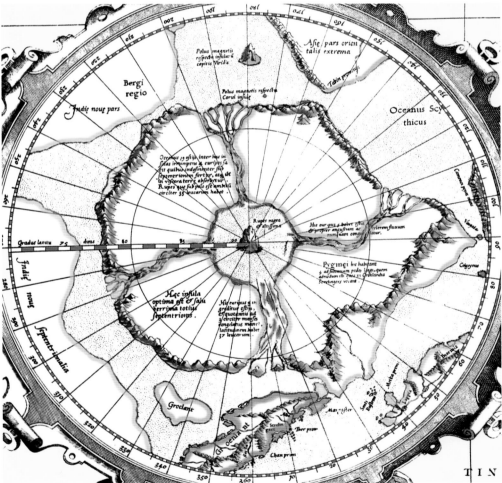

Map 17 (*above*).

The Arctic by Gerard Mercator, 1569. This map was an inset on his world map of that date, necessary because his new Mercator projection, used for the rest of the map, made the Arctic regions impossible to show. The geography reflects the classic medieval idea of a landmass divided by four outflowing rivers. A huge magnetic rock stands at the North Pole. These ideas were derived from a Carmelite monk named Nicholas of Lynn, who was said to have sailed to the Arctic about 1360. His record, a book called the *Inventio Fortunata*, was supposedly found by one Jacobus Cnoyen, a Dutch explorer. He sent it to Mercator, who promptly translated the ideas into a map showing what he took to be the very latest information.

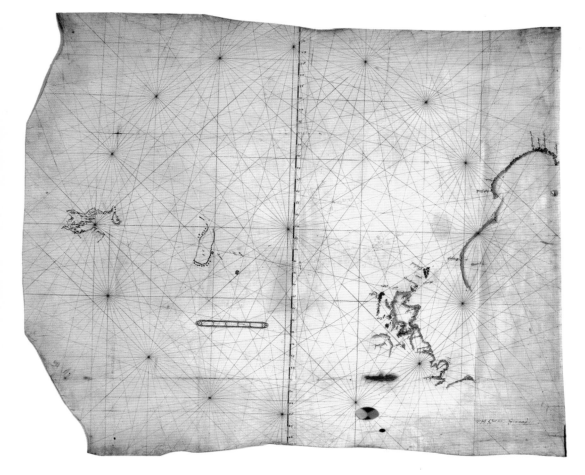

MAP 18 (above).
This was the map carried by Martin Frobisher on his first voyage in 1576. The base map was drawn by William Borough, and Frobisher has added his discoveries. The large inlet on the left is Frobisher's strait, now Frobisher Bay, Baffin Island. The peninsula at middle left is the southern tip of Greenland, although Frobisher, following the Zeno map (MAP 20, *opposite*), thought it was the mythical island of Friesland (his *Frisland*), a mistake which was to lead cartographers to misplace his strait to Greenland for a long time thereafter.

MAP 19 (below).
Part of the world map by Abraham Ortelius, published in his *Theatrum Orbis Terrarum* in 1570. It shows clear and indisputable Northwest and Northeast Passages, with not a hint of the trials and tribulations that might greet a sailor in those parts. Mercator's Arctic continent fills the northern seas. Seemingly authoritative maps such as this persuaded many to invest their money searching for the way to Cathay.

presented the Muscovy Company with a proposal for an expedition. A licence was required for any northern trading voyages, as the Muscovy Company had been granted blanket charter rights. Although initially turned down, Frobisher's proposal was eventually accepted; Lok set to work to raise the funds required, and preparations were made for a voyage in 1576. Lok would later come to rue the day he ever got involved with Martin Frobisher.

Frobisher sailed in June 1576 with three ships, *Gabriel, Michael,* and a smaller pinnace whose name we do not know. After a stormy crossing during which the pinnace was lost, on 11 July they chanced on the southern part of Greenland, which was promptly taken as the Frisland of the Zeno map (MAP 20). Thus the first achievement of Frobisher's voyages was to confirm the existence of an island that did not exist.

The master of *Michael,* distrusting the pack ice at the southern end of Davis Strait, stole back to England, leaving Frobisher in *Gabriel* to continue alone. Despite ongoing storms, Frobisher "hadde sighte of a highe lande" which turned out to be Resolution Island, at the southeast tip of Baffin Island. Sailing to the north of Resolution Island, Frobisher found the large bay now named after him, which, of course, he took to be the strait he had been seeking. The improbability of sailing across an uncharted ocean and immediately coming to the very strait he

sought did not seem to occur to Frobisher, but his personal glory required success. He named the bay "Frobisher's Streytes" after himself; Magellan had done it, so why not he?

Frobisher landed on a small island he called Hall's Island, after his sailing master Christopher Hall. There ensued a battle with Inuit with whom they had wanted to trade, and five men were lost. Now with only thirteen men, he quickly held a ceremony of possession, picking up a few stones in the process as tokens of possession—or perhaps simple souvenirs—and sailed back to England.

One of these stones was to change the future for Frobisher and his backers, for somehow it was found by an assayer to contain gold. Someone hoped to make a killing on a stock promotion perhaps; we shall never know. But this was what many wanted to hear, for after all, the Spanish had found gold in their discoveries and it had made them rich, so now, surely, it was England's turn.

As for his discoveries, Lok wrote that Frobisher "vouched to them absolutely with vehement wordes, speches and oathes, that he had fownd and discovered the straits, and open passage by Sea into the South Sea called Mar del sur which goeth to cathai." Frobisher clearly wanted another try.

All Europe was intrigued to know what Frobisher had discovered and where he had been. The Dutch mapmaker Abraham Ortelius visited England early in 1577 to try to gauge just what Frobisher had discovered. His visit was, the chronicler Richard Hakluyt recorded, "to no other ende but to prye and looke into the secretes of Frobisher's voyadge."

With the new prospect of a get-rich-quick scheme, a lot more financial support now flowed into the enterprise, with Queen Elizabeth herself making the biggest investment, but the goal of finding a Northwest Passage took a back seat to the task of finding gold. In 1577, with three ships including a much larger *Ayde*, Frobisher set off for his

MAP 20 (*above*).
The Zeno map, published in 1558, purporting to show the discoveries of Nicolo and Antonio Zeno about 1380. Although almost certainly a hoax, this map nevertheless misled navigators and mapmakers for several centuries. Greenland was mistaken for the fictitious island of Frisland or Friesland by Frobisher, who was unable to measure longitude. This resulted in his strait being placed on maps as cutting through the southern tip of Greenland, instead of on Baffin Island where his bay, if not his strait, really was (such as on MAP 27, pages 18–19). This in turn led John Davis a few years later to think that Frobisher Bay was a completely new discovery, which he named Lord Lumley's Inlet. Interestingly, despite the Zeno map's apparent denial of a Northeast Passage, this did not stop the English from continuing their efforts to find it.

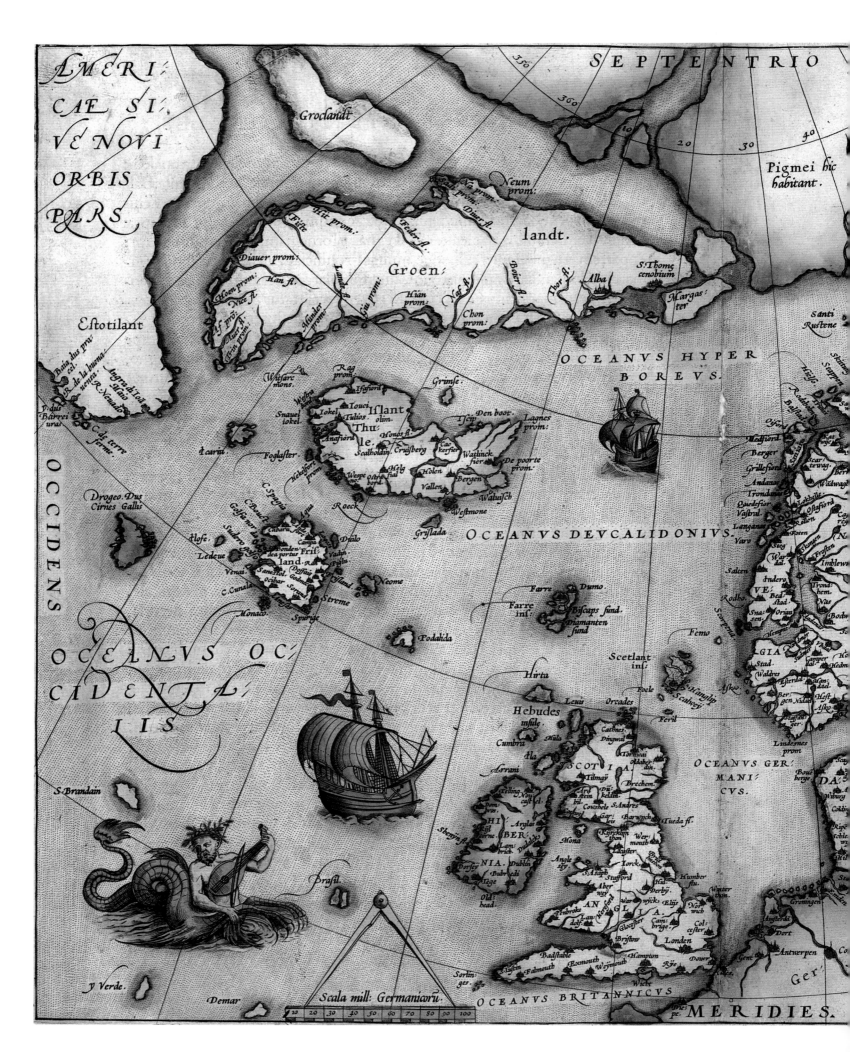

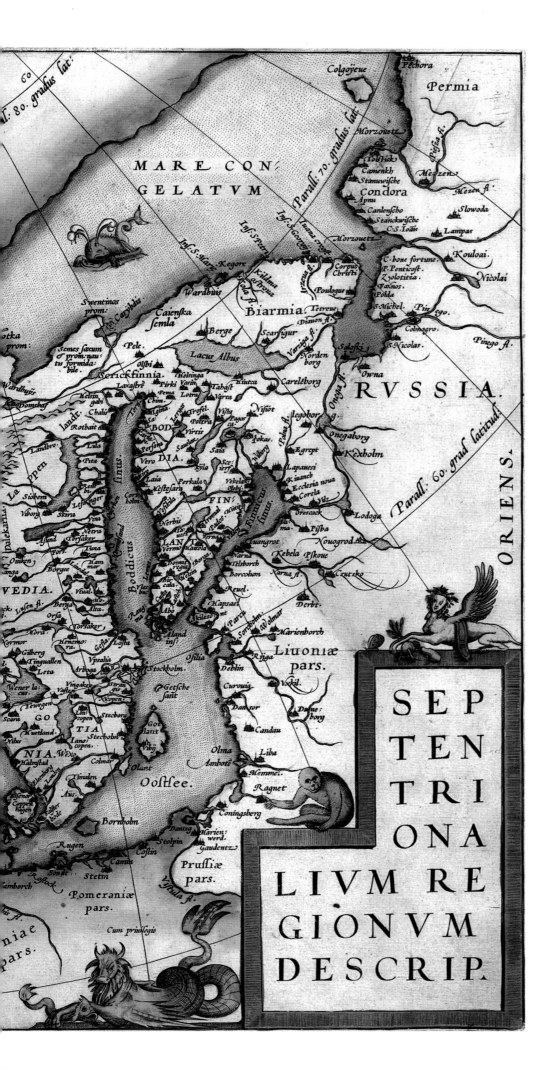

strait once more. This time the expedition returned laden with 200 tons of rock. It seems incredible, but once again assayers pronounced the rock to be gold ore, and the following year a virtual armada consisting of some fifteen ships and three hundred miners was assembled to return to Baffin Island. Also aboard were a hundred men intended to found an English colony on the new shore.

Frobisher lost his way somewhat this time, sailing into Hudson Strait, which was promptly dubbed "the Mistaken Straightes." Retracing his track, he found again his strait, his real mistake. This time an astonishing 1,350 tons of rock was mined and loaded onto the ships. Some of the ships had been lost in storms and in the ice, including half the timbers for a house for the would-be colonists, and so the colonizing project had to be abandoned. Had it been established as intended, the colony would have been the first English settlement in the New World.

Back in England, elaborate and ever more desperate measures were taken to try to extract gold from the rocks—to no avail. So worthless was the rock that when Spanish spies managed to procure some "ore" and send it for assay, King Philip II assumed his spies had been found out and worthless rock had been substituted. In the end the rock was used to build roads and walls, and the financial backers of the enterprise nursed their losses. Michael Lok, who had organized the investment and put a lot of his own money in as well, spent time in debtors' prison, unable to satisfy his creditors. As late as 1615 poor Lok was still being sued for debt.

In 1584, with the help of interested ministers of Queen Elizabeth, some merchants of Exeter and London were again persuaded that a Northwest Passage would soon be found. Led by William Sanderson, they financed another set of voyages, this time commanded by a veteran navigator, John Davis. In three separate voyages in 1585, 1586, and 1587, Davis explored the strait to the west of Greenland that came to bear his name. Over the three years, although he was able to find

MAP 21.
Another plate from Ortelius's 1570 edition of his *Theatrum Orbis Terrarum* is this superb map of the North Atlantic, Scandinavia, and the Arctic. The map does not reach far enough to the east to show Novaya Zemlya but a smaller island, *Colgoÿeue*, Ostrov Kolguyev, to the south, is shown. The ships of Arthur Pet and Charles Jackman ran aground here in 1580. The words *Mare Congelatum*— "Frozen Sea"—are prominent.

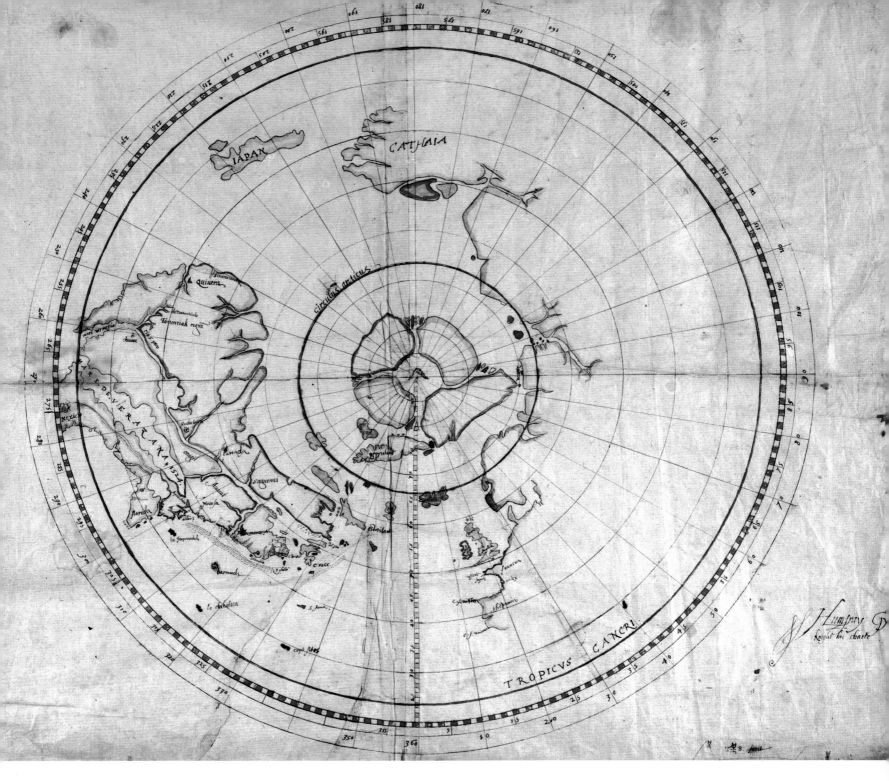

MAP 22.
This is *Humfray Gylbert knight his charte*, a rather superbly bizarre polar map drawn by English polymath John Dee for Sir Humfray Gilbert in 1583. Dee plotted Martin Frobisher's discoveries as the red-coloured two-pronged land at about 330° E and 63° N, the same longitude at which Greenland is drawn. Most later maps placed Frobisher's strait at the southern tip of Greenland. Dee's depiction of the Arctic follows that of Mercator (MAP 17, page 11), except that one of the "Arctic islands" has been stretched to become a grossly oversized Novaya Zemlya, similar to that suggested in William Borough's 1560 map (MAP 14, page 10). Clearly it was important to make the facts fit your preconceived concepts. This map also shows a superb Northwest Passage and, for good measure, another strait has been drawn through the North American continent.

financial backers each time, he became less convinced that a passage to the west existed. He explored north in Davis Strait as far as about 72° 40′ N, to a place on the west Greenland coast near present-day Upernavik he named Hope Sanderson, after his principal backer. On his way south Davis found "a very great gulfe, the water whirling and roaring as it were the meeting of tides." This later would prove to be the entrance to Hudson Strait, and it did lead to another sea—though not the one hoped for. But Davis did not investigate it; the task would await Henry Hudson a few years later.

It was not only the English who thought to circumvent the Spanish and Portuguese and arrive at Cathay. The Dutch, rapidly becoming a formidable sea power, thought so too. Their route of choice was to the northeast. The first effort seems to have been in 1584, when Oliver Brunel was sent on a voyage. Not much is known about this voyage, but it seems he reached a frozen Yugorskiy Shar, which barred his eastward progress. He returned in 1585.

Nearly a decade elapsed before another attempt was made. In 1594 an expedition of four ships was organized by Dutch merchants. Their pilot was an experienced navigator named Willem Barentsz. That year his ship managed to reach the northern tip of Novaya Zemlya before being stopped by the ice. Barentsz mapped the western coast of the islands in some detail (see MAP 26, page 18).

The following year another expedition with seven ships was sent out; those aboard included the famous Dutch chronicler Jan van Linschoten; Barentsz was the pilot. This time they tried to pass into the Kara Sea through Yugorskiy Shar, the southernmost of the three straits through Novaya Zemlya. But the ice was not cooperative, and the ships returned home without achieving much.

The merchants of Amsterdam alone refused to give up, and financed another voyage the next year themselves, but with only two ships. Jacob van Heemskerck, with Barentsz, was on one ship, and Jan Rijp and his pilot were on the other. The ships sailed in May 1596.

On 9 June, Barentsz thought he was close to the east coast of Greenland. That day the ships were confronted with pack ice. They turned to avoid it—and sighted land. It was only a bare, desolate island, but a welcome sight nevertheless. This was Bear Island (Bjørnøya) and they were nowhere near Greenland. The island did not receive that name until three days later, when the men were on their way ashore looking for food and encountered a large polar bear swimming. Despite the disadvantage of its situation, the bear put up a fight which none would easily forget.

On leaving Bear Island they sailed northwards to circumnavigate pack ice and in two days again sighted land. But this time the weather conditions were such that no attempt was made to land, and the ships continued northwards, reaching 80° 10´, farther than any previous voyage. Forced southwards again by the ice, they again sighted land, and they found a harbor at Fair Haven, at the northwest tip of this new land. After the stubborn Barentsz had made another attempt to sail northwards—and been pushed back by the ice—the land was followed southwards. Because the land "consisted only of mountains and pointed hills," wrote Barentsz, "we gave it the name 'Spitzbergen.' " Barentsz accepted that this was a new land never before drawn on charts, but many others for a long time would consider it to be part of Greenland, and it is shown on some maps as such.

Back at Bear Island, Rijp and his pilot differed with Heemskerck and Barentsz as to what they should do next, and so the ships

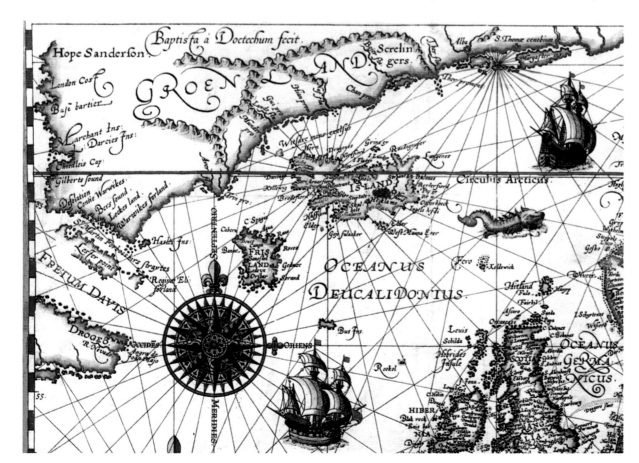

Map 23 (*top*).
Part of a map of Asia by Dutch mapmaker Gerard de Jode, published in 1578. It shows the eastern part of northern Asia as it was expected to be. The way to Cathay is south through the Strait of Anian, *Stretto de Anio* on this map.

Map 24 (*left*).
John Davis's newly discovered strait is shown on this map by Dutch mapmaker Peter Plancius in 1594, labelled in Latin *Fretum Davis*. Frobisher's strait has now been placed at the southern tip of Greenland. At Davis's northernmost point on the west Greenland coast, *Hope Sanderson*, named after his financier William Sanderson, is marked. Like many who would come after him, Sanderson received nothing for his support except a name on a map. Sanderson's name, unlike many later ones, has not survived on modern maps because it is not certain exactly where Davis was when he bestowed the name.

MAP 25 (*above*).
Gerrit de Veer's map, published in 1598. It shows the Dutch fleet unsuccessfully trying to pass through Yugorskiy Shar (here *Strtede Nassau*, Nassau Strait), to the south of Ostrov Vaygach (*Waÿgats*), on Barentsz's second voyage, in 1595.

MAP 26 (*below*).
Willem Barentsz's map of Novaya Zemlya, printed and published by Gerrit de Veer. *Het behouden Huys* is the house in which the Dutch overwintered. The ice surrounding the northern part of the islands is emphasized.

parted company. Rijp went north again, but discovered nothing new; Heemskerck and Barentsz headed east. After two weeks of sailing in dangerous waters full of ice floes, they reached Novaya Zemlya.

This time Barentsz did not examine the land closely but headed for the northern tip of the island, where he had been stopped by ice the year before. He managed to thread his

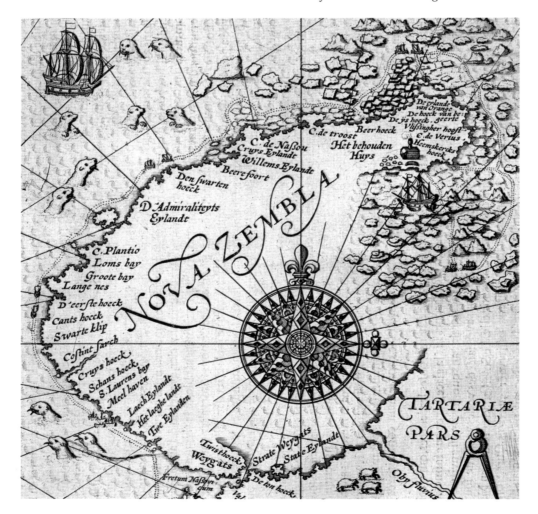

MAP 27 (*above*).
A superbly illustrated and colored summary map of the discoveries of Willem Barentsz on his three voyages from 1594 to 1597. *Het nieuwe land* is Spitsbergen, now part of Svalbard. Bear Island is here labeled *T'veere Eylandt*. As in MAP 26, *Het behouden Huys* is the location of the 1596–97 overwintering. *Waygats* is marked at the southern end of Novaya Zemlya. The track of Barentsz's ship on the third voyage is marked.

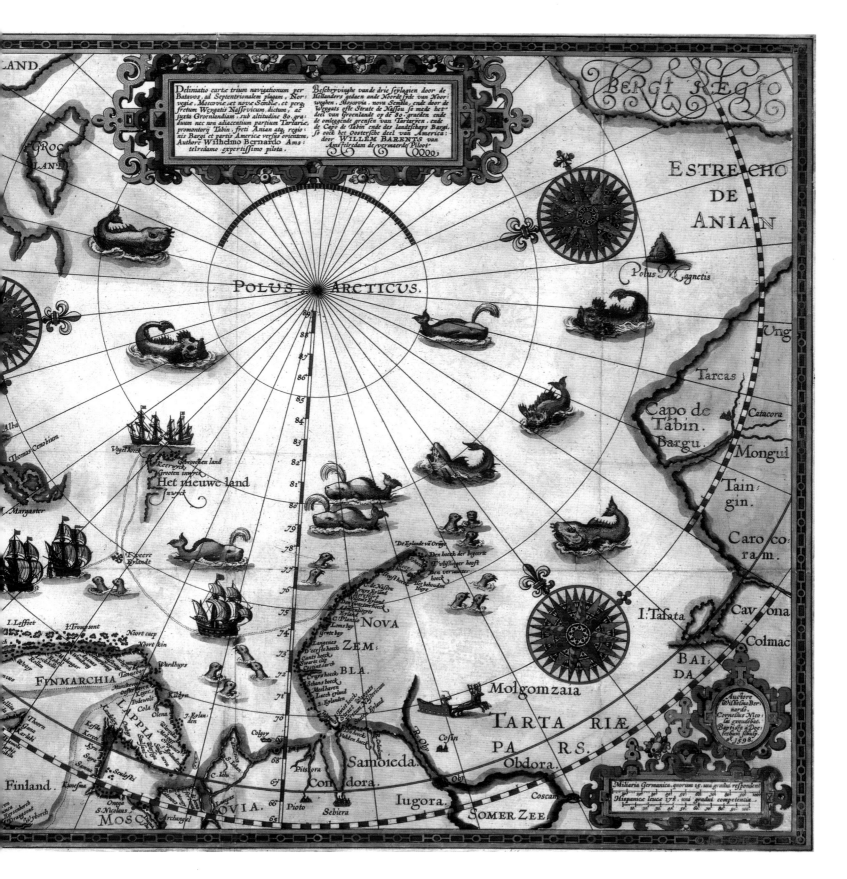

way through the ice and round the island into the Kara Sea. He intended to at least circumnavigate this land he now felt certain was an island and return through the straits he knew separated it from the mainland. But he had not reckoned on the loss of the warming influence of the Gulf Stream on the eastern coast, and by 27 August, although now trying to retrace his path to the north, the ship was trapped by ice at a place he called "Yshaven"—Ice Haven (Ledyanaya Gavan').

Timber from the ship was used to construct a house in which to spend the winter. They were relatively well stocked with food, but the intense cold—and little wood or coal to create a fire—led to much hardship. Driftwood had to be hauled from greater and greater distances to allow any semblance of comfort. And there were many encounters with fierce polar bears, from which, the men discovered, they could obtain enough grease to at least light their lamps for some time.

It seemed that the ice would never release their ship, and so on 13 June 1597, having

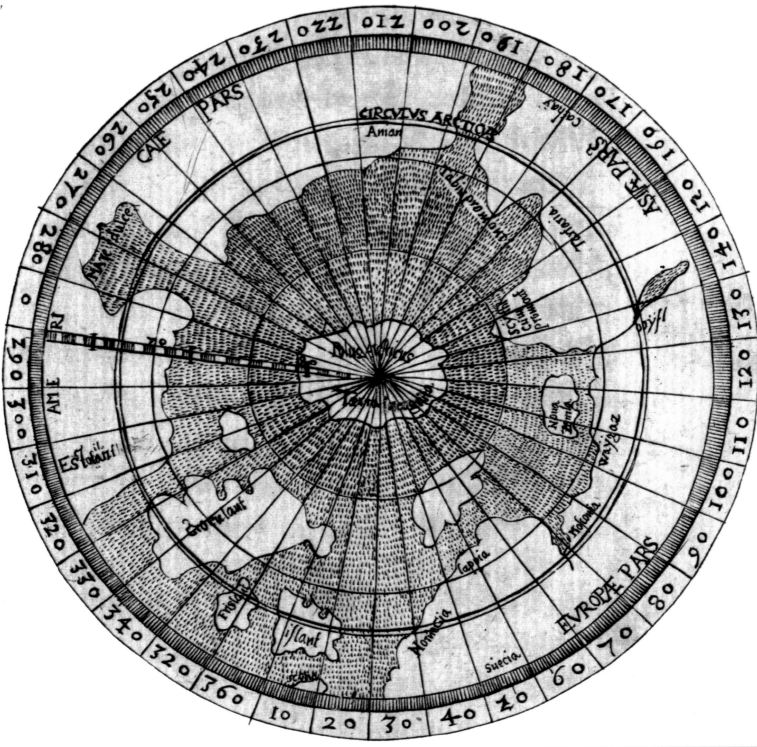

MAP 28.
This strange map was drawn in Leipzig in 1613 and shows Novaya Zemlya, albeit misshapen, and Weygats Strait (*Waygaz*). It also shows Barentsz's discovery of Spitsbergen, shown as an unnamed island mass to the north of *lappia*. The Northeast Passage is barred by the "Scythian peninsula," as on much earlier maps. To the west the map reflects the Zeno map (MAP 20, page 13) with its several islands in the North Atlantic, including Friesland—*Frislant*, and, on the mainland of North America, Estotiland—*Estotilant*. *Mare dulce* is shown as the as yet undiscovered Hudson Bay, as on Abraham Ortelius's 1570 map (MAP 19, pages 12–13). A strait between Asia and America is depicted, to the west of the land labeled *Anian*. This is the Strait of Anian, the strait which allowed transit from Northeast or Northwest Passages to Cathay, shown on this map to the south of the strait.

stocked and prepared two small boats, they set off in an attempt to escape. Before they left, Barentsz sealed a letter in a powder horn and hung it against the chimney. The letter would be found almost three hundred years later by an English sportsman, Charles Gardiner, on a hunting expedition.

It was an arduous epic—a few men sailing, rowing, and, when all else failed, dragging two small boats 2 500 km to safety.

Barentsz and two others died making the effort. But with help from Russian traders on the mainland coast, they made it to the Kola River, where they were rescued by a Dutch ship commanded by none other than Jan Rijp, out on a trading mission.

One of Heemskerck's officers, Gerrit de Veer, kept a detailed journal, later published, and produced a map (MAP 26), which was incorporated into many others thereafter.

MAP 29.
The year Barentsz left on his first voyage, Dutch mapmaker Peter Plancius drew this little map of the Novaya Zemlya region. *Sr Hugo Willoughbes land* is shown well to the west of *Nova Zemla*, and as a separate island.

Lost Colonies and Found Whales

The king of Denmark had as early as 1579 sent ships to Greenland to determine if any of his subjects—the Norse settlements—still lived; in any case, he wanted to re-establish Danish sovereignty. As part of this effort, he had several Icelandic maps redrawn (for example, MAP 7, page 7).

There were few seamen with experience of Greenland waters living in Denmark, so the king hired Englishmen. In 1605 he recruited John Cunningham and James Hall, whom, it is believed, had sailed with John Davis.

They left in May of that year, sailing into Davis Strait and as far north as 68° 35´, collecting furs and mineral samples and capturing four Inuit, but finding no Norse settlers.

The following year they sailed again, this time with five ships, and taking with them the Inuit, all of whom died on the voyage. Bad weather allowed only two of the ships to reach Greenland. Again they captured four Inuit, but found no Norse.

The next year, 1607, Hall was back, this time searching on the east coast of Greenland, but was unable to land due to the ice. So again no Norse settlements were found. Unknown to the Danes at the time, the Norse had in fact vanished two hundred years before.

About this time, England's Muscovy Company began to send out ships to Bear Island (Bjørnøya) to hunt for walruses, which because of their slowness out of water proved easy prey. A company exploring expedition

MAP 30.
James Hall was the pilot major on the king of Denmark's expedition to refind his lost colonies in 1605. Hall, an Englishman, was likely hired by the Danes because of his extensive knowledge of northern waters. He drew in his journal a number of beautiful maps of portions of the west coast of Greenland. This one, delightfully illuminated with whales and unicorns, is entitled *The coast of groineland with the latitudes of the havens and harbors as I founde them.* From his key, *D* is *Queen Sophias Cape in 67 deg 45 mm*; and *a* is *Queene anns Cape in the latitude of 66 deg.*

to Spitsbergen led by Henry Hudson in 1607 and a walrus-hunting expedition led by Jonas Poole and Thomas Edge in 1610 reported the presence of many whales in the region. The result was the dispatch of the first whaling ships in 1611. Edge led three ships, one of which was converted for whaling use and carried six Basque whalemen—the Basques at that time being the only ones with whaling experience. Only one whale was killed, but this marked the beginning of a huge and continuing industry.

The next year seventeen whales were killed, but the Dutch and the Spanish joined the hunt, and so the year after that a fleet of seven ships, including an armed vessel, was sent north. Soon seizures of catches, larger fleets, and armed clashes became the norm. The Dutch, in particular, formed the

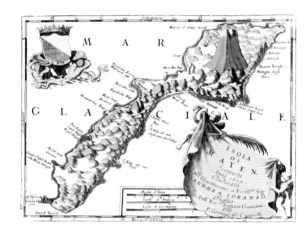

MAP 31 (*right*) and MAP 32 (*below*).
In July 1614, the Dutch whaling fleet was following the ice edge (where whales are often found) west towards Greenland when they discovered a small island with a very high mountain at its northern end. At the time they named it *Mr. Joris Eylant* after their pilot, Joris Carolus Flandro (see MAP 34, *overleaf*), but it later became known by the name of the captain of one of the ships who, it is thought, first sighted it. This was Jan Jacobsz May, and Jan Mayen Island it was. MAP 31 is a later graphic map drawn by the famous Venetian Franciscan mapmaker Vincenzo Coronelli. It shows Bear Mountain at the island's northern end. MAP 32 is part of the map of the North Atlantic drawn in 1628 by Dutch mapmaker Hessel Gerritz. It shows *Jan Mayen eyland* tucked in amongst the ice of east Greenland.

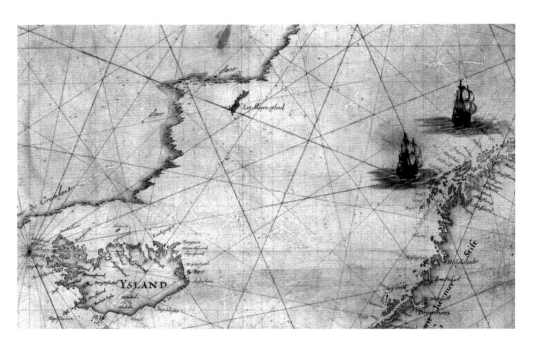

MAP 33 (*right*).
In 1630 an English whaling expedition to Spitsbergen led by William Goodlad sent eight men to hunt for deer just before sailing. They lost sight of the ships, which left for England without them. The eight men survived the winter by building a hut inside a larger shed left by Dutch whalers; it was perhaps the first-known example of the use of double insulation in the Arctic. These men were the first to overwinter in Spitsbergen. The following year the whalers returned, and the men were rescued. One of the men, gunner's mate Edward Pellham, wrote a book about his experiences, which included this superbly illustrated 1631 map of Spitsbergen, here called *Greneland*; it was thought at the time to be part of Greenland. The long book title pandered to the public taste for the gory details: *God's Power and Providence: Shewed, In the Miraculous Preservation and Deliverance of eight Englishmen, left by mischance in Green-land Anno 1630, nine monethes and twelve dayes. With a true Relation of all their miseries, their shifts [?] and hardship they were put to, their food, etc. such as neither Heathen nor Christian men ever before endured. With a description of the chiefe Places and Rarities of that barren and cold country. Faithfully reported by Edward Pellham, one of the eight men aforesaid. As also with a Map of Green-land.* To the east of Spitsbergen is *Edges Lland*, the island of Edgeøya, named after Thomas Edge and shown here attached to Spitsbergen. Still farther east is *Wiches Lande*, claimed to have been seen by an English whaler, John Ellis, in 1617 and given this name. It is likely what is today Svenskøya, the westernmost island in the Kong Karls Land group.

MAP 34 (*above*).
Joris Carolus Flandro was a pilot with the Noordsche Compagnie, the Dutch whaling company set up to compete with the English Muscovy Company. In 1614 he drew this impressively framed map of the whaling grounds and the countries surrounding them. Several glimpses of land to the east of Spitsbergen had by now been seen by some of the whalers, and Carolus took these to indicate land well to the east, even approaching the western shore of Novaya Zemlya, shown here at right. Interestingly there is nothing shown of Greenland, thought at the time to be a part of Spitsbergen. Note Jan Mayen Island shown as *Mr Joris Eylandt*, the name it was first given in 1614 when the Dutch fleet, with Carolus Flandro as pilot, discovered it. Carolus Flandro would have been expected to support this name on his own map.

MAP 35 (*right*).
This part of the North Atlantic map of Hessel Gerritz, drawn in 1628, shows Spitsbergen (*Spitsberge*), the island of Edgeøya, here again shown attached to Spitsbergen, and *Wiches Land* (likely Svenskøya, though in the wrong position). To the west is the hazily defined coast of the real Greenland, indistinguishable from the ice pack.

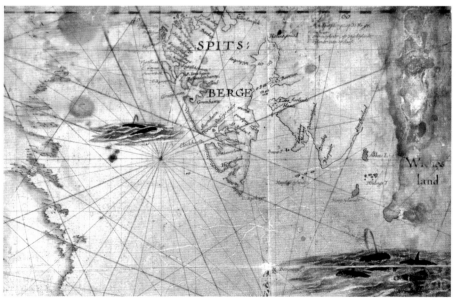

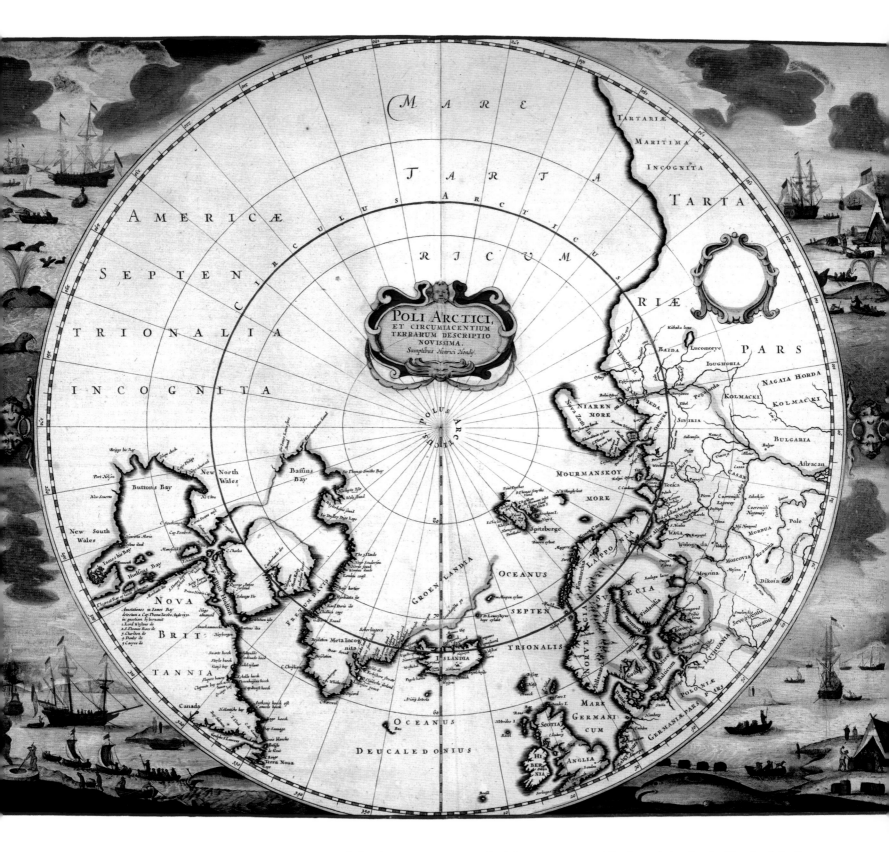

MAP 36.
Another wonderfully illustrated map is this map of the entire polar region by Henricus Hondius, published by Jan Jansson, here in a 1642 edition. From this map one can understand why Spitsbergen was long considered to be part of Greenland. In truth, ice usually stretched across the Greenland Sea, and the distinction between ice and land would have been difficult at the best of times. This map was one of the first published maps to show the explorations of Luke Foxe in 1631 and Thomas James (*James his Bay*) in 1631–32 (see pages 30–32).

Noordsche Compagnie to act as the Muscovy Company had and in 1614 sent seventeen ships, three of which were armed, to the whaling grounds.

In 1615, the Danes sent four ships to Spitsbergen, not to catch whales but to try (unsuccessfully) to assert what they considered to be Danish sovereignty over the region, because it was thought at the time to be part of Greenland.

The depredation by the Dutch of the English whalers led the Muscovy Company to sell its whaling rights in 1619, and they were purchased by a group that became known as the Greenland Company.

Whaling was by now well established, with whalers visiting the region each year; this new industry was to prosper for another three hundred years.

To the Other Ocean

With the return of the Barentsz expedition survivors in 1597, European enthusiasm for a Northeast Passage waned. There were a few efforts to sail to the northeast, but most interest was instead directed to the northwest, which seemed to hold out more hope for success. A strait to a new ocean was indeed found, but it was not the South Sea, but Hudson Bay.

And there was yet another possible route to the East—directly over the Pole. The belief that the Arctic was an open sea, an idea that would persist well into the nineteenth century, would allow a ship to sail north and then south to reach the Orient.

At the beginning of the seventeenth century, Henry Hudson, an Englishman, became one of the most experienced Arctic navigators. It was he who in 1607 reported the whales that began the whaling industry in northern seas.

His 1607 voyage, sponsored by the Muscovy Company, was aimed at sailing over the Pole. He discovered King's Bay (Kongs-Fiorden) in Spitsbergen, naming it "Whales Bay," and reached 80° 23´ N before deciding that this was not the way to Cathay. Hudson found whales, but no open polar sea.

The following year the Muscovy Company sent Hudson to try the Northeast Passage route. Hudson got as far as the west coast of Novaya Zemlya, at 72° 25´ N, but he could find no passage round the islands, and so he turned back and searched for the non-existent "Willoughby's Land" shown on some maps at this time (MAP 29, page 20); Willoughby's Land in fact was Novaya Zemlya.

Hudson had achieved little on this voyage, and the Muscovy Company washed its hands of him. But in 1609 he was invited by the Dutch East India Company to try again to the northeast. This time a mutiny prevented him from sailing beyond the Barents Sea, and he decided instead to try for a Northwest Passage. His mutinous crew were having none of it, and instead of sailing north when he reached the coast of North America, he made the best of things by sailing south, finding, in September 1609, what is now the Hudson River and the site of New York City.

On his return the English government forbade him to sail again on behalf of the Dutch. The Muscovy Company was no longer interested, but, financed by a group of interested investors, he was engaged to determine if, "through any of the passages which Davis saw, any passage might be found to the other ocean called the South Sea."

Hudson sailed from London in April 1610 and by August had sailed through Hudson Strait and into the "other ocean," which was to prove his downfall.

Hudson Bay was so large that it convinced Hudson he had found his route to Cathay, and he sailed south. A few weeks later he reached the "bottom of the bay" and was forced to again turn northwards. His ship was caught in the ice of the approaching winter, forcing him to overwinter. Unprepared, a number of his crew died.

The following year saw the outbreak of the famous mutiny. Hudson and a few others were forced into an open boat and set adrift; they were never seen again.

Command was assumed by Robert Bylot, who seems not to have had much to do with the mutiny. Due to his skill, the remaining crew made it back to England, where lack of evidence allowed them all to go free.

Hudson's journal and map of his voyage have been lost, but they seem to have been sent to Holland, where mapmaker Hessel Gerritsz published a book, with the map shown here (MAP 37). It was enormously influential, being incorporated into many other maps, because it extended geographical knowledge so much.

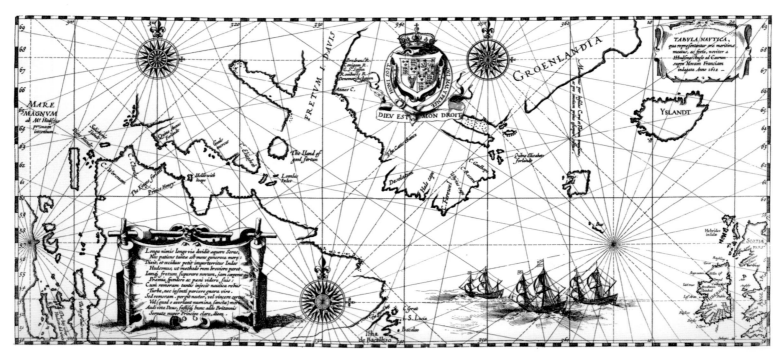

MAP 37.
This is the famous and influential *Tabula Nautica* map, drawn by Dutch mapmaker Hessel Gerritz in 1611 (this is the 1612 edition) to show the discoveries of Henry Hudson. The double bay at the south end of Hudson Bay, shown on this map for the first time, was a fiction that would show up on maps for some time thereafter. Note Frobisher's strait is shown in southern Greenland and on the west side of Davis Strait, where Davis had reported it as Lumlies Inlet.

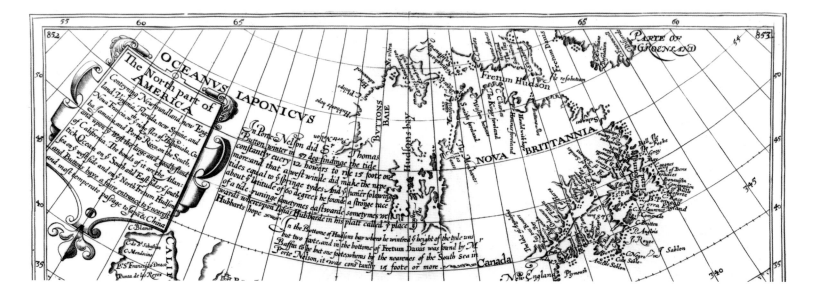

The rumour in England was that Hudson had found a Northwest Passage before the mutiny, and so a new company was formed called the Company of Merchants Discoverers of the North-West Passage, or Northwest Company. Welshman Thomas Button was charged with finding out what had happened to Hudson and completing "ye full and perfect discovery of the North-West Passage."

Taking Robert Bylot with him, Button sailed into Hudson Bay in the summer of 1612. He seems to have made little effort to search for Hudson—one wonders if Bylot had anything to do with this—and instead sailed west, expecting to reach Cathay. Instead he reached the west side of the bay at a place he called, appropriately enough, Hopes Checkt. Button explored the west coast of the bay, finding the Churchill River and reaching 65° N in Roes Welcome Sound. He wintered with considerable difficulty—men again died—at the mouth of the Nelson River and returned home the following year.

Button's explorations demonstrated the likelihood of Hudson Bay being just that—a bay, but facts were not known to deter the seekers of a Northwest Passage at this time, so more expeditions would be sent.

MAP 38 (*above*).
The first map to show the discoveries of Thomas Button was this map by English mathematician Henry Briggs, published in 1625. It is the first map to use the name *Hudson's bay*. At Button's farthest north is written *Ne ultra*—"Go no farther." Gaps in the outline of *Buttons Baie* allow for the possibility of a passage westwards, which Briggs firmly believed in; he would be one of the sponsors of Luke Foxe's voyage in 1631 (*overleaf*).

MAP 39 (*below*).
William Baffin's map of Hudson Strait and Bay from his 1615 journal. The track of his ship *Discovery* is shown by "a red prickle line" and there are flags where Baffin "made tryalle of the tide" to attempt to determine the direction of an ice-free western opening to the strait, correctly concluding that there was none. Baffin was the first, as far as is known, to measure longitude at sea.

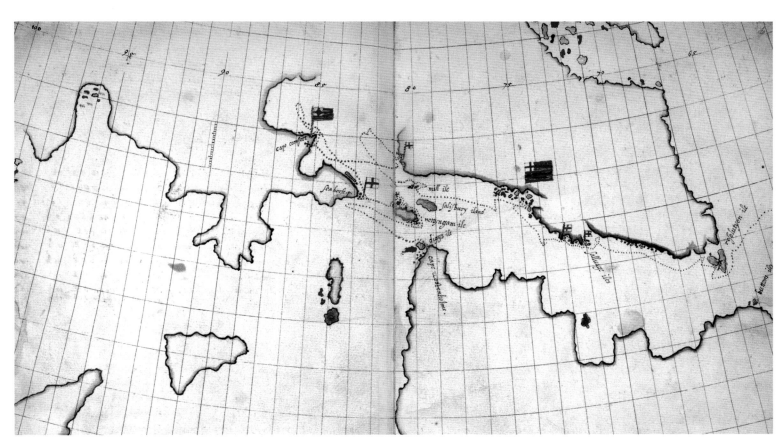

In 1615 the Northwest Company hired Robert Bylot and one of the most experienced pilots of the time, William Baffin, to try again. Baffin had sailed with James Hall (who had attempted to find the king of Denmark's lost subjects) on an English voyage in 1612 to search for silver on the west coast of Greenland.

They left England in March 1615, sailing into Hudson Strait, where Baffin measured the tides in various places (MAP 39, *previous page*) to try to find out which direction they might be coming from, and hence in which direction a western opening lay. They got as far as Frozen Strait, which joins Roes Welcome Sound to the the north of Southampton Island, but finding it living up to its name, decided there was no navigable passage in this region and returned to England in early September.

The next year the merchant investors of the Northwest Company decided that the opening to the Northwest Passage must lie to the north, up Davis Strait, and Bylot and Baffin were again dispatched.

They explored northwards, probing for openings, reaching as far north as 77° 45´, where they found an opening they named Sir Thomas Smith's Sound (Smith Sound), which is the route north that would be taken by North Pole–seekers from the late nineteenth century. But it was choked with ice. Circumnavigating the bay that would later be named after Baffin, they found one of the two major openings to the west, which they called Alderman Jones Sound (Jones Sound). It was also ice-filled and unnavigable.

On 12 July they came to another large opening to the west. They named it "Sir James Lancaster's Sound" (now Lancaster Sound), and it is the real major eastern entrance to the Northwest Passage. But Baffin did not recognize the ice-packed strait for what it was; it would be another two hundred years before Edward Parry would sail through it to the west (see page 64).

Baffin's new bay receded from the public consciousness, not to be revived until John Ross's voyage in 1818 (see page 60). This seems largely to be the fault of chronicler Samuel Purchas, who suppressed Baffin's navigation calculations and chart (which is now lost). Purchas wrote that "this map of the author's for this and the former voyage, with the tables of his journall and sayling, were somewhat troublesome and too costly to insert." By 1818, when John Barrow—second secretary of the Admiralty and the guiding light of British polar exploration in the first half of the nineteenth century—published a history of voyages to the Arctic, even the existence of Baffin Bay was in doubt; Barrow's map (MAP 104, page 60) does not show it at all. "So vague and so indefinite indeed is every information left, which could be useful," wrote Barrow, "that each succeeding geographer has drawn 'Baffin's Bay' on his chart as best accorded with his fancy" (see, for example, MAP 47, page 32).

The English now withdrew from the scene for a while. If an excellent navigator like Baffin could not find a passage, could anyone? It would take another fourteen years before English investors would be prepared to risk their money once more.

In 1619 came the next effort to find a Northwest Passage, this time by the Danes, who up till this time had been more interested in the whales of the northern waters. But they had developed expertise in sailing in ice-filled waters, and the king was concerned that another nation might find a passage to the riches of the East.

The king appointed the most experienced sailor in his navy, Jens Ericksen Munk. Munk left Denmark with two ships in May 1619 and sailed west. He entered first Frobisher Bay and then Ungava Bay by mistake before he found Hudson Strait, with the result that it was late August before he reached Hudson Bay.

Munk crossed the bay to the mouth of the Churchill River, where he became stuck in the ice of the late season and was forced to overwinter. It was a disaster; of the original sixty-five men, only three survived, one of whom was Munk himself. The three somehow managed to sail one of their ships back to Norway. The Danes went back to whaling.

It took another ten years for interest in a Northwest Passage to spawn not one, but two new expeditions. In 1629 a group of London investors calling themselves the Company of Adventurers, which included Henry Briggs (see MAP 38), petitioned King Charles I of England for a licence to send a ship to find the Northwest Passage. The king granted their request and even lent them a ship, the *Charles*. Luke Foxe was appointed as its captain.

Hearing of the London group's venture, a number of Bristol merchants hastily cobbled together their own proposal, for they feared that if the Londoners found a passage they would be given exclusive trading rights over it, thus endangering the competitive position of Bristol. King Charles was petitioned, and permission was again granted. The Bristol men appointed Thomas James to command

MAP 41 (*right*).
Part of a large map of the North Atlantic drawn by Hessel Gerritz in 1628. *Fretum Davis* (Davis Strait) leads to *Baffins Bay*, where the three major outlets found by Baffin and Bylot are marked: *Sir Thomas Smith's Sound* (barely legible), *Alderman Jones Sound*, and *Sir James Lancaster's Sound*, the latter the true eastern entrance to the Northwest Passage, unrecognized at the time by Baffin. Frobisher's strait passes through southern Greenland. Farther south the map compiles all the information from Baffin in 1615 in Hudson Strait and Thomas Button in 1612–13, with *Buttons Bay* marked at the extreme left. The southern part of Hudson Bay shows the erroneous double bay that somehow found its way onto Henry Hudson's map.

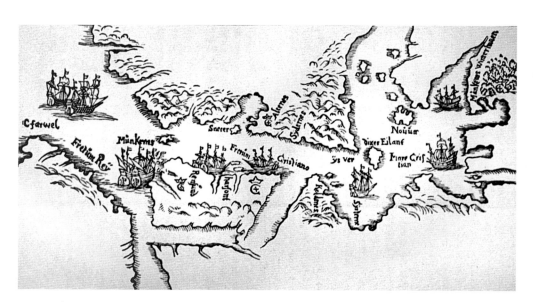

MAP 40 (*above*).
This hard-to-interpret map is a woodcut published in 1624 to illustrate an account of Munk's voyage. North is at bottom left. *C. farwel* (Cape Farewell) is the southern tip of Greenland. Munk's wintering place is labeled *Munk[ene?]s Winterhaven*. Because Thomas Button's map was not yet published, this map was the first printed map to show the west coast of Hudson Bay.

their ship and promptly named it *Henrietta Maria*, after the queen.

Both captains left England in 1631 armed with letters from the king to the Emperor of Japan. This time was going to be different!

But, of course, it wasn't. James left England two days before Foxe and headed for Hudson Bay, immediately crossing the bay to the southwest, where Briggs's map showed a large segment of undefined coastline. Instead of his expected passage to the East he found the shore of Hudson Bay, the most prominent headland of which he named Cape Henrietta Maria, after his queen and his ship. Then he turned south, hoping at least to find a way into the "River of Canada"—the St. Lawrence—thought by some to lead to a western ocean. Actually it would have led to the Great Lakes, but no farther.

At the bay's bottom he named "James his Baye," and here he was forced to winter, on Charlton Island, where he partially sank his ship to protect it from pounding surf-driven ice, a hazardous but in his case successful procedure. The following year he tried to sail northwards, reaching about 66° 30´ N

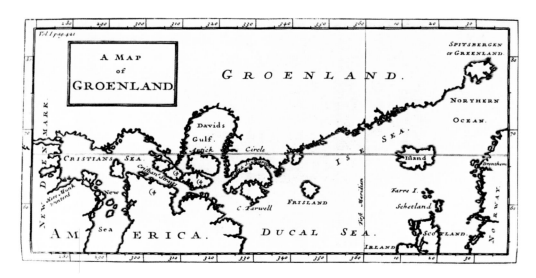

MAP 42 (*above*).
This delightful but a little confused map shows Jens Munk's explorations in 1619–20. It was published in a six-volume collection of accounts of voyages collected by John and Awnsham Churchill and published in 1732. *Christians Straits* is Hudson Strait, while *Christians Sea* is Hudson Bay; Christian was the Danish king. *New Sea* is James Bay. The mouth of the Churchill River is shown, marked *Here Munk wintered*. Frobisher's strait cuts through southern Greenland to *Davids Gulf* (Davis Strait). The *Ducal Sea* is the North Atlantic, with the fictitious island of *Frisland*. Greenland extends northeast to meet with *Spitsbergen or Greenland*, thought to be part of the larger island at the time.

MAP 43 (*below*).
Thomas James's original chart of his explorations in 1631–32, with an inset showing *James his baye*, today James Bay. On the inset, Charlton Island, where James wintered, is shown, as yet unnamed. The title of the main map reveals the purpose of the voyage—for "the discoverye of a Passage into the South Sea."

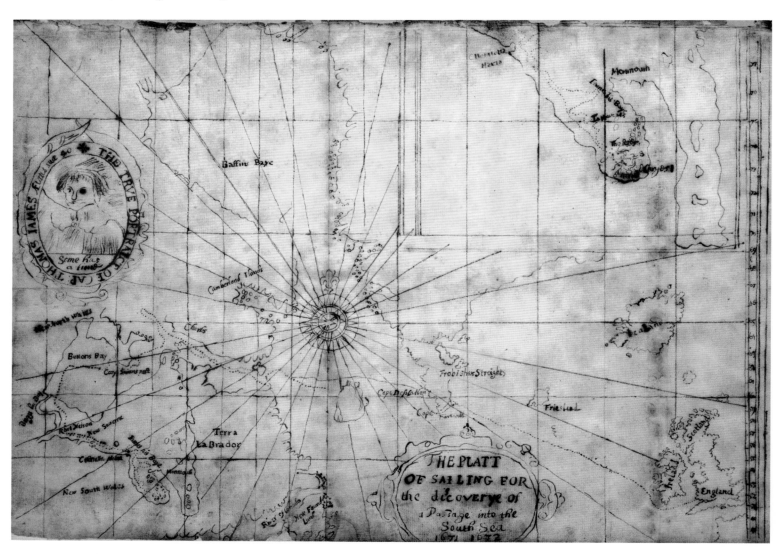

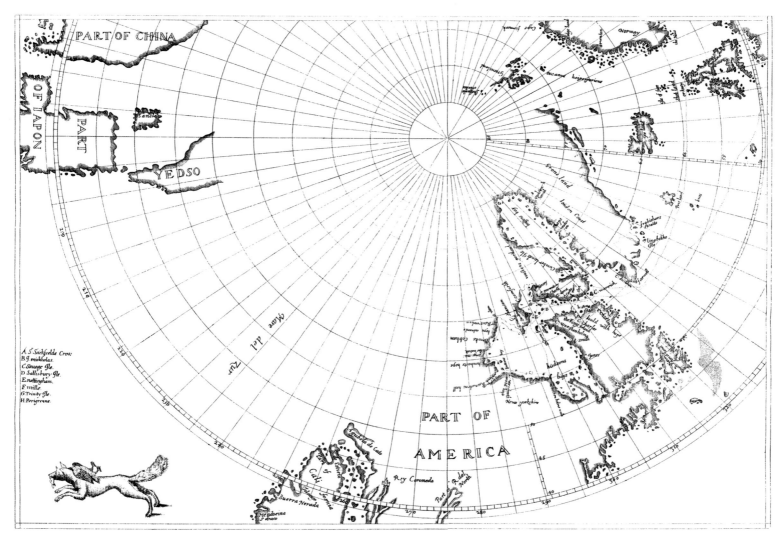

MAP 44 (above).
Luke Foxe's map of his explorations in 1631. The track of his ship *Charles* is shown, which at its northernmost point is at what is now Foxe Basin. Part of California is shown, hopelessly east of its true location. And his intended destinations, China and Japan, are drawn in, but there is a lot of blank space between the known points. At bottom left is Foxe's personal emblem.

MAP 45 (right).
The way it was supposed to be. A wide channel is shown leading westwards out of Hudson Bay to China. This 1677 edition of a 1653 map by French mapmaker Pierre du Val is based on a plate originally engraved from a map by Samuel de Champlain in 1616. Du Val has added the track of a ship going to China in 1665; there is no known record of any such voyage being attempted at that time.

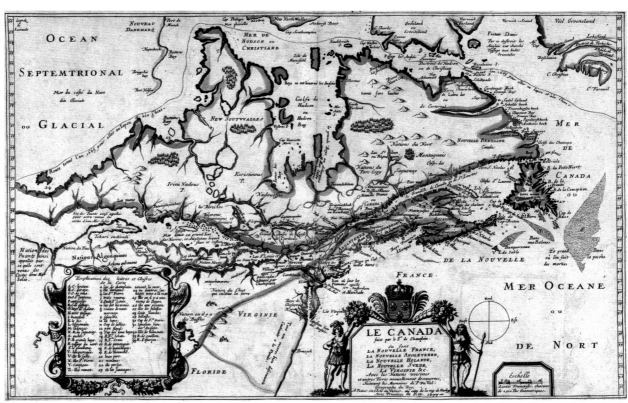

in Foxe Channel before being stopped by ice. Giving up, he returned to England in October 1632, now convinced—correctly—that there was no Northwest Passage south of 66° N. James's track is shown on his map, the original manuscript of which has survived and is now in the British Library.

Luke Foxe left England shortly after James, but returned later that same year. He also headed southwest once he reached Hudson Bay, and even met James at the entrance to James Bay. He returned northwards soon enough to escape the encroaching ice, attempting to sail northwest through Foxe Channel into Foxe Basin. He explored the southwestern shore of Baffin Island (now Foxe Peninsula), reaching 66° 47´ N before being halted by impenetrable ice.

Here he met the Arctic cold, "cold mists, thicks, and drops," he wrote, which "doth make many men droppe." And scurvy had now broken out among his crew, so perhaps wisely, Foxe decided to return to England. His track is shown on his map (Map 44).

Later Foxe concluded that if any Northwest Passage existed it must be through Roes Welcome Sound, and here on his map where Briggs had written *Ne ultra* ("Go no farther") he wrote *Ut Ultra* ("Go farther"). Unknown to Foxe, Roes Welcome Sound connects only with Frozen Strait, which leads—and that with difficulty—back into Foxe Channel. It also leads to bays later named Repulse Bay and Wager Bay, and these would be the site of later attempts to again find the elusive passage (see page 44). In the meantime the English had lost their taste for passage exploration and would not regain it for over a hundred years.

Map 46 (*left*).
This 1685 rendition of the Hudson Bay region and Greenland somehow managed to show a clear and wide channel continuing off the map to the west, despite the evidence of William Baffin, Thomas James, and Luke Foxe. A mapmaker had to sell his maps, and they no doubt sold better if they showed what the public wanted to believe. This map was created by Alain Mallet, a French engineer.

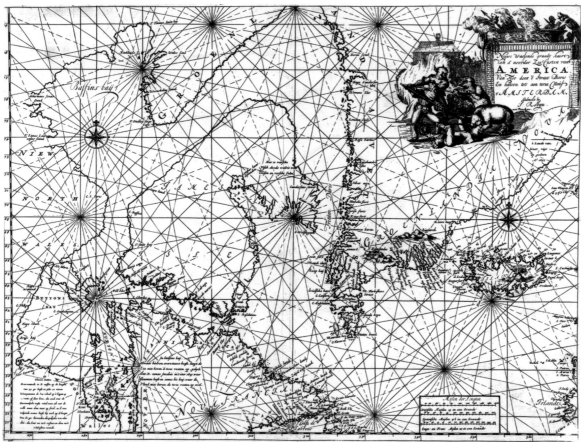

Map 47 (*left*).
A map of the Hudson Bay region by Dutch mapmaker Jacob Robyn, published in his 1683 *Zee Atlas* (Sea Atlas), a collection of maps intended for use by ship navigators. The geography shown here was predominant on maps for two hundred years. North of Hudson Strait a large *James Isle* is in the approximate position of Baffin Island but *Baffins bay* has been displaced well to the west. Note *M. Frobisher's strait* cutting through the southern tip of Greenland. The elaborate cartouche shows Jason-like characters attacking a polar bear.

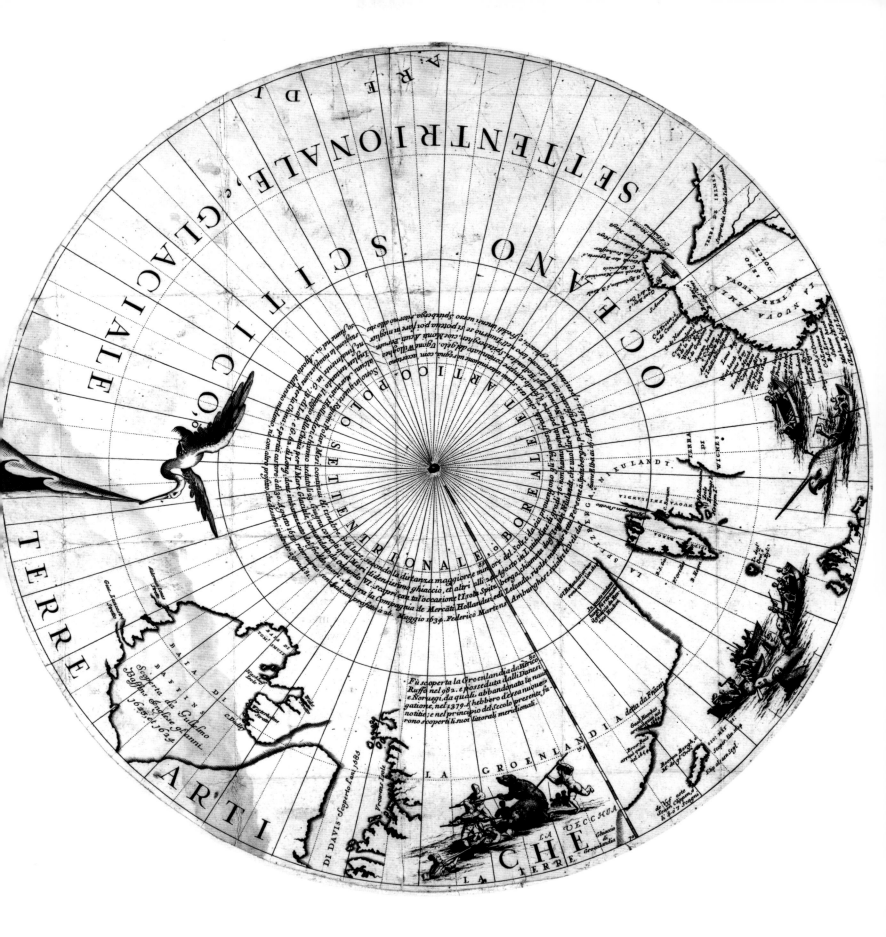

MAP 48.
This intriguing polar map was drawn by Venetian mapmaker Vincenzo Coronelli in 1692. Coronelli was well-known for his exquisite globes, some of which were huge. This map was actually designed to be stuck to the top of a globe to cover the joins of the individual globe gores. (To see the shape of a globe gore, see MAP 12, page 9.) The map is decorated with fantastic scenes of Arctic activities: whaling, walrus hunting, and bear hunting. At left a stork pulls a curtain over the geography of northern North America. Somehow Baffin Bay and Hudson Bay have become one; what is clearly intended to be Baffin Bay is displaced west to become Hudson Bay–like. The notation, in Italian, says that it was discovered by William Baffin in 1624 and 1625—Coronelli got his dates wrong. Davis Strait is more or less correctly shown. To the east, Coronelli is unsure as to whether Greenland joins Spitsbergen. Most information is reserved for Novaya Zemlya; Coronelli no doubt had seen a copy of Gerrit de Veer's map of his voyage with Barentsz almost a hundred years before.

Russian Exploration of the North

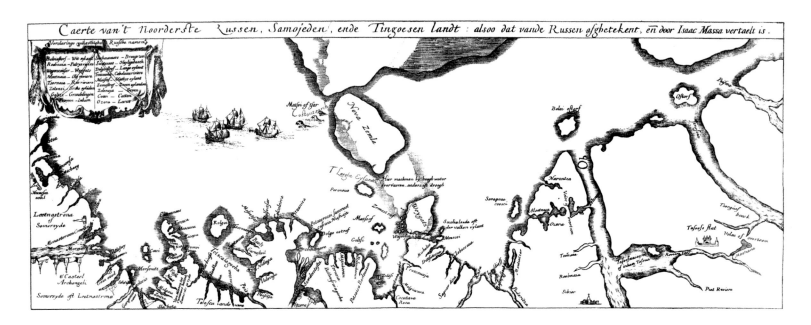

The pattern of early Russian exploration of the Arctic parts of Asia differs from that in North America because the exploration was carried out by peoples who aready lived on the continent spreading northwards and eastwards.

There were four principal reasons for northward colonization: religious—the founding of monasteries; military—the extension of empire; trade, especially for furs but also for ivory and oils; and the search for mineral wealth.

As early as 1136 a monastery was established at Arkhangel'sk, at the mouth of the Dvina River on the White Sea (Beloye More). It predated the founding of the seaport by 448 years, although one of the earliest Russian settlements in the Arctic, Kholmogory, was established 100 km upstream in 1353.

The first documented search for minerals in the Arctic was in 1491, when a large expedition was mounted with 240 miners and 100 boatmen to look for silver and copper ore in the lower Pechora River, on the coast immediately south of Novaya Zemlya. It was a success (which is probably why it was documented), and a mill was set up to process silver and gold ore.

From the twelth to the fifteenth centuries, the city of Novgorod east of Moscow had

MAP 49 (*above*).
One of the earliest surviving maps of the Russian Arctic is this map of the coast from the Kola Peninsula (at left) to the mouth of the River Yenisey (at right). It was drawn between 1604 and 1608 not by a Russian but by a Dutch merchant living in Russia, Isaac Massa. It had been copied from Russian information given to him by the brother of a Russian who had been to the Arctic. The brother did this at peril of his life, for had he been discovered slipping this information to Massa, the penalty would certainly have been death. The map is the first to show the Yamal Peninsula, with a river and lake portage across it, and an island hitherto unknown to Europeans, Belyy Ostrov (*Boloi ostorf*), at its northern tip. Clearly the Russians at this time knew a lot more about the geography east of Novaya Zemlya than the Dutch or the English, and they meant to keep it that way. The map was published by Hessel Gerritz in 1612.

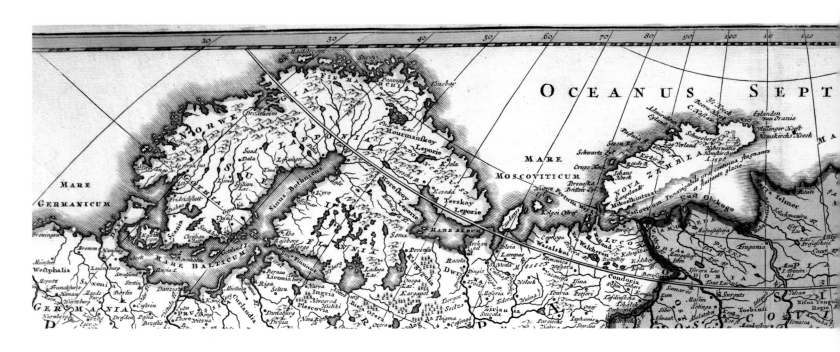

been responsible for settlements in northeastern European Russia and eastwards to Siberia just beyond the Ural Mountains as far as the River Ob'. In 1478 Novgorod fell to the armies of the Principality of Russia led by Tsar Ivan III, and from that time, Russia took up the task of colonizing farther eastwards. In 1581–85 there was a massive Russian military push into Siberia, led by Yermak Timofeyevich, and another in 1586 under Ivan Mansurov, and strongholds were established from which to subdue the rest of the country. The shores of the Sea of Okhotsk (Okhotskoye More) were reached in 1639.

The history of the Arctic in Asian Russia is bound to this slow eastward subjugation of Siberia, which reached Kamchatka and the Pacific by about 1730, after which the Russians surveyed their new dominions, including the Arctic coast (see page 38).

In 1619, Tsar Mikhail Fedorovich isued a *ukase*, or decree, closing the coast east of the White Sea to non-Russian ships; only the port of Arkhangel'sk remained open to European shipping. It seems that the tsar had heard rumours that King James I intended to claim the region for England. The ban lasted only thirty or forty years, but slowed European knowledge of the region.

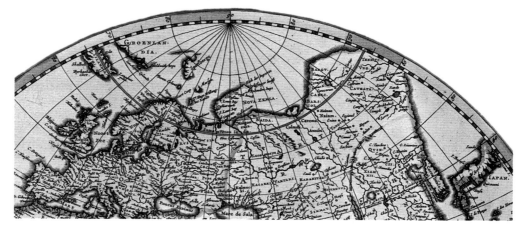

MAP 51 (*above*).
Part of the world map of Joan Blaeu, 1664. A Northeast Passage might be possible here, but what about that peninsula that goes off the edge of the map?

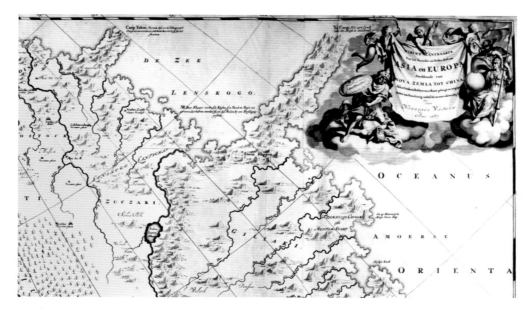

MAP 50 (*below, left*).
Part of a beautiful map of Asia drawn by German mapmaker Johann Baptist Homann and published in 1707. It shows a clear passage along the top of the continent—with one possible snag, a peninsula at the eastern end marked *Ys Caep*, Ice Cape. Widely thought to exist at this time and first shown in the maps of Nicholaas Witsen (MAP 52), it was believed to have been a land bridge to another continent—North America. Had it existed, no Northeast Passage would have been possible.

MAP 52 (*above*).
Part of Nicholaas Witsen's wall map of 1687 in which he showed his idea of an "Ice Cape" (*Ys Caep*). Cape Tabin (*Cuep Tabin*) farther north may have also had the effect of making a Northest Passage impossible. This was likely an early and misplaced representation of the Taymyr Peninsula (Poluostrov Taymyr), in fact the most northerly of all the mainland peninsulas of northern Asia. The map also seems to contain information from Semen Dezhnev's 1648 voyage (*see overleaf*). The Kolyma and Anadyr' Rivers, the beginning and end of Dezhnev's voyage, are shown in more or less their correct relative positions. How Witsen could have received this information is not known. Witsen presented a copy of this map to Tsar Peter I, who was most appreciative of the new knowledge of his dominions. Witsen went on to become Lord Mayor of Amsterdam.

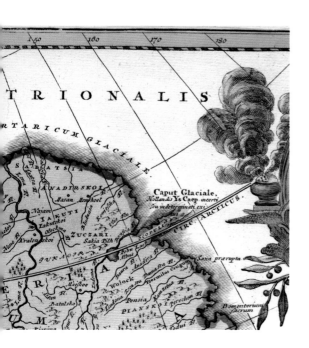

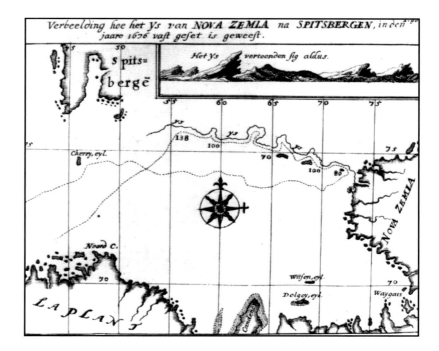

MAP 53 (*right*).
A map drawn in 1676 by Witsen showing the limit of the pack ice between Spitsbergen and Novaya Zemlya.

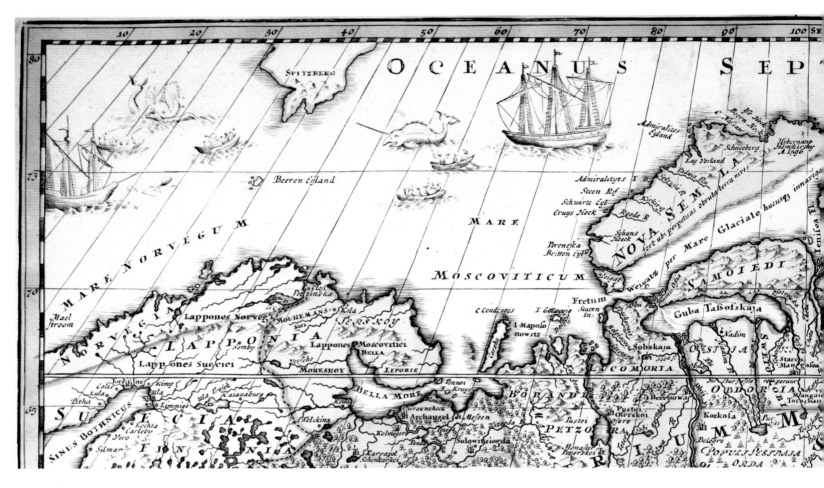

MAP 54 (*above*).
Another map by Johann Baptist Homann; this one was published in 1723. Ice Cape, *Eis Capo*, is now shown as a northward peninsula, not allowing for the possibility of being joined to another landmass. Note the land marked *Incognita* at far right; this was North America across the Bering Strait.

Russian geographical knowledge grew but slowly. To meet the demand for furs, hunters ranged over the tundra, learning much of its geography, but they made few maps, as they were generally illiterate. A military and hunting expedition under Grigoriy Semenov likely reached as far east as the Lena River in 1622–25. Semen Dezhnev and a group of hunters in 1648 sailed in small boats from the mouth of the Kolyma River (at about 160° E) round the eastern extremity of Siberia and through Bering Strait to the mouth of the Anadyr' River. This first recorded transit of Bering Strait preceded Bering by eighty years, but was largely unknown or

MAP 55 (*left*).
The map drawn by Gerhard Müller about 1736, from information from Semen Dezhnev found in Yakutsk. The huge, bull-nosed peninsula was the Chukchi Peninsula (Chukotskiy Poluostrov) that forms the extreme eastern end of the Asian continent. No doubt it appeared bigger to Dezhnev than it really was.

MAP 56 (*below*).
This map of Vitus Bering's epic journey through Asia and his voyage into the Bering Strait in 1728 was the first published record of his explorations.

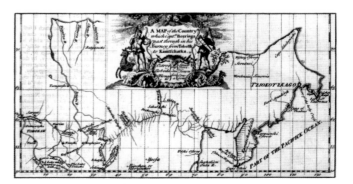

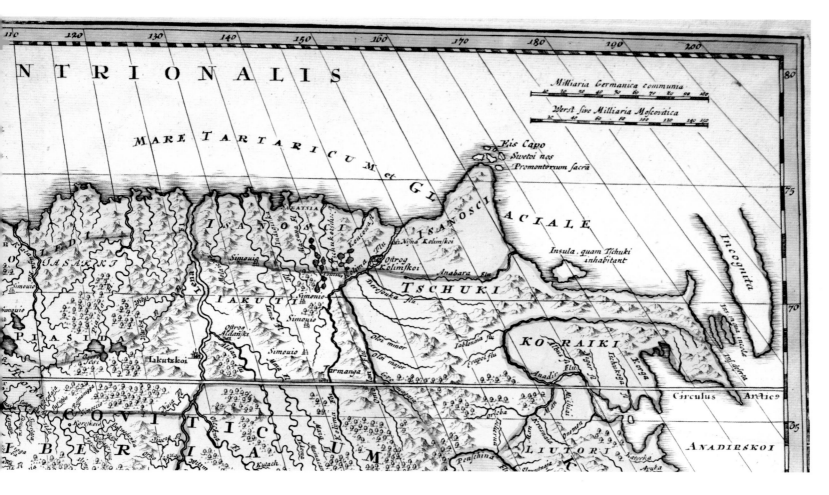

forgotten until Gerhard Müller, a scientist with the Great Northern Expedition, unearthed it and redrew it about 1736 (Map 55).

Between 1690 and 1692, Novosibirskiye Ostrova—the New Siberian Islands—were discovered by Maksim Mukhoplev while sailing from the Lena River to the Kolyma.

Finally, with Russian influence firmly spread across much of the continent, the Dane Vitus Bering, in the employ of Tsar Peter the Great, struggled across Siberia in 1725–28 and made his famous voyage through the Bering Strait. He did not see the American coast and thus was unsure that it was indeed a strait, though he was told that it was by the Chukchi he met. He did find St. Lawrence Island and the Diomede Islands. Bering's partial success in 1728 would soon spawn one of the most comprehensive Russian efforts to map the Arctic coast, the Great Northern Expedition (*see overleaf*).

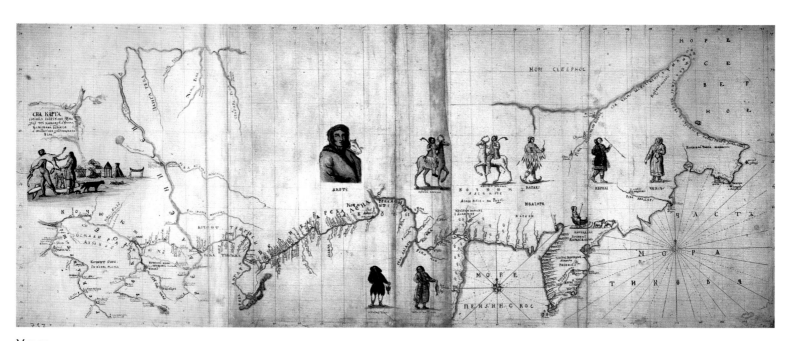

MAP 57.
One of about ten manuscript copies recording the overland journey (1725–28) and 1728 voyage of Vitus Bering into the strait that bears his name. Bering did not see the American shore, but he did sail far enough to be able to define the eastern extent of Asia. This copy of the map is from the Royal Library in Stockholm, Sweden.

The Great Northern Expedition

When Vitus Bering returned to St. Petersburg in 1730 following his voyage to his strait, he submitted a plan for the comprehensive exploration of the northern coast of Asia, which included a voyage to America.

A series of expeditions—collectively the Great Northern Expedition—were planned, all using naval personnel and all nominally to be overseen by Bering. The entire northern coast was to be mapped—to see if there was a passage to the Pacific—and a search made for any land, the "Great Land" rumored to lie to the north. As was the way in Russia at this time, communications difficulties and bureaucratic tangles meant that some expeditions took a long time to come to fruition; and most of the expeditions would take many years to complete. Bering's own expedition, called the Second Kamchatka Expedition, did not finally sail until 1741, in which year both he and Alexei Chirikov made it separately to the coast of North America at what is now the Alaska Panhandle.

The northern coast was subdivided, and the exploration and survey allocated. In the west, the coast between the White Sea and the River Ob' was the responsibility of Stepan Murav'yev and Mikhail Pavlov. Theirs was the first expedition to get going, but this did not help them, for ice prevented them from reaching the Ob' two years in a row, 1734 and 1735, and in 1736 they were relieved of command and replaced with Stepan Malygin and Aleksey Skuratov. To the government, it was irrelevant that failure was due to the conditions; failure was failure, and that was all there was to it.

Malygin and Skuratov had almost as much difficulty in 1736, resorting to mapping the Yamal Peninsula (Poluostrov Yamal) by land, but 1737 was an unusually good year for travel in the northern seas, and that year they made it to the Ob'. The return voyage on this relatively short stretch of coastline took two more years, again due to ice.

The coast between the River Ob' and the River Yenisey was assigned to Dimitri Ovtsyn, who initially had similar luck to the unfortunate Murav'yev and Pavlov. Three years in a row, from 1734 to 1736, Ovtsyn tried to sail out of the Ob' estuary (Obskaya Guba), without success. But his persistence paid off, for in 1737—the good ice year—he made it to the Yenisey. The following year, having achieved what he was assigned to do, he left his ship and traveled overland to St. Petersburg. There, however, his luck took another downturn; he was accused of friendship with an exiled prince, and for this, despite his successful expedition, he was demoted to able seaman and sent to Okhotsk to serve with Bering.

MAP 58.
Based partly on new information obtained by French geographers in Russia following the Great Northern Expedition, this new map of the Arctic coast from the White Sea to the Yenisey River was drawn by Jacques-Nicolas Bellin for an atlas in 1754. Note that the coast east of the Yenisey and the east coast of Novaya Zemlya are still shown as unknown.

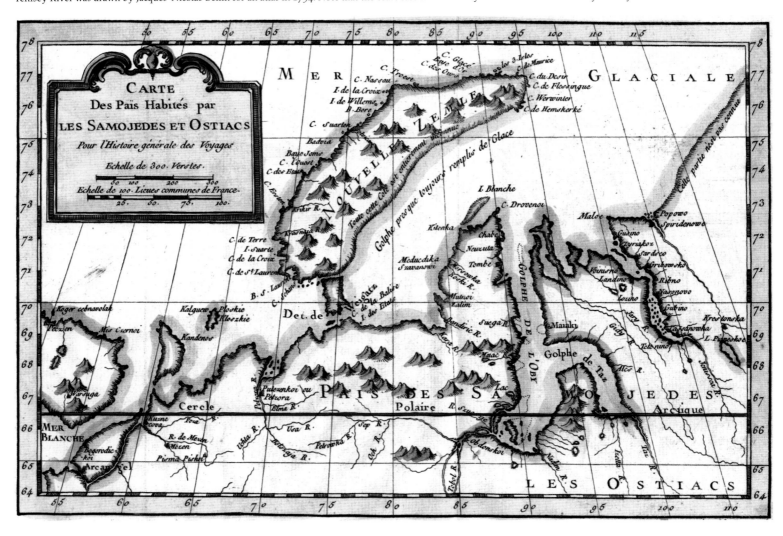

The vast, mostly uninhabited Tamyr Peninsula (Poluostrov Taymyr) lay beyond the Yenisey. To cover the area, two expeditions were assigned. One, led by Fedor Minin (who had been Ovtsyn's pilot), was to push eastwards from the Yenisey and sail around the peninsula to the Khatanga River. In 1738, Minin reached only the Pyasina River, about a fifth of the total distance. The following year he tried again, with even less success. The next year he sent Dmitri Sterlegov to map by land, but he only reached a little farther. Although Minin tried yet again in 1741, he again failed.

The other expedition, which was to survey westwards from the mouth of the Lena River, was led by Vasiliy Pronchischev and Semen Chelyuskin. They got started much earlier, in 1735, but fared only a little better than Minin. In August 1736 they reached 77° 29´ N, not far from the most northerly tip of Asia, but the ice was becoming dangerously dense, and they decided not to continue. Pronchischev died on the return trip.

The peninsula was finally surveyed by Khariton Laptev, who, after a two-year delay while new instructions were written and new stores sent out to the coast, set off in June 1739, taking with him Pronchischev's trusted navigating officer, Semen Chelyuskin.

In the second year of trying to round the Taymyr, Laptev pushed too much, and his ship was crushed by the ice and sank. At the end of 1740 Laptev reported to the Admiralty that the coast could not be charted by sea and that he was going to do it the next year by land. Three parties set off in the spring of 1741

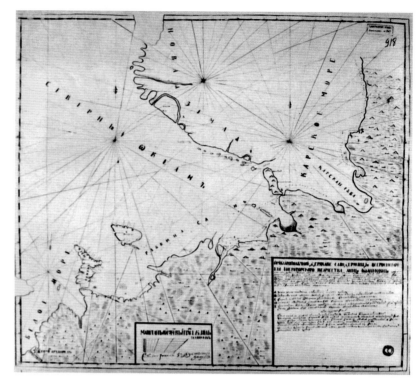

MAP 59 (*above*).
A map of the northern coast of Russia between the White Sea (Beloye More) and the western side of the Yamal Peninsula (Poluostrov Yamal), drawn by Stepan Malygin in 1736. At this stage, only part of the western coast of the Yamal Peninsula (at right) has been mapped; the expedition's geodesist, Vasiliy Selifontov, was forced to survey the western and northern coasts of this area by land in November 1736, so bad were ice conditions in that year.

MAP 60.
Khariton Laptev and Semen Chelyuskin's 1743 map of the Taymyr Peninsula, including the northern tip of the continent, Mys (Cape) Chelyuskin, named later.

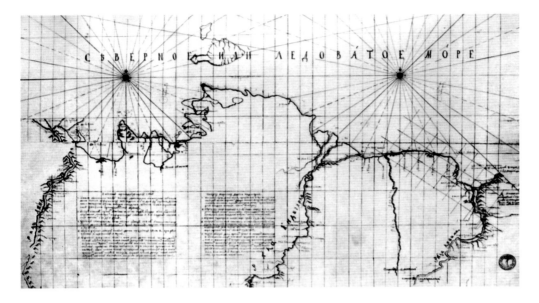

MAP 61 (*left*).
Dmitri Laptev's map of the northern coast of Russia from the Lena River (at left) to the Kolyma River (at right), drawn about 1742. The two islands of Novosibirskiye Ostrova nearest the mainland (Ostrov Bol'shoy Lyakhovskiy and Ostrov Malyy Lyakhovskiy) may have been drawn in later. Although they had been found in 1712 (*see next page*), their outline, which is generally correct, was not known until later. Mys Svyatoy Nos is the cape immediately south of the islands. The sea north of the Lena is today the Laptev Sea, named after Dmitri or his cousin Khariton, or both of them.

to map different sections, and it fell to Chelyuskin to map Taymyr's northern tip. He reached that point on 9 May 1742 and erected a cairn. This northernmost point of the Asian continent was named Mys Chelyuskin—Cape Chelyuskin—by nineteenth-century explorer Aleksandr Middendorf in 1843.

The easternmost section of the northern coast—by far the longest, as it was supposed to cover the coast from the Lena River round Bering Strait to the Anadyr' River, which flows into the Pacific—was assigned to Peter Lasinius, who in 1735, after accompanying Pronchischev's ill-fated detachment down the Lena, fared little better to the east. Stopped in mid-August by impossible ice conditions, Lasinius died while overwintering; an appalling thirty-seven died, out of a total of forty-eight men. The survivors made it back to Yakutsk, where Bering, who happened to be there at the time, on his way to the Pacific, put together a new contingent, this time led by Dmitri Laptev, a cousin of Khariton.

In 1736 Laptev could not round Mys Svyatoy Nos, the northward-projecting peninsula south of Novosibirskiye Ostrova (New Siberian Islands). He returned to the Lena Delta to winter, but the following year traveled to St. Petersburg to inform the Admiralty of his failure, thus completely missing the favorable ice year of 1737. He was instructed to try again, but it was 1739 before he did. In two years, overwintering on the way, he made it as far as the Kolyma River. But in the spring of 1741 he tried unsuccessfully to sail out of the mouth of the Kolyma and, giving up, made for the Anadyr' overland, arriving in November.

The Great Northern Expedition parties collectively mapped a huge tract of Arctic coastline, and only the Kolyma to Bering Strait stretch remained unsurveyed. It was a fine attempt, under less than ideal conditions, that achieved most of its objectives. A Northeast Passage existed, but it was not practical with eighteenth-century technology. No "Great Land" had been seen offshore. But Russia now had a much better geographical understanding of its far-flung empire-in-the-making.

MAP 62 (*below*).
A summary map of the discoveries of the Great Northern Expedition, drawn about 1742. The northern coastline of Asia is now relatively well defined.

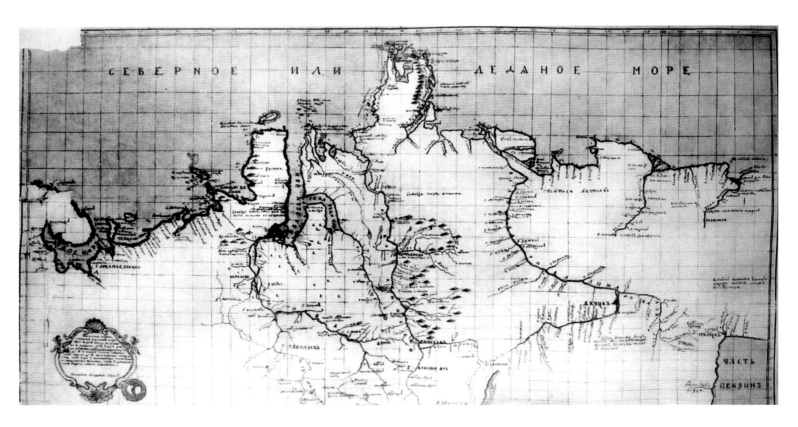

The Great Land

Rumors of land in the Arctic to the north of the mainland coast began to circulate as soon as that shore was reached, for there are indeed in the Arctic Ocean many islands, but of ice, hard to distinguish from land. In 1643 Mikhail Stadukhin, Russian discoverer of the Kolyma River, reported seeing land to the north. At least one of the small Medvezh'i Islands (Ova Medvezh'i), off the mouth of the Kolyma, was found by hunter Yakov Vyatka in 1655, and in 1711 the governor of Yakutsk sent Mercuriy Vagin to search for land. He found Ostrov Bol'shoy Lyakhovskiy, the closest of the Novosibirskiye Ostrova (New Siberian Islands), in 1712. From time to time, sightings of land would be reported, and sometimes expeditions were sent out to look for it.

In 1748 Ivan Bakhov and Nikita Shalaurov reported seeing new land while sailing south from the Anadyr' River. This might have been St. Lawrence Island and perhaps even Alaska. Certainly the ideas of land to the east—where there *is* land (Alaska)—and land to the north—where there is almost none—became confused.

In 1752 the French geographer Philippe Buache, who had access to some Russian information smuggled out of the country, published a map in which he showed a large "Great Land" to the north of eastern Siberia (MAP 63). This is also shown on his polar map on page 49 (MAP 81). His companion in cartographical mischief, Joseph-Nicolas De L'Isle, the same year published another version of the "Great Land" in which the coast of Alaska was extended westwards. This is shown in an English copy of this configuration published the following year (MAP 64).

In 1761 Ivan Bakhov and Nikita Shalaurov again, this time on a voyage from the Lena to the Kolyma, reported seeing a "vast land of mountains with seventeen peaks" beyond Mys Svyatoy Nos. Certainly this was some of the Novosibirskiye Ostrova, but this was unclear at the time. In 1763, the commander of the Anadyr' region, Fedor Plenisner, sent Stepan Andreyev to survey the Medvezh'i Islands and to search beyond them for a "Great Northern Land" supposedly lying beyond them. Andreyev exaggerated his distances, thinking that the Medvezh'i stretched 250 km or more east beyond their true location, and he reported sighting more land to the east.

MAP 63 (*left*).
Part of Philippe Buache's map of 1752, showing *Grande Terre,* "Great Land," to the north of eastern Siberia. This was in fact one of the least fantastic geographical features he showed on his map, which also shows a vast land north of North America. Buache was later known for his speculative geography (see also MAP 81, page 49), but few realized it *was* speculative at the time.

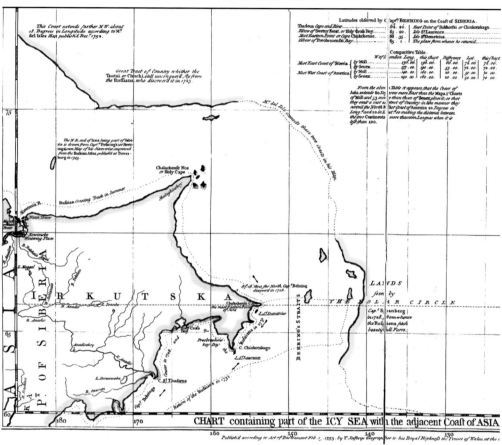

MAP 64 (*above*).
Part of a map by Englishman Bradock Mead, published under the *nom de plume* John Green in 1753. Mead was attempting to correct the unsupported speculations of Buache's map, but evidently did not think the "Great Northern Land" to be very speculative.

MAP 65 (*right*).
Nikolay Daurkin's map of eastern Siberia, 1765. It shows the mysterious "Great Land" to the north. Daurkin had likely misinterpreted native reports about Alaska—to the *east*. Or the land he showed could have simply been an elaboration from a European map of the period.

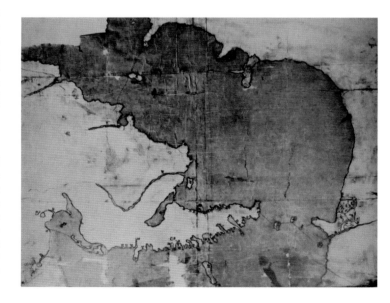

MAP 66 (left).
Fedor Plenisner's "official" map sent to St. Petersburg in 1765, reporting on what Daurkin had "found."

but Lyakhov, who was not equipped for an extended journey, could not go on. Lyakhov was later granted an exclusive hunting licence on the islands Ostrov Bol'shoy Lyakhovskiy and Ostrov Malyy Lyakhovskiy, and any others that might be found. Lyakhov is credited with the discovery of Novosibirskiye Ostrova, although he himself knew that he was not the first to find them, for he found a copper kettle on one of the islands.

In 1808, an expedition under Matvey Gedenshtrom was dispatched to survey

Andreyev was sent back to look for it the following year, and although he did not find it, "Andreyev's Land" remained on maps for a century (see, for example, MAP 221, page 144).

When he first dispatched Andreyev, Plenisner had also sent Nikolay Daurkin to report on the Chukotskiy Peninsula and its peoples, the Chukchi. Daurkin returned with a map he had drawn, which showed a vast westward extension of North America north of eastern Siberia—right where everybody expected there might be land (MAP 65).

Daurkin later participated in the expedition led by Joseph Billings and Gavriil Sarychev in 1785–93 and was probably instrumental in convincing Sarychev of the existence of a northern continent. Sarychev's later pre-eminence in the Russian navy meant a whole generation of Russian explorers also believed in the theory.

In 1770, a Russian hunter, Ivan Lyakhov, rediscovered the two nearest islands of Novosibirskiye Ostrova, off the coast of northeast Asia. He followed reindeer tracks out onto the ice and about 70 km from the mainland found an island. The tracks led farther north still and he followed them, finding another island. Still farther the tracks led,

MAP 67 (right).
This map, dating from 1775, is typical of a number of European maps before Cook in showing a "Great Land," following Daurkin's map, to the north of northeastern Siberia, as an extension of the North American continent. The inscription says that the "Great Land" was discovered in 1723, where the Tzutzy (Chukchi) escaped when they were chased by Russians, who did not capture them. Quite possibly natives who were threatened by Russian Cossacks would have escaped by fleeing out over the ice, and who was to say what was ice and what was land?

MAP 68.
A map by French mapmaker Jacques-Nicolas Bellin, published in 1766 but showing information from Gerhard Müller's map of 1754. It shows the northeastern part of Siberia and the western part of Alaska—the "Great Land" finally found. The bulbous peninsula first shown on Müller's copy of Semen Dezhnev's map from his 1648 voyage (MAP 55, page 36) has not been resolved by the expeditions of the Great North Expedition, but the relationship of America to Asia has been determined, and there is no land to the north of Siberia.

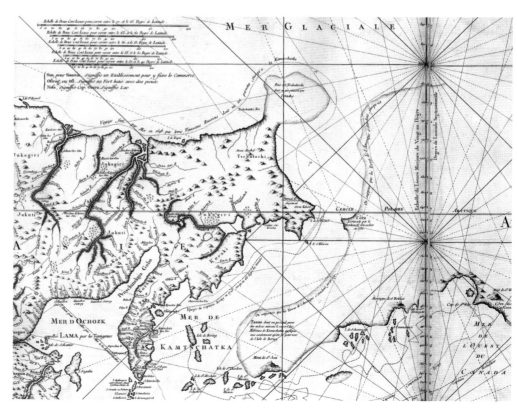

Novosibirskiye Ostrova. Most of the main islands had been discovered by this time, but there had been no proper survey and mapping. Gedenshtrom was also to search east of the islands for "Andreyev Land," the land reported by Andreyev in 1764.

During 1809, Gedenshtrom surveyed the Novosibirskiye Ostrova, although he had trouble even locating the easternmost island, Novaya Sibir'. Then in 1810 he attempted to find the land supposedly lying north of the Kolyma River. From the eastern end of Novaya Sibir' Gedenshtrom and his assistant, Yakov Sannikov, thought they saw land in the distance to the northeast. Gedenshtrom set off across the ice to try to reach the new land, failing, of course, to do so. Even though he realized that the land was an illusion, he made other attempts to reach it that year, failing each time. The mirages must have been realistic, for both he and Sannikov became convinced that there was land in the region, even though they could not find it. The "land," which subsequently became known as "Sannikov's Land," was shown extensively on their map (MAP 70, *below right*) and caused other expeditions to search for it for another century.

In 1811, Pyotr Pshenitsyn, who had been Gedenshtrom's geodesist, was sent with Yakov Sannikov to continue the exploration of the region. From the eastern end of Ostrov Faddeyevskiy, the northernmost large island in the Novosibirskiye Ostrova, Pshenitsyn saw land, this time to the north. He set out across the ice but was stopped by a lead after 25 km. But here he could still see the "land" to the north. It was probably an illusion, though it is possible that he saw Ostrov Bennetta, the discovery of which is usually attributed to George De Long and the ill-fated *Jeannette* expedition of 1879–81 (see page 135). Whatever Pshenitsyn saw or did not see, the "sighting" of this part of "Sannikov's Land" reinforced the myth of a northern land.

The tracks of Gedenshtrom and Sannikov in 1808–11, and Pshenitsyn in 1811, are shown on an 1879 map (MAP 181, page 114).

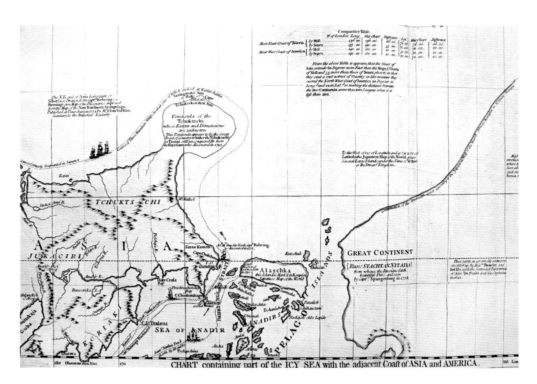

MAP 69 (*above*).
This revised edition of the Bradock Mead (John Green) map (MAP 64, page 41) was published in 1775. Now the Alaskan coast trends to the northeast from Bering Strait, much more in accordance with reality, but still too far north.

One should not be too hard on Gedenshtrom, Sannikov, Pshenitsyn, and all the others who thought they saw land in the Arctic distance. The history of the polar regions is littered with those who saw land where there was none; they were not the first, nor would they be the last.

(*Russian exploration of the region north of Asia is continued on page 112.*)

MAP 70 (*below*).
This map is by Pyotr Pshenitsyn, Matvey Gedenshtrom's geodesist on his 1808–11 expedition. Novosibirskiye Ostrova (the New Siberian Islands) are at left, while the "land" they sighted to the east is shown at right. These are, as we now know, mirages of a northern land plotted on a map. But features shown on a map tend to take on a life of their own, and others were to search for these mysterious lands for a hundred years. The "land" is also depicted on the Russian polar map of 1820–24 shown as MAP 180, on pages 112–13.

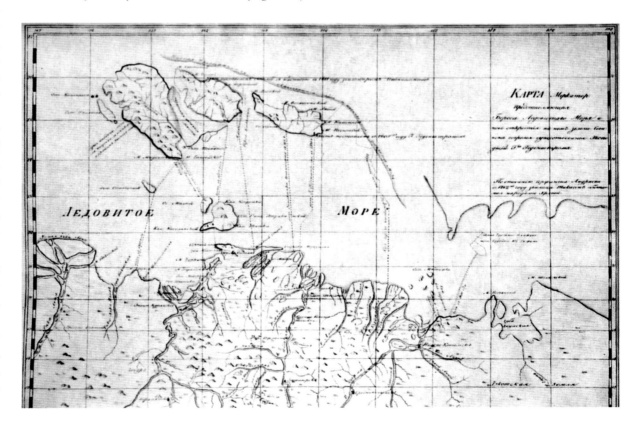

That Illusive Passage

MAP 71.
This map, part of Dutch mapmaker Joan Blaeu's world map of 1664, summarizes the assumed geography of northeastern North America to that date, which remained essentially unchanged until the advent of the passage-seekers of the mid-eighteenth century. Blaeu has added the anticipated geography to the west—a very short distance through some unspecified opening to the Strait of Anian and the western ocean. The northern tip of California, depicted as an island, is just visible here. This was the sort of map that induced investors to keep throwing their money into searches for a Northwest Passage—the Pacific was so *close*.

In the mid-eighteenth century there was a brief burst of activity from new seekers of the Northwest Passage. The concept was so appealing that unless conclusively shown to be impossible, it was bound to resurface from time to time. It was one of the "get-rich-quick" schemes of the Age—otherwise—of Reason.

In 1670, the Hudson's Bay Company had been incorporated to exploit the resources not of Cathay, but of the other sea that had been discovered seemingly on the way. The company became generally contented with the profits they could make from furs, especially beaver, which was used for making fashionable hats, and did not want to seek for a Northwest Passage.

In 1719, James Knight, Bayside governor for the Hudson's Bay Company, did persuade the company to allow him to search for a Northwest Passage, which he thought lay somewhere north of York Fort, on the west side of Hudson Bay. He had developed this idea from information given to him by natives. His voyage was disastrous; his two ships were wrecked on Marble Island, just south of Chesterfield Inlet, and Knight and all his men perished. The company's enthusiasm for exploration was understandably dimmed by Knight's demise.

It took others to renew the quest for a passage. Arthur Dobbs, an Irish landowner and member of Parliament, in 1731 began what would prove to be a long line of memorials, pamphlets, books, and the like with a memorial asserting the existence of a Northwest Passage, and by 1740 Dobbs's persistent arguments about tides and whales but more persuasively the glory and trade benefits that would fall to the nation that discovered a passage had won over the first lord of the Admiralty, Sir Charles Wager, enough that he mentioned them to King George II. The king was enthusiastic and thought that since "the expence was such a trifle," an expedition should proceed.

The Admiralty hired Christopher Middleton, an experienced Hudson's Bay Company captain, to command a voyage, and William Moor, another Bay captain. With two ships, the *Furnace* and the *Discovery*, they sailed in June 1741 after a struggle to find enough men to man the ships, for Britain was at this time at war with Spain. The lateness of the season meant that the two ships had to overwinter, which they did at Churchill River. After a difficult winter, the ships sailed north on 1 July 1742.

Starting his search for a westward-leading channel at about 65° N, Middleton found one that led deep inland, which he

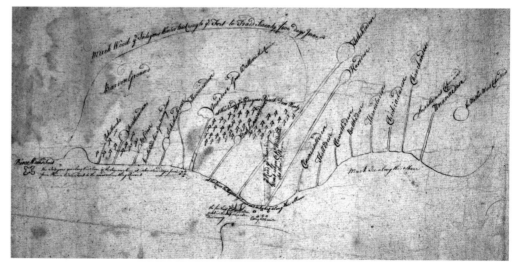

MAP 72.
This map, which shows the rivers draining into Hudson Bay north of Prince of Wales Fort, was originally drawn by James Knight around 1719, the year he left on his fateful voyage north. Clearly he did not take it on the voyage, or it would have been lost with him. A later hand has added information: Prince of Wales Fort, which was named in August 1719, after Knight had left; and the drawing of the little ship with the notation *the furthest Capt Middleton went in the ship when on* [a Voyage of?] *Discovery*. Christopher Middleton discovered Repulse Bay in 1742.

named after Sir Charles, Wager River (Wager Bay). Middleton was fairly certain it was no strait, but controversy would follow from his failure to follow the inlet to its western conclusion.

Continuing north, Middleton found that the land trended to the west beyond a cape; he wrote that it gave him "great Joy and hopes of it's being the extream Part of America, on which account I named it C. Hope." But again he was disappointed, for careful examination revealed that he was in a large bay with no openings. He named it Repulse Bay.

Returning southwards, he found Rankin Inlet (named after one of Middleton's officers, John Rankin) but not Chesterfield Inlet, missed because of the screen of small islands that covers its entrance. One reason Dobbs felt that this area was where the Northwest Passage lay were reports of strong tides in Roes Welcome Sound. Middleton's investigation correctly concluded that they were actually coming from Frozen Strait, at the north end of Southampton Island.

Middleton arrived back in England in October 1742 to a furious Dobbs, who was certain that Middleton must have falsified his log to hide his discovery of a passage, and where else could it be than in his Wager River? No doubt after some money passed under the table, John Wigate, who had been Middleton's clerk, produced a map which showed Wager "Straits" (MAP 77, page 47).

The following year Dobbs was presented with a map of North America reputed to have been drawn on the dirt floor of a London tavern by a disaffected *coureur de bois* named Joseph La France (MAP 76, *overleaf*). Transferred to paper and with the features discovered on Middleton's voyage added, the map showed a remarkable west coast of North America coming so close to Hudson Bay that it looked like Rankin Inlet led straight to the sea. Plied perhaps with gin, La France had given Dobbs's supporters just what they wanted.

There followed a war of words, with printed vituperations flying back and forth in the form of remarks, replies, and rejoinders for all the world to see. One title, published by Dobbs, was *Remarks upon Capt.*

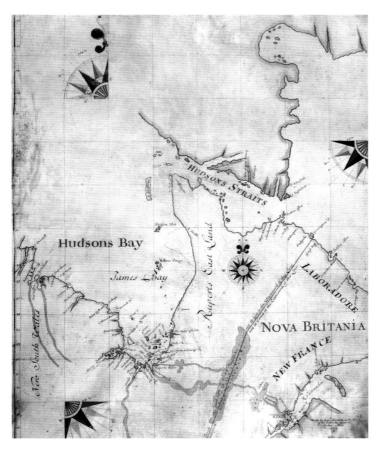

MAP 73 (*above*).
A superb map of Hudson Bay drawn in 1709 by Samuel Thornton for the Hudson's Bay Company. Company activity, and hence geographic information, was concentrated at the "bottom of the bay" at the time.

MAP 74 (*below*).
The Hudson Bay part of Christopher Middleton's map, 1743. North is to the right. Middleton's track in 1741 and 1742 is shown by the dotted lines. *Churchill R[iver]*, where he overwintered, is shown at left, and his major discoveries, *Wager River* (Wager Bay) and *Repulse Bay*, at right. Note C[ape] Dobbs at the entrance to Wager Bay.

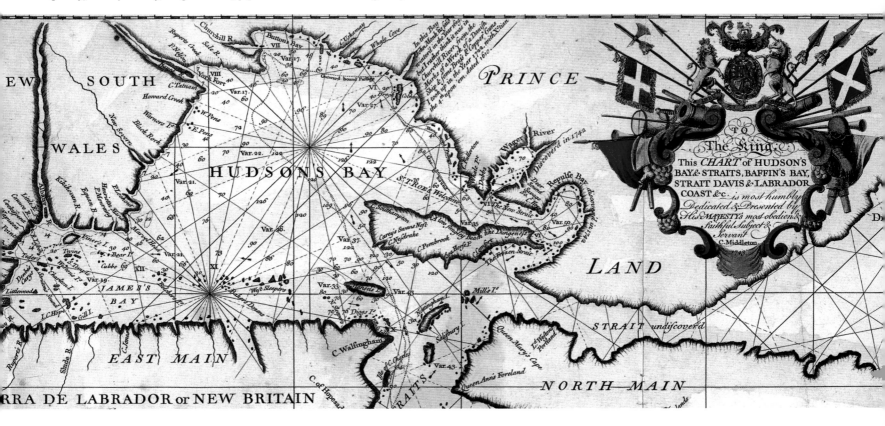

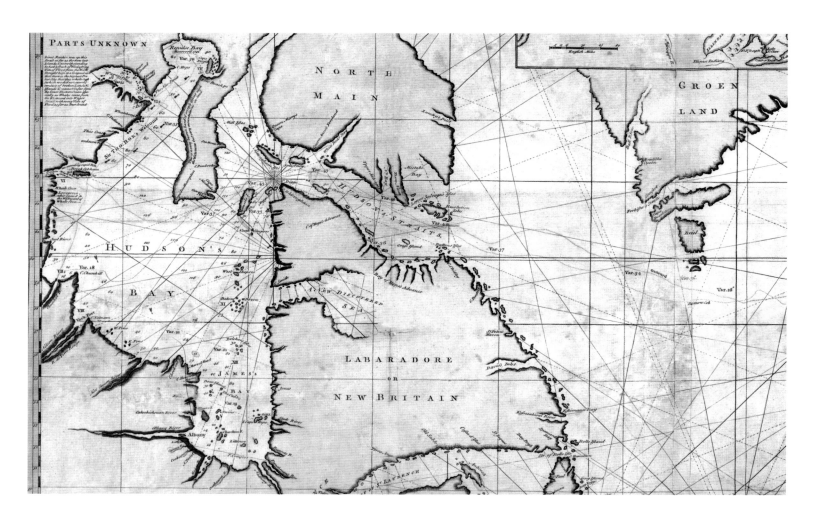

Middleton's defence wherein his conduct during his late voyage for discovering a passage from Hudson's-Bay to the south sea is impartially examin'd, His Neglects and Omissions in that Affair fully Prov'd; the Falsities and Evasions in his Defence Expos'd; the Errors of his Charts laid open, and His Accounts of Currents, Streights and Rivers, Confuted; Whereby it will appear, with the highest Probability, That there is such a Passage as he went in search of—and that was just the title. Middleton had, wrote Dobbs, "wilfully misbehaved, by neglecting to look into those places where he had Reason to expect a Passage, by falsifying facts, by making Currents and Tides contrary to Truth, and by forging a large Frozen Streight to bring in Tides and Whales, in order to support the scheme he had laid to conceal the Passage." Dobbs was utterly convinced that, in cahoots with the Hudson's Bay Company, Middleton was trying to hide his discovery of a Northwest Passage.

MAP 75 (*above*).
A map published by John Wigate in 1746, incorporating his more detailed map of Wager Bay (MAP 77). Wager *River* has become Wager *Strait*, with its western end conveniently disappearing behind the cartouche. This map would have been the principal one with which Dobbs managed to interest investors in the voyages of William Moor and Francis Smith in 1747.

MAP 76.
Joseph La France's 1742 map of North America, with details added from Middleton's voyage. The entrance to the Western Sea seems to be at *Rankin's Inlet*. La France combined what he knew of western geography and added those his benefactors wanted to see.

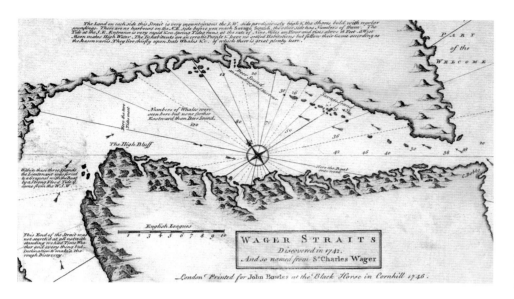

MAP 77 (above).
A detail map of Wager Bay as a strait published by John Wigate in 1746 as an inset on MAP 75. The label at the western end of the bay states *This End of the Strait was not search'd at all notwithstanding we had Time Weather and every thing but inclination to make a thorough discovery*, referring to Middleton's failure to search the inlet to its western end.

Moor and Smith returned to Britain in October 1747. Dobbs, naturally enough, was unhappy with their finding and spent the next while attempting to discredit them for their alleged "Timidity, ill Conduct or bad inclinations," maintaining still that the Hudson's Bay Company was behind it all. In 1749 Dobbs even managed to get a parliamentary inquiry into the company, for which a famously bizarre map was produced (MAP 79, *overleaf*). A motion to end the company's trade monopoly—Dobbs's real object—was defeated, and with it the search for a Northwest Passage effectively ended until the next century.

MAP 78 (below).
Henry Ellis's map, published in 1748. Ellis was Arthur Dobbs's agent on the voyage led by Moor and Smith. At the western end of Wager Bay is the notation *Thus ends Wagers Bay after the warmest Expectations of a Passage*. But Dobbs's hopes were still alive after seeing this map, for now the newly discovered Chesterfield's Inlet had not been explored to its western end and, on the map, conveniently disappears behind a cartouche. Clearly *this* was where the Northwest Passage must lie!

Today it seems amazing that an armchair geographer with no experience of the area he was talking about would even be given the time of day. But Dobbs was persuasive, of that there is no doubt at all, for in May 1745 a petition to Parliament presented on Dobbs's behalf resulted in an Act of Parliament authorizing a reward of £20,000 for the finder of a passage. Armed with this, the tenacious Dobbs was able to get enough subscribers to a private fund to finance another voyage.

Two ships were purchased, the *Dobbs Galley* and the optimistically named *California*. Middleton's second-in-command, William Moor, was appointed to lead the expedition, and Francis Smith, another Hudson's Bay Company captain, was also hired. Doubtless Dobbs would have preferred men not associated with the company, but there were none with the requisite Hudson Bay experience.

Moor and Smith sailed in May 1746 and wintered near the Nelson River, on the west side of the bay. In 1747 they sailed north but could not do even as well as Middleton. The only new discovery they made was another long inlet, which Smith named Bowden's Inlet, but Henry Ellis, Dobbs's agent with the expedition, named it Chesterfield Inlet after their principal investor, the Earl of Chesterfield, and this is the name that prevailed.

They did, however, explore to the western end of Wager Bay, where, as Henry Ellis wrote, they "had the Mortification to see clearly that our hitherto imagined Strait ended in two small unnavigable rivers."

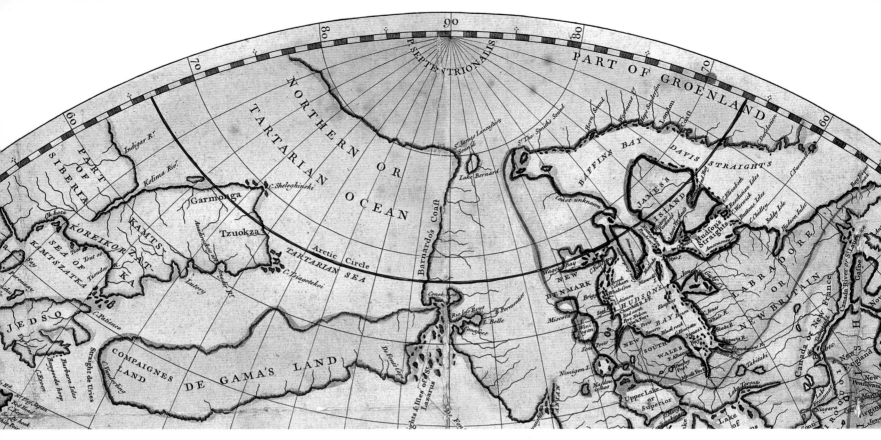

Map 79 (*above, left*).
This wonderfully speculative map, one of the the strangest maps of the western Arctic and northern Pacific ever drawn, was produced at the instigation of the Hudson's Bay Company for a parliamentary inquiry in 1749. Somehow this map was intended to prove that the company had probed deep into the continent and thus should be allowed to retain its trading monopoly. No one knows why such bizarre geography was shown, but it is known that the company had second thoughts and did not show the map to the committee. It is not even clear who drew the map, although it was engraved by a normally reputable London mapmaker, Richard Seale. Many maps must have been destroyed, for only ten copies are now known to exist, and most of those are in the company archives.

Map 80 (*left*).
The geography of the Hudson Bay region as it was believed to be after the voyages of Christopher Middleton in 1741–42 and William Moor and Francis Smith in 1746–47. The map was drawn by John Roque in 1762.

Map 81 (*above*).
Also vying for the "strangest map award" is this polar map by French geographer Philippe Buache, drawn in 1752. In 1747 Joseph-Nicolas De L'Isle returned to France from Russia, where he had spent twenty years at the Academy of Sciences in St. Petersburg. He smuggled out maps and manuscripts pertaining to Vitus Bering's voyage of 1741–42 and other maps of the Great Northern Expedition (see page 38). De L'Isle, together with his uncle, Philippe Buache, then perpetrated a number of maps on which they drew various interpretations of the new Russian discoveries in light of another "discovery" which had recently been revived, that of the apocryphal voyage of Bartholomew de Fonte in 1640. This mammoth hoax, nevertheless believed by many at the time, told of a voyage from the northern Pacific through channels that led to Hudson Bay or Baffin Bay—a Northwest Passage (see page 66). First published in 1708, the account was revived by De L'Isle and Buache to merge on their maps with the latest discoveries. No doubt they were hoping to make a name for themselves. To the north of eastern Siberia is an island named *Great Land 1723*, a reflection of the several Russian accounts of land in this region (see page 41). Greenland is connected to a huge Alaskan peninsula. A number of channels lead to Hudson Bay from the Western Sea, and one almost reaches Baffin Bay.

Cook Probes the Ice

Bering Strait, both the western exit from a Northwest Passage and an eastern exit from a Northeast, was discovered and recorded at least as early as 1648, by Semen Dezhnev (see page 36). Vitus Bering, after whom the strait is named, sailed into it in 1728 (see page 37), but it was not until 1778, when Britain's famous navigator James Cook explored it, during his third voyage, that the strait was finally mapped accurately and the true relationship of the continents of Asia and America revealed to the world.

Bering Strait was first referred to as the Strait of Anian—a name derived from the travels of Marco Polo—in an apocryphal story told by English seaman Thomas Cowles in 1579 of a Portuguese captain named Martin Chake in 1555. It was first shown and named on a map by Giacomo Gastaldi in 1562, but whether the idea of a strait in this location was derived from facts or simply wishful thinking, no one knows. Like the early depictions of what could be Hudson Bay on a map (MAP 19, page 12), the Strait of Anian/Bering could have been a lucky guess or based on real information now lost.

James Cook was by the time of his return from his second voyage a navigator of considerable repute. On that voyage he had demolished once and for all the idea that there was a huge Southern Continent. He was the obvious choice to command a new naval expedition in 1776 to search for an entrance to the Northwest Passage from the western end.

The Admiralty even planned to have ships waiting at the eastern end to assist Cook when he emerged. Richard Pickersgill, who had been with Cook on his first two voyages, was dispatched in 1776 to Davis Strait to carry out preliminary explorations. He discovered little due to heavy ice, and was court-martialled on his return for alleged drunkenness and discharged from the navy; the task of waiting for Cook in 1777 was assigned to Walter Young. In an almost useless voyage, Young turned back for England by the end of June, again due to heavy ice conditions. The Admiralty had not yet realized the virtue in assigning Greenland whalemen as ice pilots; easing a ship through the ice was a perilous and tricky task. How then was Cook to pass right through such a passage?

The Admiralty by this time knew that there could be no passage south of about 70° N, because it had the reports of Samuel

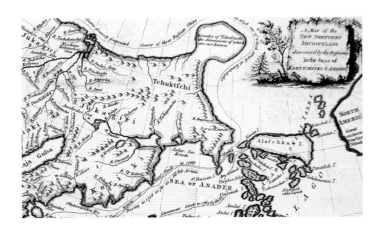

MAP 83 (*right*).
Part of Jacob von Stählin's 1774 *Map of the New Northern Archipelago*, ostensibly showing new Russian information. This map misled Cook and caused him to waste much time in 1778 before concluding that the map was false. Alaska is shown as an island.

MAP 84 (*below*).
Track of *Discovery* in 1778 in Bering Strait as plotted by William Bayley, Cook's astronomer.

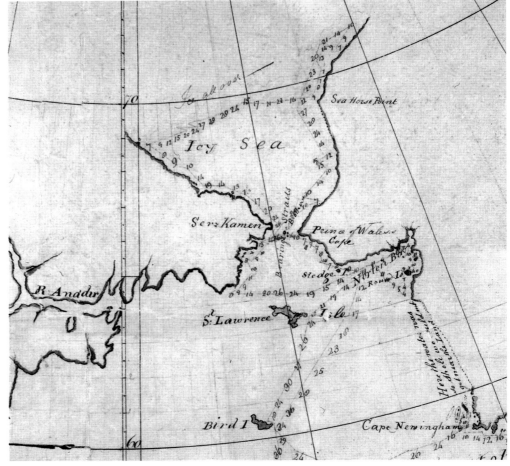

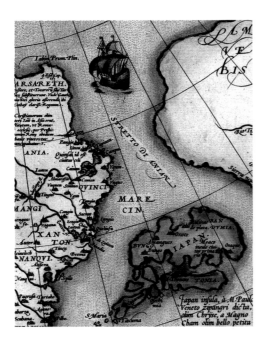

MAP 82.
A *Stretto di Anian* (Strait of Anian) is colorfully shown on this 1570 map by Abraham Ortelius.

Hearne's overland trek (*see next page*). But by extrapolating the latitude where Constantine John Phipps had reached the ice pack in 1773 (page 55), it was thought that between the sea Hearne had discovered at about 70° N and the pack that stopped Phipps at about 80° N, there should be an open sea. It was not realized at the time that climatic conditions vary depending on many factors, not least the influence of the Gulf Stream in Phipps's case and the continental influence in Hearne's. However fallacious the reasoning, Cook's ships were not strengthened to push through ice.

Cook left England in 1776 with two ships, HMS *Resolution* and *Discovery*, the latter commanded by Charles Clerke. Delays—during which he discovered the Hawaiian Islands—meant that he did not reach Bering Strait until August 1778. His instructions were to begin looking for an eastward-leading strait north of 65° N, partly because of the knowledge of Hearne's explorations and partly because of Cook's reliance on a map published in 1773 by Jacob von Stählin purporting to show the results of the voyage of a Russian naval officer, Ivan Sindt (MAP 83). Cook wasted valuable time trying to find the way through the Aleutians using this map, which, incidentally, showed Alaska as an island. Finally Cook realized that the map was wrong. "What could induce him to publish so erroneous a Map?" wrote Cook. "A Map that the most illiterate of his Sea-faring men would have been ashamed to put his name to."

On 17 August, sailing northeast (to Baffin Bay), Cook met the ice. He was near a promontory he named Icy Cape. It would be his "farthest north." He turned west, probing the pack as he went, but with an unstrengthened ship, he could do nothing. He turned south, to winter in his new-found Hawaii, where, the following February, he met his death at the hands of natives.

Charles Clerke assumed command, and in the following season he made many attempts to find a way through the ice, to no avail. It was five weeks earlier than in the previous year yet the ice formed an impassable barrier—to the unstrengthened ships—at about 67° N. Clerke, who was ill, wrote in his last instructions that "the amazing mass of ice [is] an insurmountable barrier... a passage, I fear, is totally out of the question." He died at sea and was buried in Kamchatka; John Gore, the first lieutenant on *Resolution*, returned the expedition to Britain. No passage had been found and the famous Cook was dead. All Britain was saddened.

But Cook had established the general trend of the Northwest Coast of America—though proving that there were no western entrances to a Northwest Passage lurking under the fog would have to await the survey of one of his midshipmen, George Vancouver, in 1792–94. And Cook for the first time mapped Alaska accurately as far north as his Icy Cape, setting the stage for later probes farther into the Arctic Ocean from the west (see page 75).

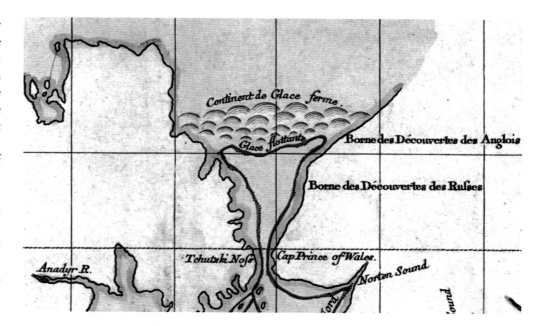

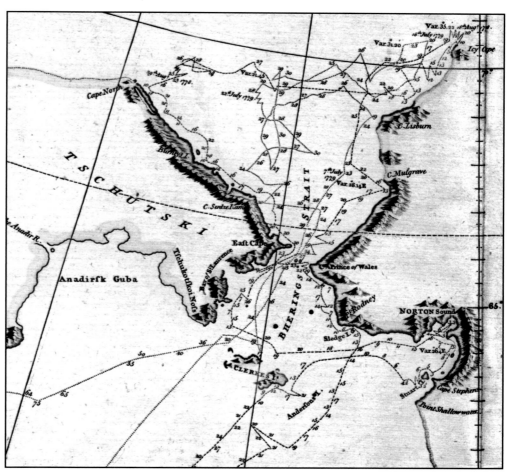

MAP 85 (*above top*).
Part of a "scooped" map of Cook's third voyage drawn by German geographer Conrad Lotter in 1781. Lotter had used information from a member of Cook's crew. Two books by crew members were published before the official book in 1784. Lotter graphically depicts the barrier of ice that blocked farther northward passage.

MAP 86 (*above*).
Part of the "official" map from Cook's book, published posthumously in 1784. This shows the tracks of the ships in 1779, under the command of Charles Clerke, as well as those of 1778 under Cook.

The Arctic by Land

Moses Norton, the governor of the Hudson's Bay Company's Prince of Wales Fort at Churchill River, had for many years collected from native sources information about the region to the north. In 1768 he proposed what he called "an inland journey," to promote an extension of trade "as well as for the discovery of a North West Passage, Copper Mines, &c." This tall order was assigned to a promising young employee, Samuel Hearne.

After false starts in 1769 and 1770, Hearne set off in December 1770 on an unprecedented trek northwest, accompanied by Matonabbee, a Chipewyan native chief, and a large entourage of his wives and followers. Hearne would never have survived, let alone found his way to the shores of the Arctic Ocean, without them.

After enduring many hardships and witnessing a massacre of Inuit people by the Chipewyans, Hearne reached the sea on 17 July 1771. He was certain it was the sea because of the whalebones and sealskins left by fleeing Inuit, and indeed it was. But Hearne miscalculated his latitude, thinking it 71° 54´ N when it was in fact 67° 49´ N; he was out by 240 km. This error would lead to disputes about whether or not he had reached the Arctic Ocean, especially from the East India Company hydrographer, Alexander Dalrymple, whose map (MAP 90) showed the Arctic shore at the correct latitude.

But Hearne had roughly fixed the position of the southern shore of the Arctic seas and the northern coast of North America. He was the first European to reach the Arctic Ocean. His trek established that there could be no Northwest Passage below 71° 54´ N—or 67° 49´ N—and that any such passage would have to lie in latitudes likely to lessen its utility due to the cold.

The Hudson's Bay Company, with its usual secrecy, did not publish the results of Hearne's expedition, and his book was published in 1795, after his death.

In the 1780s the company was beginning to receive some serious competition in the fur trade from a group of traders who formed

MAP 88 (*right*).
Samuel Hearne's map of the Coppermine River at the point where he reached the shores of Coronation Gulf on 17 July 1771. He has marked his incorrectly calculated latitude, 71° 54´ N. Also noted is the *Fall of 16 Feet* [this] *is where the Northern Ind*[ia]*ns killd the Eskamaux*.

MAP 87 (*above*).
This map published by British mapmaker Aaron Arrowsmith, dated 1 January 1795, was immediately outdated, for Samuel Hearne's book and map of his trek to the shores of the Polar Sea was published that year. Arrowsmith solved this problem by pasting a patch containing the new information onto his map. There is absolutely nothing under the patch. In fact, it seems that Arrowsmith, usually privy to Hudson's Bay Company information, missed the boat entirely with his mapping of Hearne's geography, for an edition of this map "with additions to 1796" still does not contain the Hearne information, although it does show Alexander Mackenzie's Arctic expedition of 1789, as does this map.

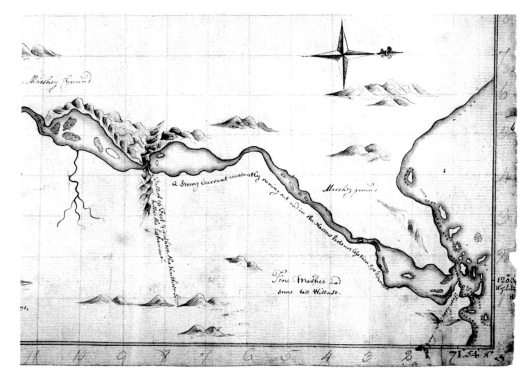

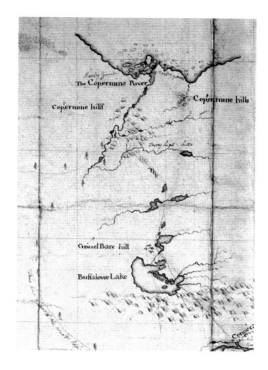

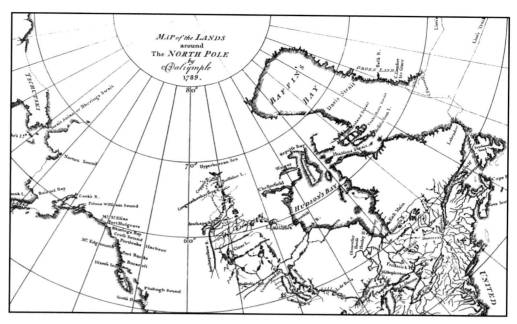

MAP 89 (*above*).
Part of Samuel Hearne's map showing the route he took via the Coppermine River to the Arctic Ocean. It was drawn in 1771 following his return from his trek.

MAP 90 (*above, right*).
East India Company hydrographer Alexander Dalrymple drew this map of the polar regions in 1789, placing the mouth of the Coppermine River at more or less its correct latitude. Whether it was at this latitude or where Hearne had placed it, the possibility of a Northwest Passage anywhere to the south of it was effectively removed.

MAP 91 (*below*).
One of North West Company trader Peter Pond's maps, first drawn in 1787. Like all of Pond's surviving maps, this is a copy, the original having been lost. This one is a copy of a map that Alexander Mackenzie took with him in 1789, intending to present it to Catherine, the Empress of Russia. Of course, he got nowhere near Russia, ending up on an island in the Mackenzie Delta. Pond's information was derived from native accounts, which were in essence correct, but which perhaps lost a little in the translation. The route that Mackenzie presumed to take, west from Great Slave Lake to Cook Inlet (off the map) or Prince William Sound (here Sandwich Sound) is indicated, and the route he did take, greatly foreshortened, is also shown. At the top of the map is the Arctic Ocean: *Here according to the account of the Natives the water ebbs and flows and they know of no Land further to the Northward*. Samuel Hearne's route via the Coppermine River seems to be shown at right, marked (partly off this portion of the map) *Mr Haring's rout to the North Sea*. The notation about *Ochipawayans* who *kill'd a number of Eskimaux* must be a misplaced and misdated reference to Hearne's expedition derived from native report, although we know from a later map that Pond found out that the expedition was Hearne's. For all its shortcomings, this pioneer map was a valiant first attempt to map native knowledge of the Northwest, and the geographical ideas it embodied were certainly the motivation for Mackenzie's efforts to find the Pacific in 1789, a venture that would not prove successful until another attempt was made in 1793, when Mackenzie became the first to cross the North American continent north of the Gulf of Mexico.

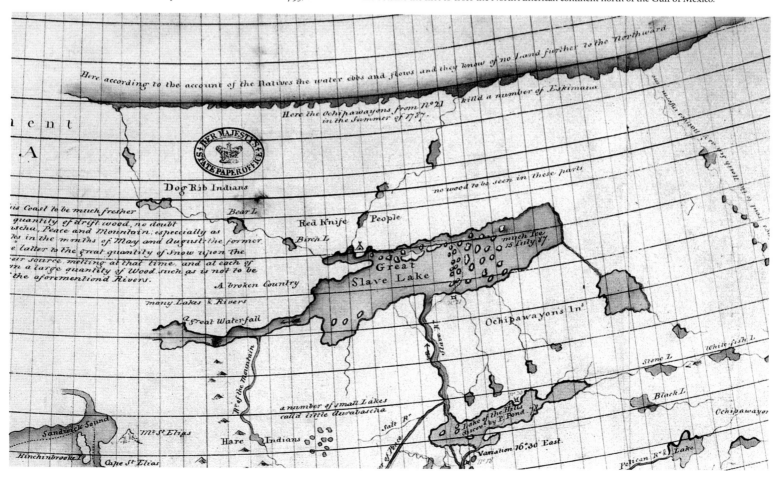

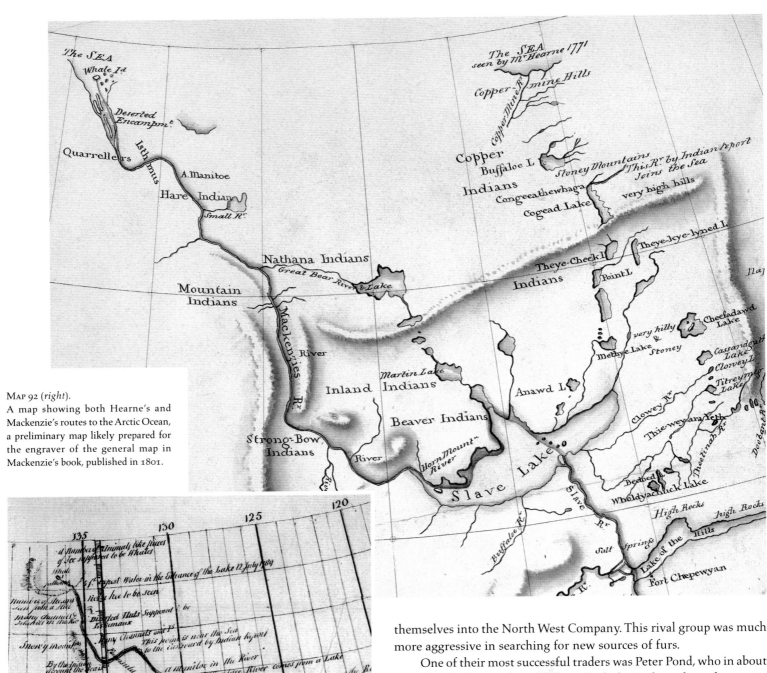

MAP 92 (right).
A map showing both Hearne's and Mackenzie's routes to the Arctic Ocean, a preliminary map likely prepared for the engraver of the general map in Mackenzie's book, published in 1801.

themselves into the North West Company. This rival group was much more aggressive in searching for new sources of furs.

One of their most successful traders was Peter Pond, who in about 1780 found the Athabasca River, which drained north to the Arctic. Pond produced a series of maps, one of which is shown here (MAP 91), in which he showed what he thought was the route to the Pacific Ocean, via Cook's River (Cook Inlet), found by James Cook in Alaska. In 1788 he showed this map to an ambitious young trader, Alexander Mackenzie, who was so taken with Pond's ideas that he set off in 1789 to find the route to the Pacific. But Pond had not got his geography quite right, and Mackenzie ended up instead at the shores of the Arctic Ocean, canoeing down the mighty river that now bears his name. In July 1789 Mackenzie became the second European to reach the Arctic Ocean. His map (MAP 93, *left*) now allowed a second point of its southern shore to be defined. Both Hearne's and Mackenzie's river mouths were part of a real Northwest Passage, but were for now but points on what would prove to be a long and circuitous line.

MAP 93 (*left*).
A copy of Mackenzie's 1789 field map showing his route down his eponymous river to the ocean. Mackenzie seems not to have initially realized that he had reached an ocean, for the notation dated 12 July refers to the *Entrance of the Lake*.

To Sail over the Top of the Earth

Sailing directly over the Pole was the third option of an easy shortcut to the East. It was proposed as early as 1527 by Robert Thorne, an English merchant. Henry Hudson had tried it in 1607 (see page 26) to no avail. The idea was that the continuous daylight in the summer months would prevent the formation of any impenetrable barrier of ice. And it was inconceivable to many right up until the 1860s that a whole ocean, constantly in motion, could actually freeze. Peter Plancius, the Dutch mapmaker, wrote influential papers in the early seventeenth century promoting these ideas. In the next century the Russian scientist, poet, and polymath Mikhail Lomonosov advanced the idea that an ice pack could only form if it were attached to land, and that if one sailed far enough north an open-water passage would be found. In 1763 Lomonosov drew the polar map shown here (Map 94), with the northern passages indicated, somewhat ambiguously, by dotted lines.

In 1759 and 1760 Lomonosov wrote papers in which he proposed a program of Arctic exploration and scientific research from a base to be established in northern Spitsbergen, with an attempt to be made to sail over the Pole to Bering Strait. Because of his pioneering scientific research in the Arctic Ocean, the major undersea ridge in the Arctic Basin, discovered by Russian scientists in 1948, was named after him the Lomonosov Ridge (khrebet Lomonosova; see page 190).

In 1764 six ships under Mikhail Nemtinov sailed for Spitsbergen to set up the base, and in 1765 Vasiliy Chichagov was instructed to sail from there to Bering Strait. He reached 80° 26´ N before being convinced that there was no way through the ice. The next year he tried again, with, predictably, the same result, attaining 80° 28´ N, about the same latitude that was by now regularly reached by whaling ships hunting for whales at the edge of the ice pack.

The theories of an open polar sea were expounded again by famous French explorer Louis-Antoine de Bougainville in 1772 and by Swiss polymath Samuel Engel in a book the same year. Engel thought that ice could only form from fresh water, and if a ship kept away from land, with its riverine sources of fresh water, the sea would be open—"une mer vaste et libre" in the far north. Bougainville presented detailed plans for a polar voyage to his government, but they were rejected. His proposal became known to the Royal Society in Britain, where its new vice-president, Daines Barrington, championed the cause. Like Lomonosov before him, Barrington gathered information on all claimed voyages to high latitudes. In January 1773 Barrington proposed to Britain's Royal Society that an expedition be mounted to search for the passage across the Pole he was convinced must exist. After all, Barrington wrote, in a treatise first published in 1775, "if the ice extends from N. Lat. 80½ to the Pole, all the intermediate space is denied to Spitsbergen whales." This argument, coupled with the others, convinced the British Admiralty that sailing over the Pole to the East just might be possible.

An expedition was organized "to try how far navigation was practicable towards the North Pole." Constantine John Phipps was appointed to command two ships, HMS *Racehorse* and *Carcass*, which were strengthened bomb ships (sturdy ships used for bombardment). The expedition was outfitted very carefully, with lots of relatively warm clothing for the men and stockpiles of food well beyond that normally provided on naval vessels at the time.

Map 94 (*above*).
Mikhail Lomonosov's polar map, drawn in 1763. Possible high-latitude passages to Bering Strait are indicated. The land shown to the north of eastern Siberia is labeled "doubtful."

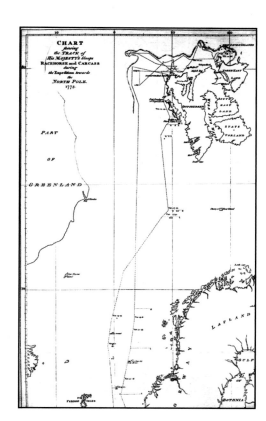

Map 95.
The route of Constantine John Phipps's 1773 expedition "towards the North Pole," from his book, published in 1774.

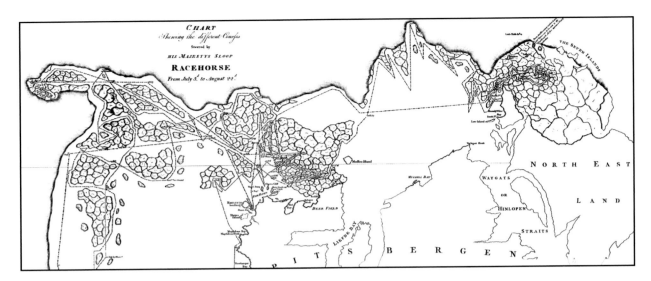

MAP 96.
The tracks of *Racehorse* and *Carcass* at the edge of the ice pack to the north of Spitsbergen in 1773, from Phipps's book. The most prominent feature on the map is the edge of the pack. At the eastern end is a small island, with the notation *From this Mountain neither land nor Water was to be seen to the Northward, the Ice appeared Flat & unbroken.* This was Phipps Island (now Phippsøya), where some crew members from the *Carcass* climbed a hill to try to see a passage through the ice.

The voyage is thought by many to be the first purely geographical and scientific mission to the Arctic. Instructions were received from John Hunter, considered to be the founder of modern surgery, and the famous naturalist Joseph Banks.

The ships sailed for Spitsbergen in June 1773, reaching it at the end of the month. All was going well. On 5 July Phipps was able to write to Banks: "I am in 79° 30´ & have not seen a single bit of ice." But this was to change, for the same day the pack was met.

Now the two ships probed and shoved, trying to find a way through the ice, managing to reach 80° 36´ N, thus beating the previous record—that of Chichagov in 1766—although this was unknown to them at the time. But that was as far north as they were going to get. Open sea or not beyond the vicinity of land, the ships could not get far enough away from Spitsbergen to test Engel's theory. Six weeks were spent in the loose ice in front of the pack, often anchored to floes, making scientific observations and trying to find a clear passage north. But none existed.

It was during this period that one of the midshipmen, a fourteen-year-old named Horatio Nelson, had a famous encounter with a polar bear, trying to kill it with the butt of his gun after it misfired and only being saved by the firing of a gun from the ship to scare the bear away. The incident would doubtless hardly be remembered were it not for Nelson's subsequent illustrious career, which was to end as the victor of Trafalgar in 1805.

From a hill on a small island north of Spitsbergen, Phippsøya, they could see far to the north—and could see nothing but unbroken ice (MAP 96, *above*). They narrowly escaped being trapped in the ice on 6 August. Phipps ordered the boats out on the ice when they were saved by a change in the direction of the wind, which blew the ice away from the ships. With the season now being "so very far advanced," it was time to leave.

Phipps's voyage was bound to fail in its objective and for this reason has been generally overlooked in the annals of exploration history, but it was an important first step in scientific exploration of the Arctic.

The idea of an open polar sea would not disappear, however. Samuel Engel found ways to incorporate Phipps's experience into his theories, and in 1818 the last British attempt to sail over the Pole to the East (as opposed to attempts to get to the Pole itself) was initiated. This was the voyage of David Buchan and John Franklin.

Conceived by John Barrow, second secretary to the Admiralty, and part of the two-pronged effort that also saw John Ross and Edward Parry try to find a Northwest Passage, sailing over the Pole was thought possible again because of recent reports from whalers of clear seas north of Spitsbergen (see page 58).

Buchan and Franklin reached 80° 34´ N, not even as far north as Phipps, before being stopped by the pack. Good year or not, there was still no passage over the Pole. The ice had beaten the theories; in future polar attempts, the British navy would use boats and sledges, not ships (see page 124).

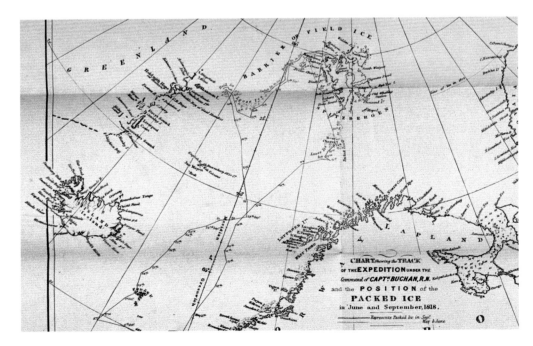

MAP 97.
The track of David Buchan in the *Dorothea* and John Franklin in the *Trent* in 1818 on their attempt to find a passage over the Pole. The ships were first stopped near Spitsbergen and later off the eastern coast of Greenland. No map or account of this voyage was published at the time, and this map comes from a much later book, published in 1843, by Frederick Beechey, a lieutenant on the *Trent* at the time of the voyage.

MAP 98.
A map of the region north of North East Land (Nordaustlandet), the large island in the east part of the Svalbard group, and south of the Seven Islands (Sjuøyane) and the edge of the ice (shown at the right of MAP 96, *above left*). It was drawn and illustrated by Philip D'Auvergne, a midshipman on the *Carcass*. The tip of Nordaustlandet at bottom left is Nordkapp.

For Glory Not Gold

The British seemed unsurprised and undismayed by Phipps's failure to sail over the Pole, and in 1776 a £5,000 parliamentary reward was offered to any ship that sailed to within a degree of the North Pole. But this was a small sum compared to the £20,000 reward that was still offered from the 1745 Act for the finding of a passage west from Hudson Bay. This reward was now also offered to naval vessels—and for discovery of *any* passage north of 52° leading to the Pacific. But the rewards had little effect, largely because the cost of a private voyage would likely eat up most of the reward, and the risk was too high. It would take expanded rewards and the use of naval vessels to produce some results. But war with France would intervene.

A number of maps of the period doubted the existence of Baffin Bay (Map 100 and Map 103) while others showed it not as a bay at all but as an open-ended continuation of Davis Strait (Map 102 and Map 104). This was largely due to the loss of William Baffin's maps from 1616. Thus it seemed that Baffin Bay might offer an entrance to a Northwest Passage.

With the defeat of Napoleon in 1815, British interest again turned to the Arctic. The end of the war had left thousands of naval officers on half pay; what better way to employ at least a few of them than to find the Northwest Passage, if only for the glory of the British navy? This view was advanced by John Barrow, the second secretary to the Admiralty—the highest-ranked civil servant, in a key position to influence naval affairs.

And there were reports of Russian incursions. Russian traders had by now established

Map 99.
A map of Baffin Bay by British whaling captain William Scoresby Jr., published in 1815. Despite Scoresby's frequent visits to Davis Strait, Baffin Bay is still shown in a similar fashion to maps of seventy-five years before. The true entrance to the Northwest Passage, Lancaster Sound, is shown as *Sir James Lancaster's Sound*, found and named by Robert Bylot and William Baffin in 1616. What Scoresby considered *The most probable Communication with the Sea seen by Hearne in 1772* [1771] *and by McKenzie* [Mackenzie] *in the year 1780* [1789] is shown to the south of Lancaster Sound.

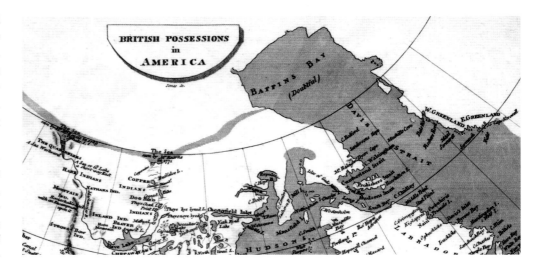

Map 100 (*above*).
A map published in 1804 by American mapmaker Samuel Lewis in cooperation with Britain's Aaron Arrowsmith. It shows conjecture as to how Baffin Bay might be connected to the seas seen by Samuel Hearne in 1771 and Alexander Mackenzie in 1789.

Map 101 (*below*).
This German map published in 1805 shows a speculative peninsula which would have been a problem had it really existed.

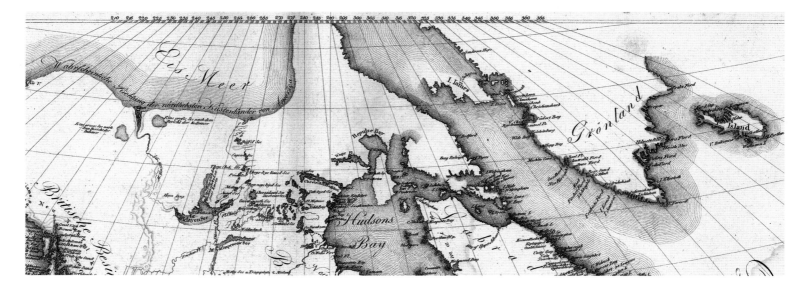

themselves in Alaska, and in 1816 Otto von Kotzebue, sent to search for a Northwest Passage, found Kotzebue Sound, on the American shore of Bering Strait.

Barrow was prepared to concede that the Northeast Passage was now the preserve of the Russians, but that was all, for he perceived the Northwest Passage to be what he called "a peculiarly British object." He planned a double probe, of the Northwest Passage and a route over the Pole. He defined the objectives in a letter to the prime minister: "To endeavour to correct and amend the very defective geography of the Arctic Regions, especially on the side of America. To attempt the circumnavigation of old Greenland, if an island or islands as there is reason to suppose. To prove the existence or non-existence of Baffin's Bay; and to endeavour the practicability of a Passage to the Pacific Ocean, along the Northern Coast of America."

Reports of ice-free seas encouraged the plan. Veteran whaling captain William Scoresby Jr. wrote to Joseph Banks in 1817 that he had found on his last voyage "about 2,000 square leagues [about 46 000 km²] of the surface of the Greenland Sea, between the parallels of 74° and 80° N, perfectly void of ice which is usually covered with it." Scoresby met with John Barrow, proposing himself to lead an expedition, but Barrow would have none of it. Only naval officers would do.

Joseph Banks, on behalf of the Royal Society, lobbied the Admiralty, and Barrow came up with his two-pronged plan. David Buchan and John Franklin were to sail northwards (see page 56), and at the same time John Ross and Edward Parry, in HMS *Isabella* and *Alexander*, were to sail into Baffin Bay, find any channels leading west, and sail through the Northwest Passage to the Pacific, where they were to meet Buchan and Franklin. Of course, it did not quite work out that way.

Parry met with Sir Joseph Banks in December 1817 and heard his ideas about the state of the ice around eastern Greenland. Banks theorized that recent decreases in the summer temperatures of the temperate zones—backed up by accounts from Boston of corn not having ripened—were due to the breakup of polar ice, resulting in icebergs floating southwards.

In 1818, before the four ships left, by way of encouragement the £5,000 reward was expanded to include incremental amounts for sailing to 83°, 85°, 87°, and 88° N, and £15,000 for sailing to 150°, £10,000 for 130°, and £5,000 for 110° W.

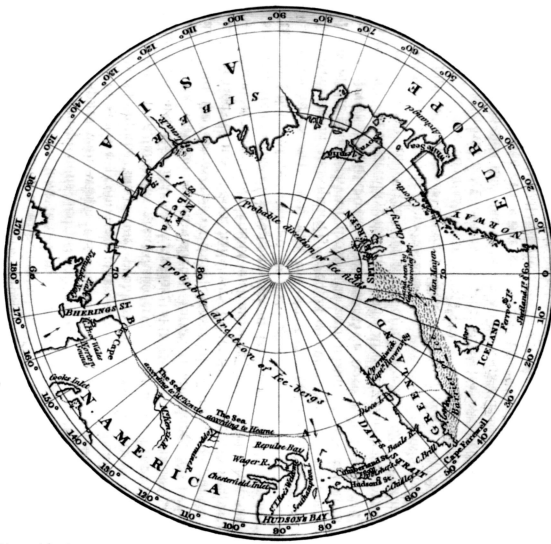

MAP 102 (*above*).
John Barrow's 1817 anonymously published polar map, showing his theory that Baffin Bay was open at its northern end, allowing a stream of icebergs from the Bering Strait region to flow across a presumed open polar sea, exiting southwards (as indeed they do) through Davis Strait. The map has been rotated 90° clockwise from its original orientation.

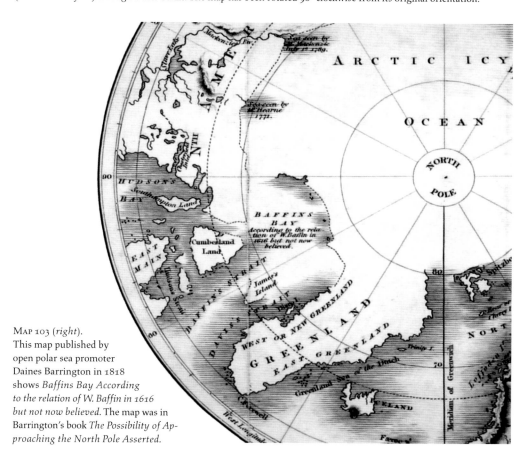

MAP 103 (*right*).
This map published by open polar sea promoter Daines Barrington in 1818 shows *Baffins Bay According to the relation of W. Baffin in 1616 but not now believed*. The map was in Barrington's book *The Possibility of Approaching the North Pole Asserted*.

MAP 104.
John Barrow published an account of Arctic exploration while the four ships were away in 1818. This is the map from that book. As in MAP 101, Baffin Bay is shown open ended. There is a considerable contrast in geographical knowledge between the northern coast of Asia and that of North America. This map was periodically updated to add information as the geographical puzzle was solved (see MAP 136, page 81, for an 1846 map, and MAP 217, page 139, for one dated 1879) and it is interesting to compare them.

Ross's instructions were very similar to those given to Walter Young in 1777 (see page 50)—survey Baffin Bay, find a westward opening, and sail through it to the Pacific—but Ross would at least survey Baffin Bay, which Young had failed to do while waiting for James Cook. Ross's voyage is chiefly remembered, however, for what he failed to do rather than what he did.

Ross and Parry sailed in April 1818 and followed the west coast of Greenland north, where, at 75° 55´ N, in Melville Bay (named by Ross, now Melville Bugt), he encountered native people he called "Arctic Highlanders." They were an isolated Inuit group that had never before seen Europeans. With the help of their interpreter, John Sackheuse (an Inuit who had stowed away on a whaler sailing for England in 1816), the natives asked questions: "Do they [the ships] come from the sun or the moon," for "they are alive, we have seen them move their wings."

Continuing north to about 78°, Ross could not enter Smith Sound because of ice, but he did name the capes on either side of it after his two ships. He then sailed south along the Ellesmere Island shore, finding the entrance to Jones Sound, but again it was blocked by ice. Continuing south he came to Lancaster Sound, the true eastern entrance to the Northwest Passage. But it was here he made one of the more famous errors in the history of exploration, for, sailing west, he decided that it was just another inlet, with its head blocked by a range of mountains he named the Croker Mountains

This view of the western "end" of Lancaster Sound, complete with mountains, was published with MAP 106 by Ross in his book to show what he thought he saw. The Admiralty was not convinced.

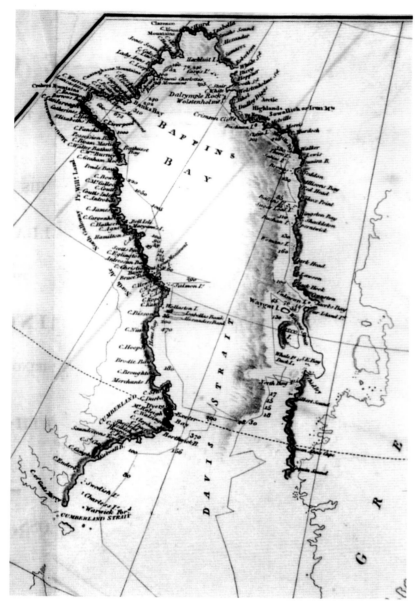

after the first secretary to the Admiralty, John Croker. His insistence, even against the views of most of his officers that there was no passage here, was later to lead to ridicule. His map shows clearly his perception of the sound, depicting it as a bay (MAP 106, *below, left*).

Alexander Fisher, surgeon on the *Alexander*, published an anonymous book on his return to England in which he described what happened. "The breadth of the inlet was estimated to be from ten to twelve leagues [48 to 57 km] . . . alas! the sanguine hopes and high expectations excited by this promising appearance of things were of a short duration, for, about three o'clock in the afternoon, the *Isabella* tacked, very much to our surprise indeed, as we could not see any thing like land at the bottom of the inlet."

We can only assume that Ross was deceived by a wonderful mirage, for the range of mountains he saw blocking the sound appears in his book along with his map of the "bay."

The ships proceeded southwards, and a good part of the east coast of Baffin Island was surveyed, but neither Cumberland Sound nor Frobisher Bay was mapped.

On his return Ross sent a letter to the Admiralty from Lerwick reporting their safe arrival, "after fully ascertaining that no passage exists from the Atlantic to the Pacific Oceans through Davis' Strait and Baffins Bay which were found to be bounded by land extending to the North as far as the 84th Degree, in Latitude 74° N., and generally surrounded by a Chain of Mountains."

MAP 105 (*above, left*).
John Ross's map of his survey of Baffin Bay from his 1819 book. It shows as bays all three of the major waterways leading from Baffin Bay. Lancaster Sound is at top left (but not labeled), and *Crokers Mountains* are marked.

MAP 106 (*below, left*).
Ross's map of the eastern entrance to Lancaster Sound, shown as a bay, with the path westward blocked by *Crokers Mountains*. Here *Lancasters Sound of Baffin* has been relegated to a small bay.

MAP 107 (*below*).
The map published anonymously by Alexander Fisher, surgeon on the *Alexander*, reflected the views of the majority of officers that Lancaster Sound was just that, a sound, and not a bay as Ross claimed. It could be dangerous to one's career to go against your superior officer's opinion, and this was why Fisher's book was published without his name being on it. The Admiralty soon came to the same conclusion, however, and dispatched Edward Parry the following year to confirm it.

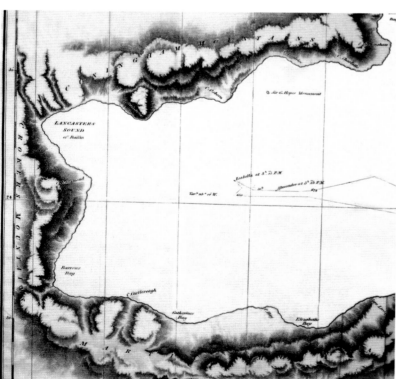

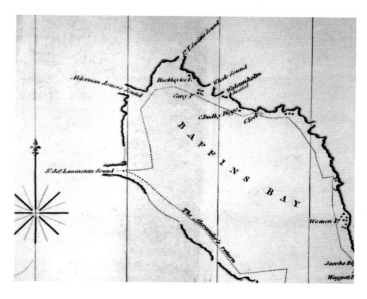

61

Ross's senior officer, Edward Parry, was one of those who disagreed with his assessment of Lancaster Sound, and he, without Ross, would the next year be assigned to investigate the sound further.

Parry had written of Lancaster Sound that "the swell of the sea, the breath of the opening, and the depth of the water, are all flattering appearances," supporting the view that it was a strait, not a bay. The Admiralty agreed, and thus began a long series of expeditions that would, over the next forty years, transform the map of the Canadian Arctic.

Parry was given command of HMS *Hecla*, and Matthew Liddon was appointed captain of HMS *Griper*.

There were to be no more surveys of Baffin Bay. Parry was instructed to

proceed in the first instance to that part of the coast [Lancaster Sound] *and use your best endeavours to explore the bottom of that Sound; or, in the event of its proving a strait opening to the westward, you are to use all possible means, consistently with the safety of the two ships, to pass through it, and ascertain its direction and communications; and if it should be found to connect itself with the northern sea, you are to make the best of your way to Behring's Strait.*

Stores and provisions for two years were loaded on board. They included "a large supply of fresh meats and soups, preserved in tin cases, by Messrs Donkin and Gamble, of Burkit's essence of malt and hops, and of the essence of spruce." The invention of canned food early in the nineteenth century (to feed Napoleon's armies) was to be a significant factor in facilitating overwintering in the Arctic.

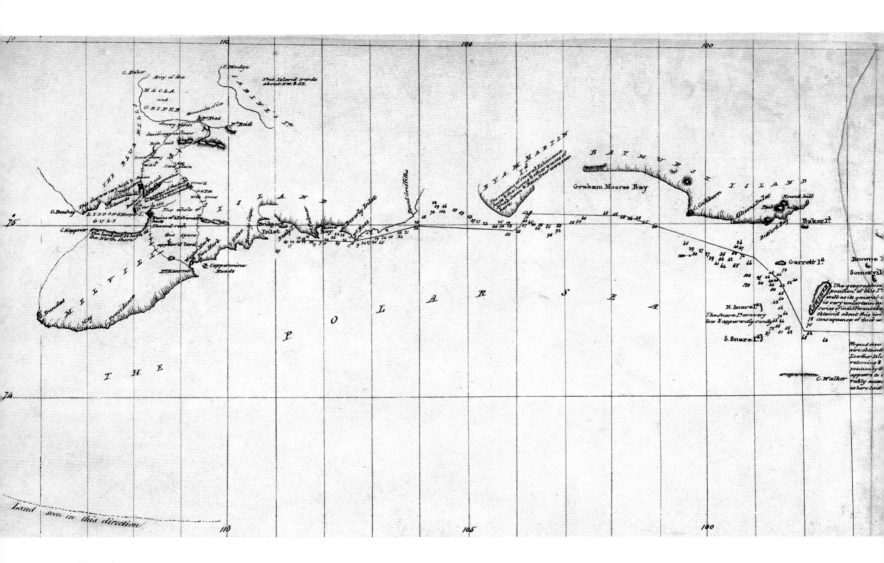

MAP 108.
Edward Parry's map of his discoveries in 1819 and 1820, drawn on two pieces of paper joined together. The track of the ships in 1819, but not in 1820, is shown. On the western part of Melville Island the overland expedition of Parry and Edward Sabine, his astronomer, in June 1820 is shown, while at bottom left is the land seen by Parry while at his farthest west, which he named Banks Land after Joseph Banks. Parry named geographical features for his superiors, sponsors, and friends. Devon Island was named for Matthew Liddon's home in England; Cornwallis Island after Admiral Sir William Cornwallis, under whom Parry had previously served; Bathurst Island after the colonial secretary, the earl of the same name; Byam Martin Island after Vice Admiral Sir Thomas Byam Martin, comptroller of the British navy; and Melville Island after Viscount Melville, first lord of the Admiralty. Originally in the files at Trinity House, Britain's authority for the regulation of shipping, this map is now in the British Library.

Parry and Liddon sailed in May 1819. Sailing through Ross's mountains, they entered territory unknown to them. Keeping to the north side of the channel, they sailed westwards in a channel Parry named—what else?—Barrow Strait. With the going becoming difficult, Parry saw "a broad and open channel"—Prince Regent Inlet—to the south. He followed this, thinking it might provide a clearer passage to the west. The ships penetrated about 100 km before being stopped by ice, whereupon they retraced their path back to Barrow Strait.

Parry found he could continue his path westwards, ice conditions having improved. (His track can be followed on MAP 108.) Now he encountered a problem that would afflict all travelers to these parts. He found his compasses next to useless and realized that the magnetic pole was now to the southwest. So useless were the compasses that he had the binnacles removed "as so much useless lumber." At times he resorted to unorthodox methods of navigation, and for some time steered the two ships on each other so as to maintain a straight path. *Hecla* had to keep the *Griper* dead astern, and *Griper* had to keep *Hecla* dead ahead. When the weather closed in, only the constancy of wind direction could be used. Parry discovered the coasts of a string of islands now collectively known as the Parry Islands—Cornwallis, Bathurst, Byam Martin and Melville Islands, all named by Parry. Off the latter island, on 4 September, the ships crossed the 110° W meridian and became entitled to the £5,000 prize offered by the British government. Parry, whose personal share was £1,000, named the nearest conspicuous promontory Cape Bounty. It is *Bounty Cape* on MAP 108.

With the advancing season, ice prevented the ships from progressing much farther west, and a harbor was found on the south coast of Melville Island. Even now a channel had to be cut in the ice to get the ships to safety. Parry named the harbor without much creativity—Winter Harbour.

So began the first deliberate overwintering by British naval ships. They were prepared. The ships were covered with a framework and canvas and banked with snow to insulate them as much as possible. The problem of keeping the men happy and occupied through the dark winter was solved by preparing theater presentations and writing a newspaper, called the *North Georgia Gazette*; by ordering physical exercises; and by running a school, all in addition to more normal naval duties. After what Parry called the "death-like stillness of the most dreary desolation" of the deep winter, hunting was organized once game returned; with impeccable fairness, Parry ordered that "every animal killed was to be considered as public property" and be evenly distributed to all the crew, including the officers.

In June, Parry, with his astronomer Edward Sabine and ten others, carried out a fifteen-day, 200 km exploration by land from Winter Harbour, visiting the north side of the island at Hecla and Griper Bay and the west at Liddon Gulf. The track of this foray is shown on MAP 108.

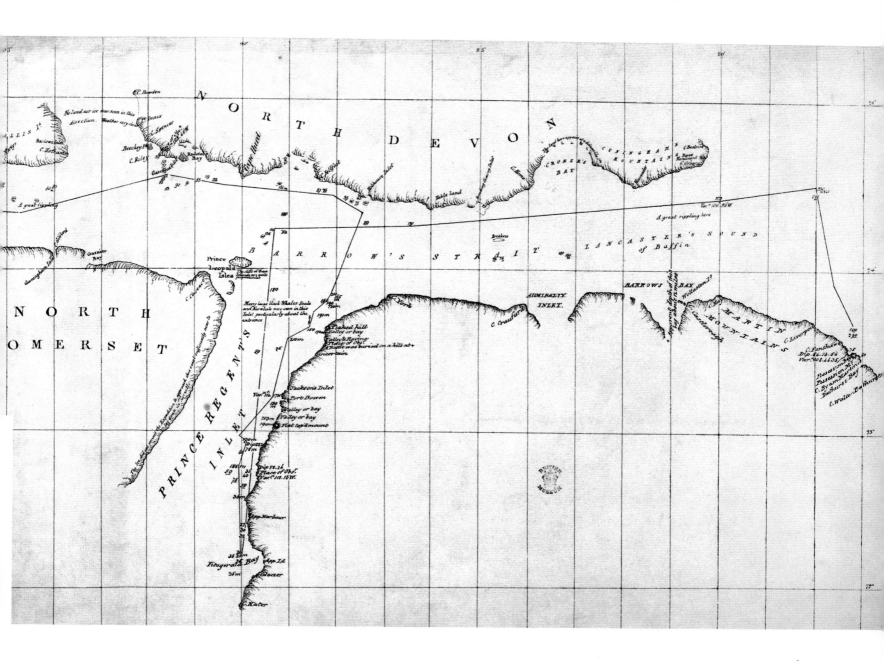

Finally, there was the work of preparing the ships to sail, but it was not until 1 August that the ships could escape from Winter Harbour's icy grip.

Parry now attempted to complete the voyage ordered in his instructions, so he sailed westwards, hoping to complete a transit of a Northwest Passage. But, unknown to him at the time, he was now at one of the locations in the Arctic where the multi-year ice pack presses in between the islands. Parry reached 113° 46′ 43.5″ W (as corrected; all Parry's positional fixes west of Prince Regent Inlet had later to be corrected by a few minutes to the east, due to faulty astronomical tables, not Parry's inaccuracy). From here, however, he could see another island to the southwest, which he named Bank's Land (Banks Island). It is shown as a tentative partial coastline on Map 108 and Map 109.

It now remained to escape their icy domain before the onset of the next winter. But Parry need not have worried; the distance from Winter Harbour to the entrance to Lancaster Sound was covered in only six days, as compared to five weeks in the other direction. They returned to England in November 1820 to widespread acclaim, having lost only one man out of ninety-four.

Parry had at a single stroke added a great swath of geographical knowledge to the map. Where once there was a blank, there were now islands and inlets, coastlines and straits. He correctly concluded that

of the existence of . . . a [Northwest] Passage, and that the outlet will be found at Behring's Strait, it is scarcely possible, on an inspection of the map, with the addition of our late discoveries, and in conjunction with those of Cook and Mackenzie, any longer to entertain a reasonable doubt.

At the same time as Parry began his voyage, John Franklin, newly returned from his northward voyage on the *Trent*, was given another assignment, which was not really a naval expedition at all, although it was sponsored by the Admiralty. Franklin was to travel overland to the mouth of Hearne's Coppermine River and explore the coast to the east. This, it was hoped, would link Hearne's dis-

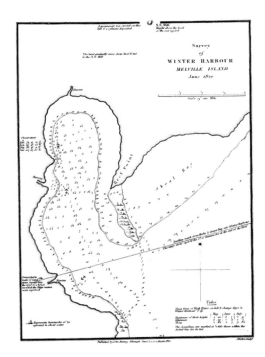

MAP 110 (*above*).
Parry's map of Winter Harbour, on the south coast of Melville Island, where *Hecla* and *Griper* wintered in 1819–20. Parry carried out the soundings of the harbor on the day the ships left, 1 August 1820, while equipment was being loaded onto the ships from the beach, where it had been stored. Parry and his men spent ten months trapped in this bay.

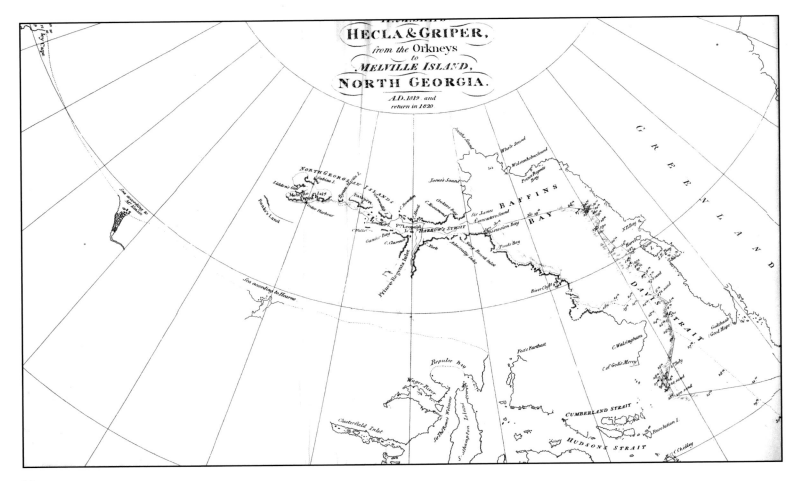

MAP 109.
The general map from Edward Parry's book about his voyage of 1819–20 shows the sudden increase in geographical knowledge where before there was only a blank on the map. Note Hearne's mouth of the Coppermine River, with its tentative coastline drawn to the coast north of Repulse Bay. The knowledge of the Hudson Bay region remains much as it was by the middle of the previous century.

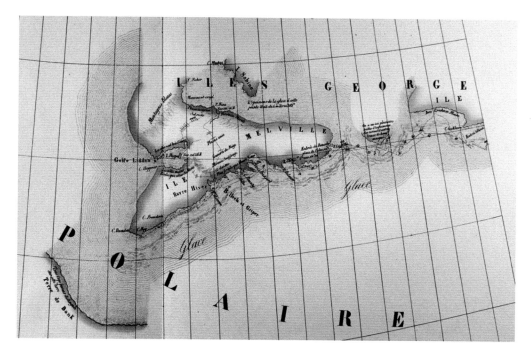

Franklin's canoes are tossed about in Coronation Gulf in this illustration by George Back, from his 1821 sketchbook.

The following June, Franklin traveled north, accompanied by a group of natives and twenty voyageurs. On 18 July 1821 he reached the mouth of the Coppermine River at the point Hearne had attained fifty years before. The natives left, and with the voyageurs, Franklin began his survey of the coast to the east in two canoes.

coveries with Repulse Bay and "amend the very defective geography of the northern part of North America." One detects the voice of John Barrow here.

Accompanying Franklin were John Richardson, a naval surgeon, and two midshipmen, George Back and Robert Hood. They sailed on a Hudson's Bay Company vessel in 1819 and in October reached Cumberland House, a fur trade post established by Samuel Hearne on the Saskatchewan River.

The next season was taken up in getting their supplies north, and a base for the winter of 1820–21 was established on Winter Lake, north of Great Slave Lake. This they called Fort Enterprise.

MAP 111 (*above*).
This map of the western part of Parry's discoveries appeared in an atlas by French mapmaker Phillippe Marie Guillaume van de Meulen in 1825. Parry's North Georgia Islands have become *Îles George*.

MAP 112 (*right*).
Franklin's map of his route from Great Slave Lake to the mouth of the Coppermine River. The "strip" style of the map makes it appear as if there is one river connecting the two, but this is an illusion; in reality the system of lakes and rivers in the region is very complex. Note *Fort Enterprise*, on *Winter Lake*, where Franklin's party wintered in 1820–21.

The elaborate cartouche from one of Franklin's maps, of part of his journey to the Coppermine River in 1820. This was drawn by Robert Hood at Fort Enterprise in March 1820.

MAP 113 (right).
Armchair geographers don't give up easily. When Parry's discoveries in 1819–20 became known (his book was published in 1821) they did not fit well with maps illustrating a number of apocryphal voyages through straits linking Hudson Bay or Baffin Bay with the Pacific. Pierre Lapie, one time geographer to the French king, drew this map to reconcile, in his mind, Parry's discoveries with these preconceived geographical notions. And he did quite a good job of it.

Parry's discoveries are shown as islands (which in fact they are) in the top center of the map. Banks Island, which shows on Parry's map (MAP 108, pages 62–63) as *Land seen in this direction* and as *Bank's Land* on his published map (MAP 109, page 64) has been made conveniently small so that the track of Captain Lorenzo Ferrer Maldonado, a voyage reported to have taken place in 1588, can be shown. This report first showed up as a memorial to the Spanish king in 1609. A copy of a drawing of the Straits of Anian is shown on Lapie's map in the bottom left corner. The track of Admiral Bartholomew de Fonte supposedly in 1640 (and first reported in 1708) is shown through a *Lac de Fonte*—Great Slave Lake—and a passage to Hudson Bay through *Ent[rée] de Chesterfield* (Chesterfield "Entrance"). North of Parry's island discoveries is the track of *Capitaine Bernarda*, reported to be one of De Fonte's captains, Pedro de Barnarda, whom he had sent to investigate regions farther north. Barnarda was said to have reported that Davis Strait was frozen solid. Both Barnarda's and Maldonado's western entrances to the enormous *Mer Polaire* depicted here are shown in somewhat convoluted form; clearly Lapie is not quite sure of his "facts." These pathways between the two oceans are also shown—after a fashion—on Philippe Buache's 1752 map (MAP 81, page 49), which is the polar version of another 1752 map of Asia and America that has the passages in greater detail.

Lapie's map also shows a journey of a Hudson's Bay Company man, Alexander Cluny, that supposedly took place in 1745, and which was published in a book called *The American Traveller . . . By an old and experienced Trader* in 1769. Though clearly serendipitous, the book annoyed the company, newly sensitized to any suggestion that it did not do enough exploring. Cluny's Northwest Passage crosses from Repulse Bay—Christopher Middleton must have missed it—to the northern *Mer Polaire*. Amazingly, this is very similar to the route across the neck of land between Repulse Bay and the Gulf of Boothia taken by later travelers, but overland (see, for example, MAP 176, page 110). Following the Buache model (MAP 81, page 49) Greenland is miraculously connected to, and becomes part of, northern Alaska.

Lapie no doubt spent many happy hours concocting this map, which lives on for our enjoyment.

MAP 114 (below).
This is a 1774 map showing the De Fonte geography, included here for comparison with Lapie's map.

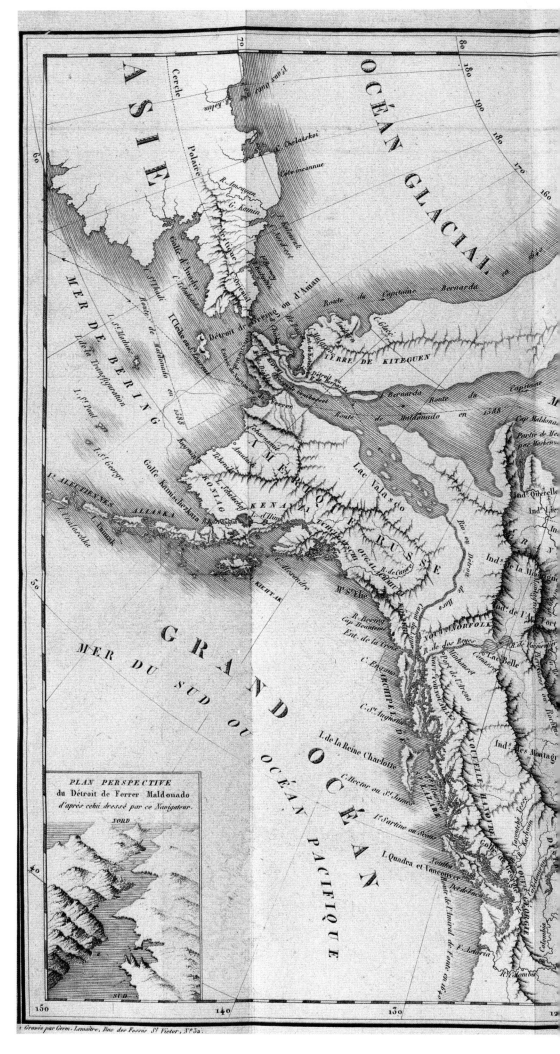

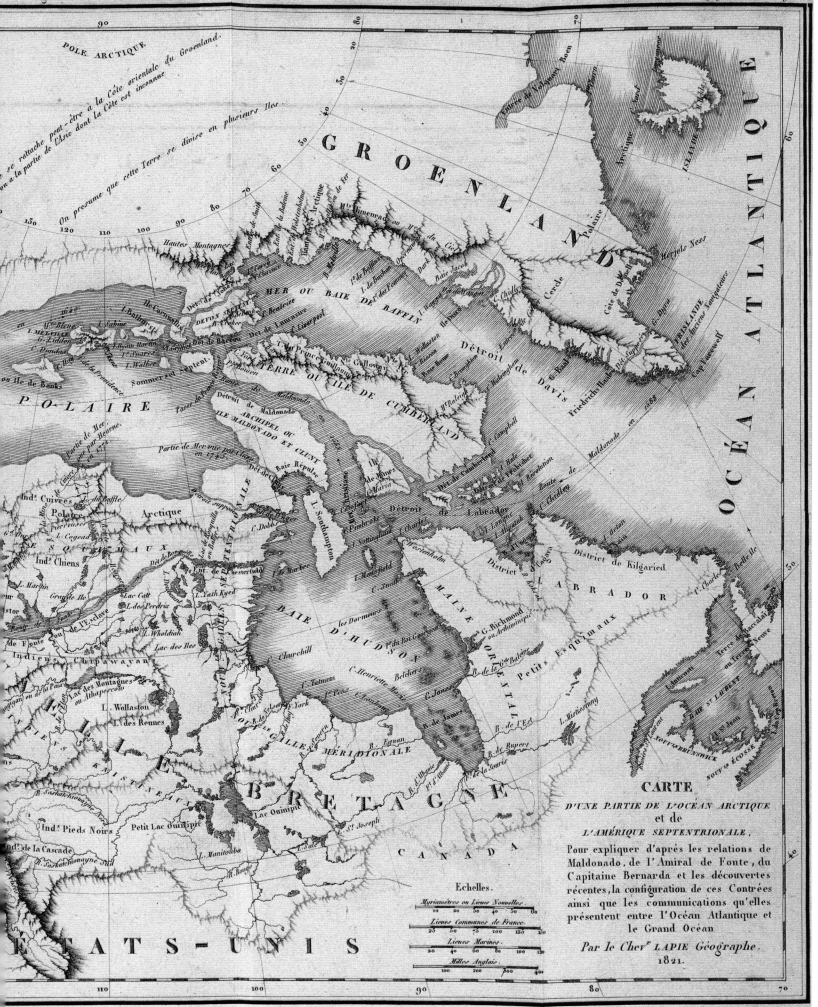

Franklin mapped the coast as far as the Kent Peninsula, to a place he named, appropriately enough, Point Turnagain.

But now Franklin began to run into trouble. The canoes were literally worn out, and so he turned inland, with the intetnion of reaching the Hudson's Bay Company's post of Fort Providence. An ongoing feud between the company and the rival North West Company had resulted in Franklin being short of supplies he had been promised, and food began to run out. Ten of his voyageurs were to die.

At this time one of the more bizarre incidents of Arctic exploration history occurred. When midshipman Robert Hood became ill he was murdered by a voyaguer named Michel Teroahauté. Richardson, in turn, thinking that they might all be next, and "thoroughly convinced of the necessity of such a dreadful act," killed Teroahauté.

On their return to Fort Enterprise, they found it bare. George Back went off to try to find the natives, who would be able to hunt for them. Reduced to eating anything remotely edible—including moccasins—they were saved only by the return of Back and the natives. Franklin finally made it to Fort Providence on 11 December 1821, and in the summer of 1822 he returned to England on the Hudson's Bay Company supply ship.

Franklin's book, published in 1824, was an immediate best seller, and its Gothic tales of hardship—Franklin became known as "the man who ate his shoes"—were lapped up by an admiring British public. Franklin achieved his objective, mapping over 300 km of the Arctic coastline of North America. MAP 115 (*below*) was drawn to show the way in which the discoveries of Ross, Parry, and Franklin expanded knowledge of the Arctic.

On his 1819–20 voyage, Parry had written that he considered a future expedition would "act with greater advantage, by at once employing its best energies in the attempt to

A view of Fort Enterprise, one of Robert Hood's cartouches.

penetrate from the eastern coast of America, along its northern shore." Parry had studied the accounts of Christopher Middleton and the other eighteenth-century voyages and had concluded that a Northwest Passage lay north of Repulse Bay. He was actually right; the southernmost possible Northwest Passage does indeed lie partly through Fury and Hecla Strait, which Parry was to find and explore, but unfortunately for him it is impassable to sailing vessels. Modern high-powered vessels strengthened against the ice can pass this way. Parry was simply ahead of his time.

MAP 115 (*below*).
This *Outline to Shew the Connected Discoveries of Captains Ross, Parry, and Franklin* covers the period of Ross's voyage to Baffin Bay in 1818, Parry's first voyage, to Melville Island, in 1819–20, and Franklin's first overland expedition in 1819–22. It thus represents the state of geographical knowledge by 1822. The map was published in 1823.

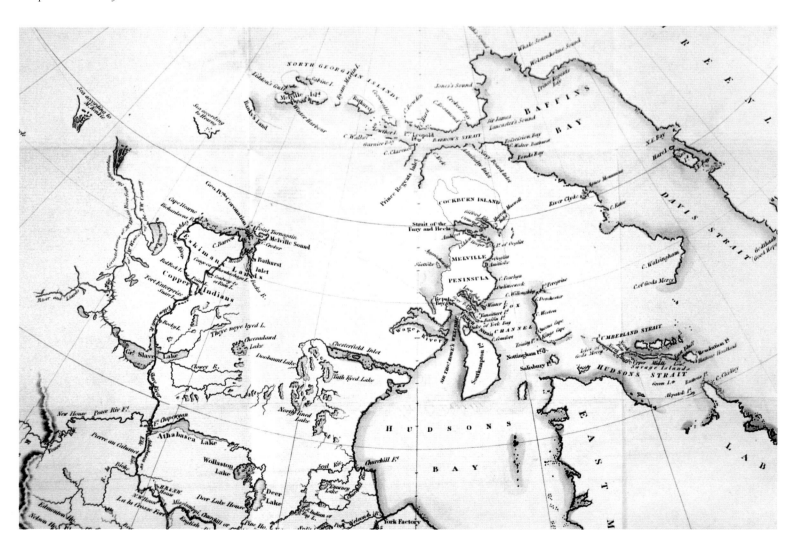

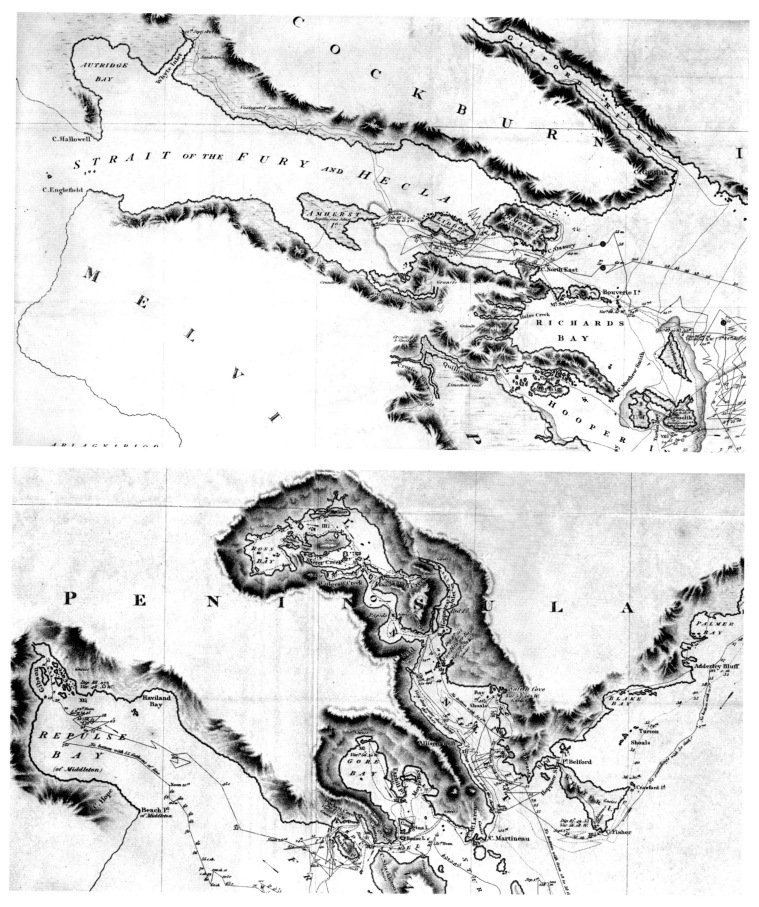

Map 116 (*above, top*).
Parry's map, from his second voyage, of Fury and Hecla Strait, named after his ships. The strait lies at the north end of Foxe Basin between the Melville Peninsula to the south and Baffin Island to the north, here called *Cockburn Island*. Parry named Baffin Island on this voyage, but thought the land to the north of this strait to be a separate island. Parry's wintering place in 1822–23, Igloolik Island, is shown at bottom right. Ship tracks are shown, as are tracks on the frozen strait and on its northern shore.

Map 117 (*above, bottom*).
Repulse Bay and Lyon Inlet, explored by George Lyon in 1821. Winter Island, where Parry wintered in 1821–22, is at bottom right.

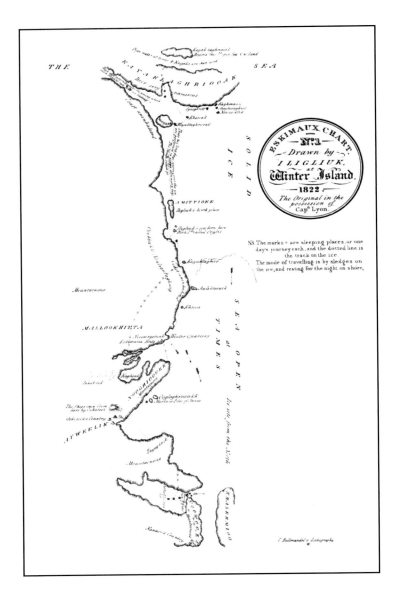

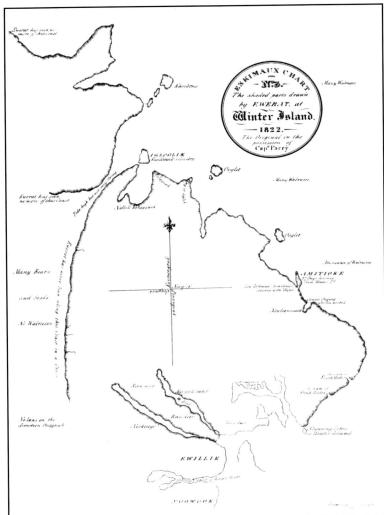

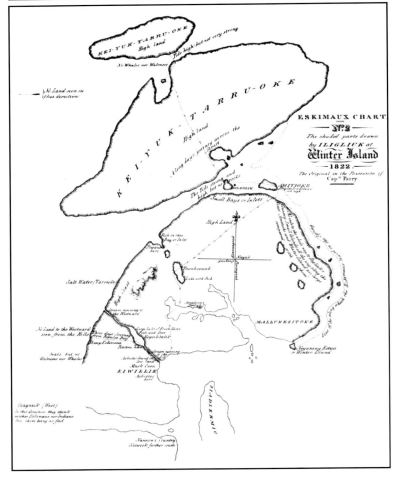

Parry sailed again on 29 April 1821, having been in England for only five months since his return from his first voyage. He took two ships, HMS *Hecla* and *Griper*, commanded, respectively, by George Lyon and Henry Hoppner. On the *Hecla* was a young midshipman named James Clark Ross, John Ross's nephew, on his first Arctic voyage. He would in due course become the most experienced polar navigator of his time, with voyages to both the Arctic and Antarctic. Even more cans of meat were taken aboard; the ships were so well supplied that a transport ship had to accompany them across the Atlantic.

Parry's instructions were not only the usual—to find a Northwest Passage—but also to determine the northern boundary of North America. Franklin's whereabouts were unknown at this time, but he had been instructed to survey eastwards, so perhaps the two could join.

Parry decided to sail north of Southampton Island into Foxe Basin and into Middleton's Frozen Strait, belatedly exonerating Middleton (see page 45). Repulse Bay was thoroughly explored, then inlets to the north were probed, including Lyon Inlet (MAP 117). Then the season ran out, and they wintered at a place Parry named Winter Island; clearly, he liked that rather obvious name.

During the winter, contact with a group of Inuit was made, and, once they had got to know each other, some were induced to draw maps for Parry, showing where they were and what lay to the north and, more importantly, the west. Some are shown here. The Inuit had an excellent sense of geography and all bear a good relation to reality.

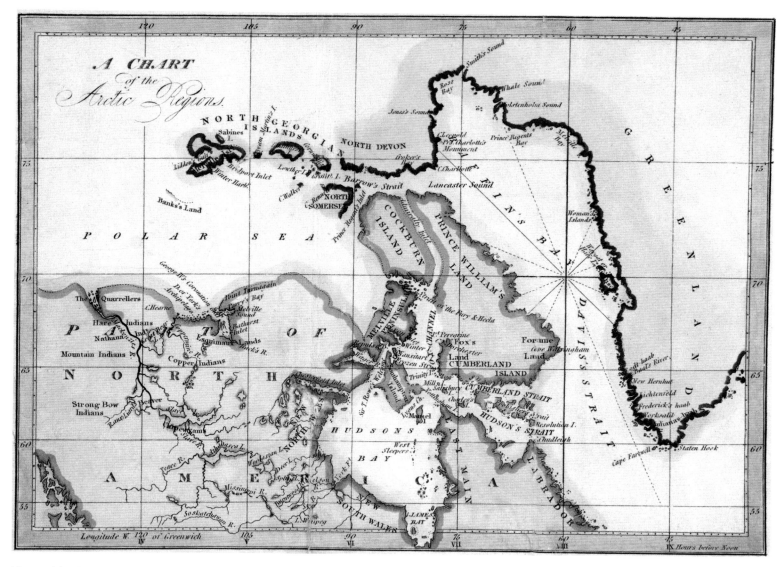

MAP 121 (above).
This summary map drawn by British mapmaker John Harris in April 1825 (before the return of Parry from his third voyage) records the discoveries of Edward Parry's second voyage in 1821–23, added to those of Ross, Parry's first voyage, and Franklin's first overland expedition. The coast between Fury and Hecla Strait and Franklin's Point Turnagain is shown, tentatively, but it had not been mapped at this time. Admiralty Inlet is shown connected to Foxe Basin, leaving *Cockburn Island*, actually part of Baffin Island, as a separate island.

MAP 118 (above, far left), MAP 119 (above, left), and MAP 120 (left).
Copies of Inuit maps drawn for Parry at Winter Island in 1821–22. All show the coast of Melville Peninsula north of Winter Island, for Parry asked the Inuit to draw what lay to the north and west. In MAP 118 just the eastern coast of Melville Peninsula is shown, with Fury and Hecla Strait at the top. MAP 119 shows what lies to the west of the strait as well and a small part of the coast to the north of the strait, which is that of Baffin Island. MAP 120 is the smallest-scale map of all, showing a wide area. Not only is the entire northern end of Melville Peninsula shown, with Fury and Hecla Strait at center, but all, or at least a large part, of Baffin Island is depicted—this is the large island called *Kei-yuk-tarru-oke* on this map—and at the top is Bylot Island. The indentation in the coast of Baffin Island south of Bylot is Pond Inlet.

These maps show a remarkable degree of Inuit awareness of the geography of the areas over which they roamed, and certainly gave Parry a good feel for what to expect. The notation *No land to the Westward seen from the Hills*, on the west coast of the peninsula on MAP 120, convinced Parry that this was the location of the Polar Sea and all he had to do was to sail through the strait the Inuit all drew. Parry did indeed find the strait, but was unable to sail through it, for it was frozen.

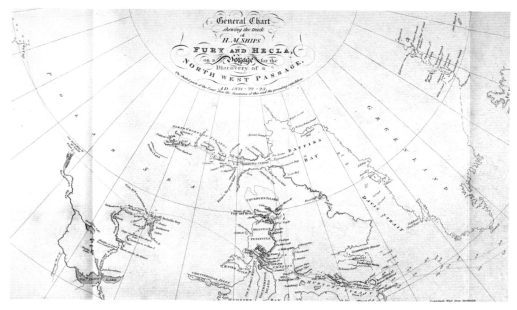

MAP 122.
The general map from Parry's book about his second voyage, published in 1824. Franklin's coast to Point Turnagain is connected with a dotted line to Parry's latest discoveries. This map was likely the source for some of the information on Harris's map, above.

Parry had learned some lessons from his previous overwintering in 1819–20 and had installed the new "Mr. Sylvester's coal-burning stove and heat distribution system." This, it was found, worked quite well to prevent condensation, which of course rapidly turned to ice, on the walls of the cabins.

The next year Parry sailed north, fully expecting to find the strait to the Polar Sea the Inuit maps had seemed to show. He found his strait, which he named after his ships the Strait of Fury and Hecla (now Fury and Hecla Strait), but it was frozen. Traveling on foot across the ice and then along the north side of the strait, Parry reached a point on 12 September where he could see the western end of the strait. His track is shown on MAP 116, page 69.

Thinking he had finally found a Northwest Passage, Parry decided to try again the next year, and so retreated to Igloolik Island (also shown on MAP 116) for a second overwintering. On 12 August 1823 he began a second attempt to sail through the strait, to no avail. Worrying now about his provisions and the possible onset of scurvy, he gave up and returned to England. He had learned that it wasn't enough to find a passage; it had to be a practical one. Nevertheless, a considerable addition to geographical knowledge of the Arctic had been made.

In 1824, the Admiralty initiated no less than four roughly concurrent expeditions; by "blitzing" the north part of America it was hoped to delineate a Northwest Passage once and for all. John Franklin was again to be sent overland, this time down the Mackenzie, exploring the coast both east and west of the Mackenzie Delta; Frederick Beechey was to sail through Bering Strait and then east, to meet with Franklin going west; George Lyon was to sail into Repulse Bay and then strike out west, overland, until he reached Franklin's Point Turnagain; and finally Edward Parry was to sail down Prince Regent Inlet and find any westward-leading channels.

George Lyon set off first. Unfortunately for him, as it turned out, he was given HMS *Griper*, a poor sailor. Lyon later wrote that it was "a vessel of such lubberly, shameful construction as to baffle the ingenuity of the most ingenious seaman in England." In fact

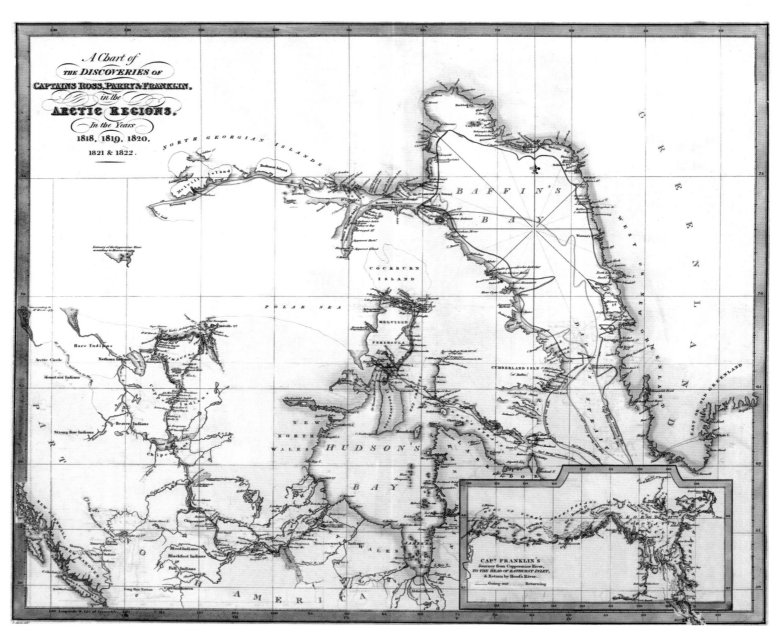

MAP 123.
This summary map of the explorations of Ross, Parry, and Franklin up to the return of Parry from his second voyage, in 1823, was published to feed the growing British public interest in Arctic exploration. Suddenly commercial mapmakers—this map was by John Thomson—found they had a market for people who wanted to know where the explorers had been and what they had discovered. This map also shows the originally reported positions for Mackenzie's mouth of the Mackenzie River and Hearne's mouth of the Coppermine.

it had to be towed across the Atlantic by another ship. *Griper* had just returned from a voyage to Spitsbergen the previous year under the command of Douglas Clavering, who had taken Parry's astronomer, Edward Sabine, there to carry out scientific observations.

Lyon decided to approach Repulse Bay through Roes Welcome Sound to the south, thinking that this would be less likely to give him problems with ice. But he had not reckoned on the weather, which affected *Griper* because it could not work off a lee shore (where the wind is blowing onshore), a very dangerous situation. When the ship was still about 80 km from Repulse Bay, a storm forced Lyon to cut the ship's anchor cables, and after a "most anxious night," they survived the storm only because of a change of wind direction. After this, because they had lost their cables, all the ship's officers agreed that they should return to England.

Edward Parry sailed from England shortly after Lyon, in May 1824, again after having been back in England barely seven months. He had the same ships as on his second voyage, *Hecla*, and *Fury*, under Henry Hoppner. Even getting to Lancaster Sound was a problem this year, such were the ice-choked waters. It took eight weeks to get from Greenland to the sound, using warping (pulling on an anchor imbedded in the ice ahead) or towing by boats. They did not enter Prince Regent Inlet until September and had to overwinter at Port Bowen, on Baffin Island's Brodeur Peninsula.

Parry, contrary to Barrow's belief that ships should stay away from land to avoid ice, had concluded (correctly) that free passage was more likely close to shore. The next year, therefore, after having initially to sail north, he managed to ease southwards close to the western shore of the inlet, along the coast of Somerset Island. After countless collisions with ice, the ships reached about 72° 30′ N, somewhat less penetration into Prince Regent Inlet than Parry had achieved in 1819, on the opposite shore. His track is shown on MAP 125.

Returning northwards, *Fury* was badly holed and then, during the night of 2 August, forced onto the beach by surrounding ice. Attempts to repair her were thwarted by bad weather and ice, and the decision was made to abandon the ship. Both ships had been specifically designed to be able to receive the other's crew in such a situation, and so *Fury*'s crew were taken aboard *Hecla*, and they sailed for home while the chance presented itself.

This time, the veteran Arctic explorer Parry had failed to make any significant new discoveries and, to boot, lost a ship. The Arctic ice had won this round. But *Fury*'s supplies and boats would stay in their refrigerated storage on Fury Beach and would prove to be a lifesaver for a later shipwrecked crew, that of John Ross in 1832 (see page 76).

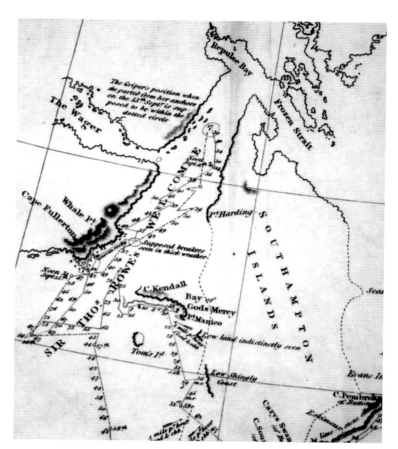

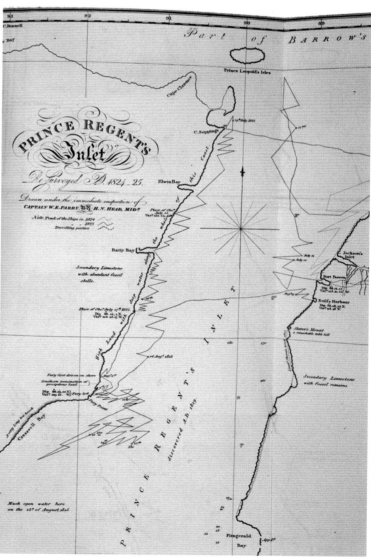

MAP 124 (*above, right*).
Part of George Lyon's map of his abortive mission in 1824. He was attempting to reach Repulse Bay, shown at top, via Roes Welcome Sound, but was prevented from attaining his goal by a tremendous storm in which the ship lost its anchors. Three years before, Parry had reached Repulse Bay via Frozen Strait.

MAP 125 (*right*).
Edward Parry's map of Prince Regent Inlet, explored (farther south than in 1819) during his third voyage, 1824–25. The place where *Fury* was driven ashore is shown, at a place later called Fury Beach. Here it is marked *Fury left*. The stores and boats left by Parry when he abandoned *Fury* were to prove a lifesaver for John Ross's expedition in 1832 (see page 76). Parry has been careful to note on the map that Prince Regent Inlet was *discovered* A.D. *1819*, for it was his discovery.

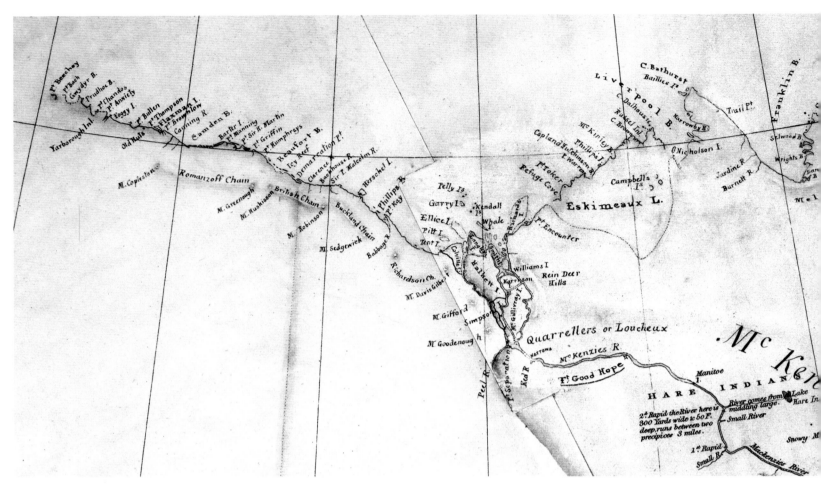

MAP 126.
This interesting map is from a map of North America published in 1824 by British mapmaker Aaron Arrowsmith, but only the portion on the right is the printed map, which shows the results of John Franklin's first expedition, east of the Coppermine River in 1821. The rest of the part shown here is not that map at all but hand-drawn patches applied to update the map with the results of Franklin's second overland expedition, when he mapped the coast west of the Mackenzie Delta and Richardson mapped the coast to the east, in 1826. At left is *Gwydyr B*[ay], the location of Return Reef; while they were waiting for squalls and surf to die down, the coast 24 km farther west, to *P*[oin]*t. Beechey*, was surveyed. This unique map belonged to the Hudson's Bay Company, and so it was probable that someone in that company drew and applied the patches containing the new information.

Early in 1825 the third component of the Admiralty's Arctic blitz left England: John Franklin, accompanied by John Richardson, plus George Back and Edward Kendall. They sailed for New York and traveled to Cumberland House, where they arrived in June. Here Franklin learned that his wife, ill from tuberculosis when he left England, had died.

Advance parties of men and equipment had been prepared the previous year, so that Franklin, learning from his "Gothic" experiences on his earlier expedition, was much better equipped this time. And, importantly, some naval men, marines, were to be employed. Franklin wanted no more murders.

Franklin and Kendall carried out a preliminary survey of the Mackenzie Delta as the season drew to a close (MAP 128, page 76), unfurling "on the shores of the Polar Sea" a flag made by his now deceased wife, doubtless an emotional moment for Franklin, though as a British naval officer in front of his men, he probably did not reveal this. The winter was spent at a base, again constructed in advance, on Great Bear Lake. This they called Fort Franklin (now Déline).

In June 1826, they went down the Great Bear and Mackenzie Rivers to Point Separation, where the delta begins, and there split into two parties. Franklin and Back, with boats they named *Lion* and *Reliance*, went west; Richardson and Kendall, with boats *Dolphin* and *Union*, went east.

From 4 July to 8 August, Richardson and Kendall traveled 1 450 km from Point Separation to the mouth of the Coppermine and mapped the coast. Kendall, the surveyor, had only a pocket watch, for two chronometers had been broken, but he was able to map very well because he knew how to calculate his position using an observational method called lunar distances. Across a strait they named Dolphin and Union Strait after their boats, they saw part of Victoria Island, which Franklin named Wollaston Land after a British scientist; it is now the Wollaston Peninsula. Richardson was of the opinion that the strait would hold "greater prospects of success" for a Northwest Passage, as progress was possible close to the land. He pointed out a "strong current of flood and ebb" which kept the ice in motion. The strait would in fact be used by Roald Amundsen in 1905, during the first transit by ship of the Northwest Passage (see page 158), and later by many others.

After reaching the Coppermine River, Richardson and Kendall returned to Fort Franklin overland.

Franklin and Back, meanwhile, explored to the west. In the delta, they encountered Inuvialuit who, by sheer numbers, managed to take a significant amount of their supplies, in a fracas only stopped by threats to shoot. Franklin named the spot Pillage Point.

On 9 July they reached the open sea, finding it frozen, not clear of ice as it had been the previous year, and progress westward was made by keeping close to shore, where there was a lead in the ice.

They reached and named Herschel Island (after the astronomer William Herschel) and struggled farther west intermittently, sometimes with their way being blocked by

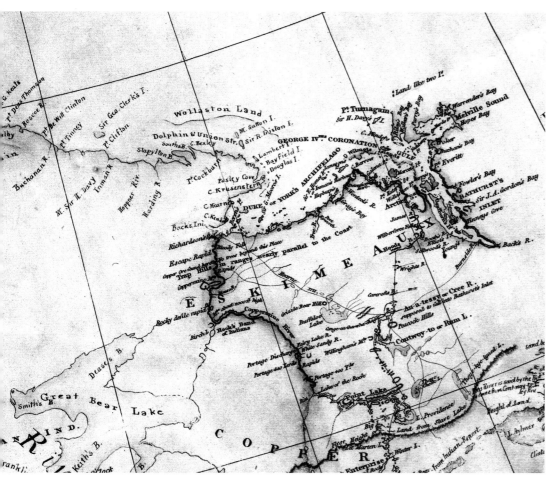

MAP 127 (below).
Frederick Beechey's map of the north coast of Alaska, showing the part he mapped for the first time, between Icy Cape and Point Barrow. Note the differing positions of the edge of the ice pack shown for 1826 and 1827.

ice, when the boats had to be dragged. One of Franklin's place names, Point Anxiety, reflects the struggle. By 6 August, having made only half of the distance from the Mackenzie to Icy Cape, where he had hoped to meet up with Frederick Beechey, dispatched to sail through Bering Strait to connect with Franklin (*see below*), Franklin realized that he would have to turn back. On 18 August, having reached 149° 37′ W, at a place he called Return Reef (just west of *Pt. Beechey* on MAP 126), Franklin turned back. He had been instructed to turn back no later than 20 August in any case.

Franklin's second land expedition achieved much of its purpose; although Franklin did not connect with Beechey, some 1 500 km of coastline was mapped for the first time. This was a considerable addition to knowledge of the Arctic.

The fourth element of the Admiralty's 1824 plan was the voyage of Frederick Beechey. The idea was that he would approach from the west through Bering Strait and, it was hoped, connect with Franklin coming east. Beechey had been a lieutenant with Franklin on his northward voyage on the *Trent* in 1818 and with Parry on the *Hecla* in 1819–20.

Beechey sailed in HMS *Blossom* in May 1825 and reached Chamisso Island in Kotzebue Sound, just north of Bering Strait,

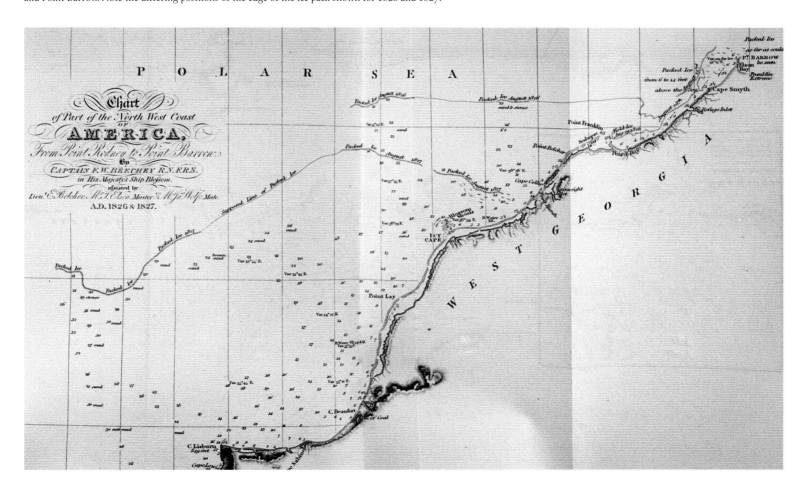

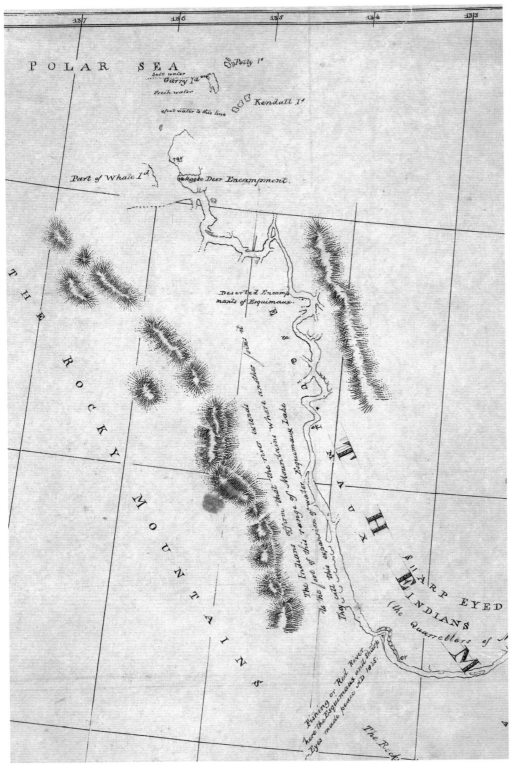

MAP 128.
Part of John Franklin's map showing the lower reaches of the Mackenzie River, the Mackenzie Delta, and some of the islands in the Beaufort Sea opposite the delta. Alexander Mackenzie's Whale Island has been renamed *Garry I[slan]d*, while part of *Whale I[slan]d* is shown farther south. The map was drawn by Edward Kendall in 1825.

Beechey did not find Franklin, but he did put 230 km of the northern Alaska coastline on the map, leaving only a 300 km gap in the known north coast of North America, west of Franklin's Point Turnagain. In fact, the Admiralty's 1824 plan had been quite successful due to the Franklin and Beechey part of it; in Arctic exploration two out of four isn't bad.

The next expedition to the Canadian Arctic was not arranged by the Admiralty, however, but was the first that was privately financed. It was commanded by none other than John Ross, who after his long hiatus following his mistake in thinking Lancaster Sound was a bay (see pages 60–61) was beginning to despair that he would never again be employed by the navy. Now he was to try something completely new, and hopefully recapture his reputation by finding the Northwest Passage.

Ross had long been convinced that navigation of Arctic seas would be easier using a steam-powered vessel. "When the ice is open or the sea navigable," he wrote, "it is either calm, or the wind is adverse, since it is to southerly winds that this state of things is owing: so that the sailing vessel is stopped exactly where everything else is in her favour, while the steam boat can make a valuable progress."

And so Ross fitted out a shallow-draft paddle steamer, *Victory*, lately used for the transport of packets from Liverpool to the Isle of Man. But right at the beginning of his book published after his voyage, Ross refers to the makers of his "execrable machinery," Messrs. Braithwaite and Erickson—giving an ominous hint of troubles to come.

The financing for his innovative voyage was found by Ross with the Lord Mayor of London, Felix Booth, who had made a fortune in the spirits trade. Thus it would come about that many places in Canada's North today bear the name of a famous brand of gin.

Ross sailed—for the steaming was to be mainly reserved for the ice—in May 1829, taking with him his nephew, James Clark Ross, now a commander following his voyage to the north of Spitsbergen with Edward Parry in 1827 (see page 124).

John Ross thought that the Northwest Passage would be found at the southern end of Prince Regent Inlet, and after steaming through his own "mountains" at the entrance to Lancaster Sound—which must have been a moment of reflection for Ross—he proceeded south down that inlet. Ross intended

in July 1826. This was where they had optimistically arranged to meet with Franklin, but of course he was not there and so they continued northwards, then eastwards, passing James Cook's Icy Cape, after which point Beechey was on an unmapped coastline.

Although *Blossom* was soon stopped by ice, Beechey sent a small schooner ahead under his sailing master, Thomas Elson. By sailing very close to shore, Elson managed to get to a major headland which was named Point Barrow, after "that distinguished member of our naval administration." Point Barrow is the northernmost point in North America west of the Boothia Peninsula. Returning to the *Blossom* was difficult, with the ship having to be tracked (pulled by men on shore).

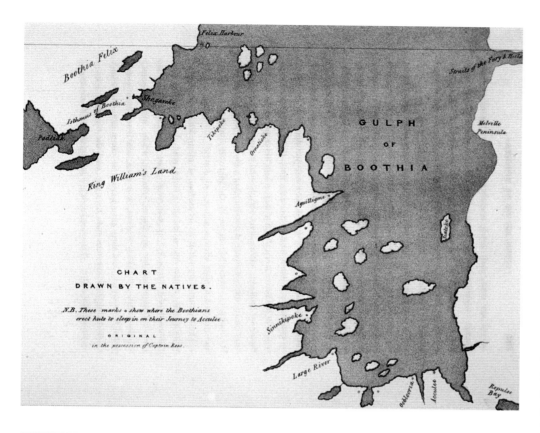

Map 129 (*above*).
The map of the southern end of the Gulf of Boothia drawn by Ikmalick and Apelaglu. They were Inuit; in perhaps the ultimate act of Eurocentrism Ross called them "Boothians," for they lived in the country he had named after the gin magnate.

This illustration from Ross's book shows Ikmalick and Apelaglu in the cabin of the *Victory* drawing a map.

Apelaglu, to draw a map of what they knew about the geography of their surroundings. What they drew must have disappointed Ross immensely, for their map showed correctly that his Gulf of Boothia was closed at its southern end (MAP 129).

Deciding to make the best of it, Ross undertook explorations by land. James Clark Ross crossed the Boothia Peninsula, discovered King William Island, and mapped its northern coast, believing it, however, to be part of the mainland, and it is shown as such on MAP 131, *overleaf*.

The ice would not release *Victory* that summer; they managed only 5 km before being stopped, and another winter was spent in another small bay, which Ross named Sheriff Harbour (Felix Booth being also the Sheriff of London). From here the following spring James Clark Ross made another foray across the peninsula and located the North Magnetic Pole at a "calculated place" just north of Kent Bay, on 1 June 1831. It was an unremarkable place: "Nature had here erected no monument to denote the spot which she had chosen as the centre of one of her great and dark powers," he wrote for John Ross's book. "We fixed the British flag on the spot, and took possession of the North Magnetic Pole and its adjoining territory, in the name of Great Britain and King William the Fourth." Its position was confirmed by a dipping needle showing 89° 59´.

The following summer the ice again prevented their escape; truly it was a prison. *Victory* moved only 25 km this time. Ross now realized that desperate measures were required, and the following spring, 1832, all

to sail from Cresswell Bay, the last point mapped by Parry in 1825 (see MAP 125, page 73), to Franklin's Point Turnagain, and thus along the northern coast of America mapped by Franklin and Richardson. It was a reasonable plan, but overlooked the fact that there was land in the way, a peninsula Ross soon found, which he named Boothia Felix after his sponsor the gin merchant.

By 7 October the "prison door," as Ross called it, had shut, and they were entombed by ice in a bay he named Felix Harbour, on the northern side of the much larger Lord Mayor Bay. The steam engine had by this time given Ross so much trouble that he decided to remove it; henceforth *Victory* was to be a sailing vessel only. Parts of the engine are still on the beach at Felix Harbour today.

In January 1830 a group of Inuit arrived, and Ross persuaded two of them, Ikmalick and

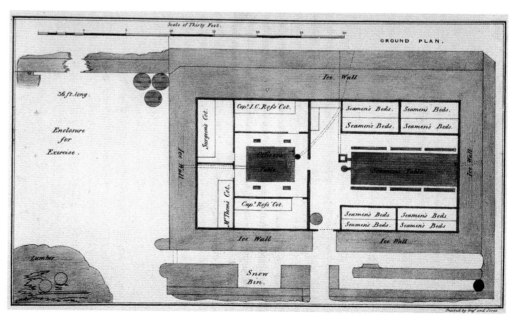

MAP 130.
Somerset House, built of wood and canvas, in which John Ross and his men spent the winter of 1832–33 at Fury Beach.

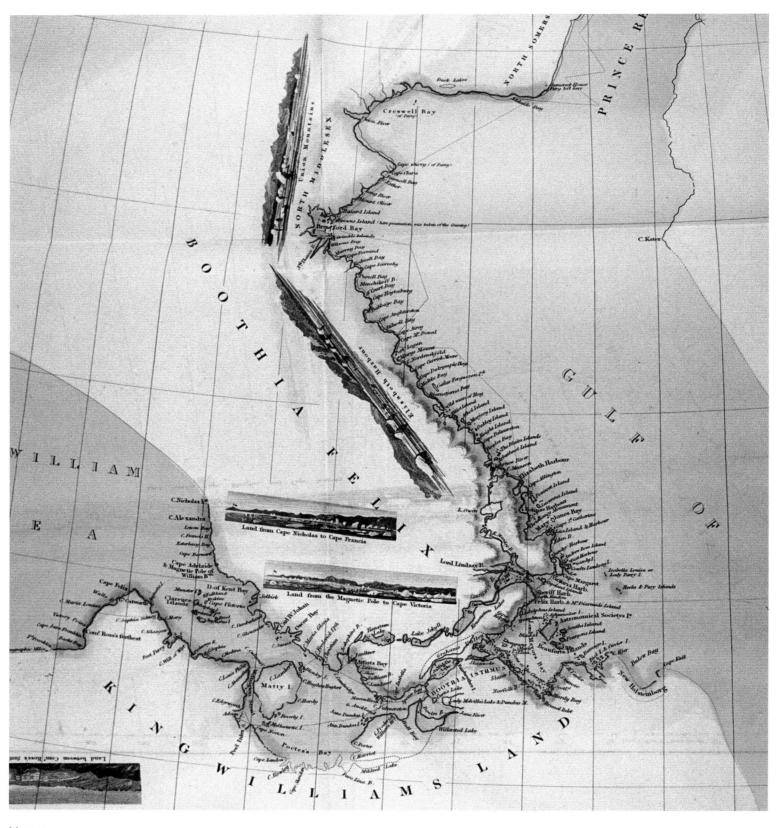

MAP 131.
A rather beautiful map of John Ross's *Boothia Felix*, today the Boothia Peninsula, except for the northern part. Where Ross has marked *North Middlesex* is a separate island, separated by the narrow Bellot Strait, discovered in 1852. The North Magnetic Pole, found by James Clark Ross in 1831, is shown just north of *D. of Kent Bay*, Kent Bay, on the west side of the Boothia Peninsula. *King Williams Land*, now King William Island, is shown as part of the mainland. In reality the island is separated where Ross has noted *Poctes's Bay*, albeit with a dotted line; it does not exist.

supplies were put on land and the ship was abandoned. There followed a laborious journey north to Fury Beach, where Ross hoped to find plentiful supplies and, just as importantly, boats that had been left when *Fury* was abandoned by Parry in 1825. Luckily for them, everything was still largely intact; the food was edible and the boats repairable.

Even then, attempts to leave in *Fury*'s boats in August met with frustration; the ice that year was just too heavy. Nothing remained but to hunker down for yet another winter—their fourth—in the Arctic. Although they were able to build a reasonably comfortable shelter, by Arctic standards at any rate, and although food was in good supply, the canned soups and meats being still in good condition, scurvy was now beginning to rear its ugly head and several of the crew died. Their shelter they named Somerset

House, a plan of which is shown as MAP 130 (page 77).

Luckily for them all, the ice the following summer was not so bad. They transported supplies and the sick north to where the boats had been stopped the previous season and on 15 August tried again. By arduous rowing and some sailing they reached Lancaster Sound, where, on 25 August, they were seen by a whaling ship and taken aboard. By an enormous coincidence it was the *Isabella*, Ross's old ship, now converted into a whaler sailing out of Hull, England. The crew of the whaler were astonished; the mate, wrote Ross, "assured me that I had been dead for two years." Ross, like Mark Twain, might well have remarked that the reports of his death were greatly exaggerated.

Ross's reputation was certainly rehabilitated by the venture, for he received a knighthood, the usual accolade for successful British Arctic explorers. As if reflecting the length of his record-breaking stay in the Arctic, his book, published in 1835, runs to 740 pages.

In February 1833, with the realization in Britain that Ross must be in trouble, George Back had left on a rescue mission. He planned to descend the 974-km-long Great Fish River, now the Back River, which, it was felt, must flow to the general vicinity of where Ross was likely to be. His proposed route, and the expected geography, is shown very well in MAP 132 (*above, right*).

While at Great Slave Lake the following winter, he received news of Ross's return, but decided to continue on his intended route anyway, planning to reach the sea and then explore west until he reached Franklin's Point Turnagain. Chantrey Inlet, into which the Back River flows, was blocked by ice, so Back was unable to carry out his plan fully, but he did get as far as a promontory he named Point Richardson, where he detected a strait, thus suggesting (correctly) that Ross's *Poctes's Bay* did not exist and that his *King Williams Land* was an island. MAP 133 was the map Back drew on his return and is thus an interesting comparison with MAP 132. He had not achieved his objective but he had added something significant to the geographical puzzle. The prestigious journal of the new Royal Geographical Society was moved to concede that "the return of this expedition has thrown a new and extended light over the geography of the north-eastern extremity of America."

Back was not so successful with his next attempt at Arctic exploration, however. In 1836 he was instructed to sail to Repulse Bay

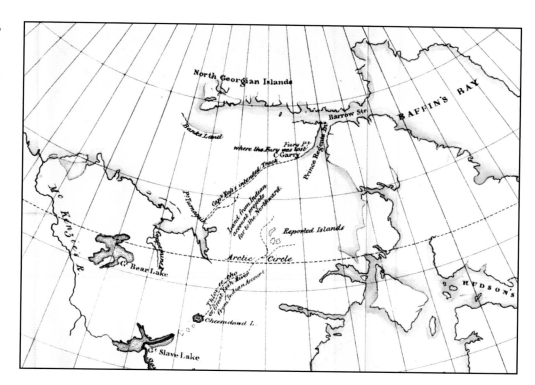

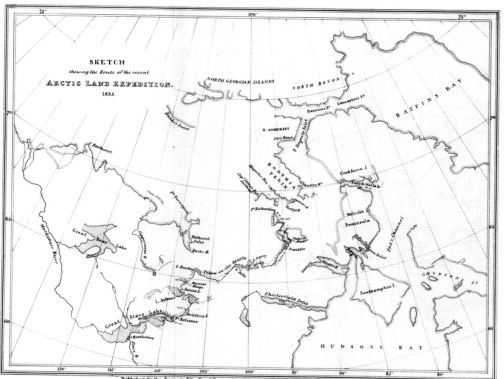

MAP 132 (*above, top*).
George Back's proposed route, shown on a map drawn in 1833 at the time he left England. It shows John Ross's proposed track from the known point from Parry's expedition in 1825 to Franklin's Point Turnagain, and of course does not include information from Ross. The improbability of finding Ross in such a huge and frozen land would have been immense. And the *Great Fish River* (Back River), shown close to Great Slave Lake, was itself difficult to find in a country laced with lakes and a multitude of rivers.

MAP 133 (*above*).
An equivalent map published in 1835 from information sent ahead by Back. Ross's explorations are now shown, but *Poetess B*[ay], Ross's *Poctes's Bay* (see MAP 131), is shown open, suggesting the strait around King William Island. Back's route from Great Slave Lake to the ocean is shown in red. Note *Capt. James Ross farthest* at the western end of the northern coast of King William Island as mapped to this time. Both MAP 132 and this map were published by the Royal Geographical Society.

and from there carry boats overland to chart the remaining unknown coast, from Point Turnagain to Fury and Hecla Strait. It was a tall order, but Back never even got to Repulse Bay. With HMS *Terror*, he sailed into Foxe Channel, intending to approach Repulse Bay through Frozen Strait, as Parry had done in 1821. The ship was caught in the ice north of Southampton Island and drifted back along its coast for almost a year, after which, badly damaged, the ship escaped the ice in July 1837. Back had to give up all thoughts of continuing, and he struggled across the Atlantic, finally beaching the ship on the west coast of Ireland.

In the late 1830s the Hudson's Bay Company, aware of the necessity of renewing its exclusive trading licence in 1842, decided to try to enhance its reputation by completing the mapping of the northern coast of North America. This was done between 1837 and 1839 by two company traders, Thomas Simpson, who was a cousin of the company governor George Simpson, and Peter Warren Dease, who had assisted John Franklin during his second land expedition, in 1825–27.

Thomas Simpson was the more enthusiastic explorer, it seems, likely because he was out to advance his career. The first "gap" in the explored coastline the pair tackled was that between Frederick Beechey's Point Barrow and Franklin's Return Reef. This was explored and mapped in 1837, using two small boats they named *Castor* and *Pollux* after the twins of legend, with Simpson walking the last four days to Point Barrow when the going for the boats became impossible.

After a winter at a base they constructed on Great Bear Lake, which they named Fort Providence, the two set off again, this time to the east. They explored the coast from the Coppermine River east, but were stopped by ice at Franklin's Point Turnagain. From here the redoubtable Simpson continued on foot, reaching Cape Alexander and, two more days beyond, the Beaufort River at 106° W, before turning back. During this trek, Simpson saw land across the water, which he named Victoria Land for the new queen. It is in fact a large island, today Victoria Island.

After another winter at Fort Providence, Simpson and Dease were back the following year. This time the coast beyond Cape

MAP 134 (left).
Back's map of the track and drift of *Terror* in Foxe Channel, 1836–37. Despite their predicament, the crew managed to survey the northeast coast of Southampton Island.

MAP 135 (below).
A Hudson's Bay Company map of Peter Warren Dease and Thomas Simpson's explorations in 1838 and 1839, when they filled in the gap in the mapped coastline between Franklin's Point Turnagain and the coast mapped by James Clark Ross and George Back. Just east of C.[ape] *Alexander* is the small Beaufort River, the point at which Simpson finished his survey for 1838 and began it the following year. The *Riv. twice as large as Copp.rmine Riv.* is the Ellice River.

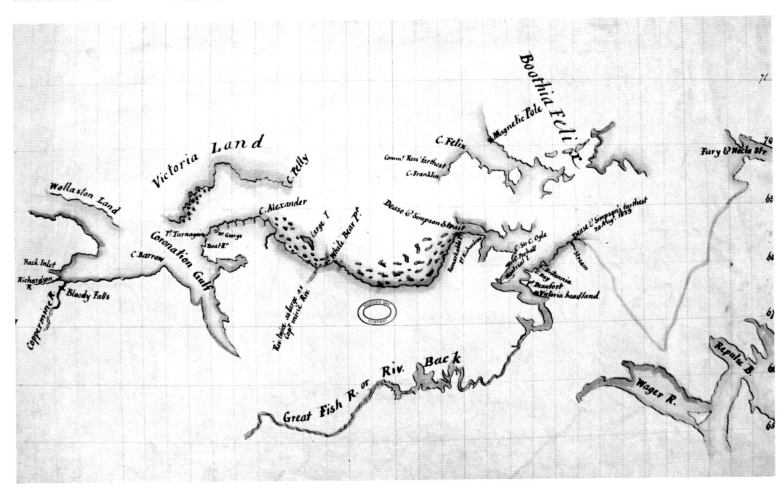

MAP 136 (above).
In 1846 Sir John Barrow published a book updating his 1818 history of polar voyages of discovery to include those he had himself had a major role in organizing. In the book he included an update of his polar map of 1818 (MAP 104, page 60), and this is the map, beautifully colored in this rare example. Although the notation in the margin gives 1847 for the map date, it was published in 1846. The southern end of the Gulf of Boothia is taken from the Inuit map drawn for John Ross in 1830 (MAP 129, page 77), but the Boothia Peninsula is still shown as an island, with a strait to its south. The notable blank on this map is the region north of the North American mainland as far north as Parry's 1819–20 discoveries, and the Franklin expedition had just been dispatched to find its way through it. But it was not Franklin but the numerous searches for him that were destined to fill in much of this blank on the map.

MAP 137 (right).
John Rae's survey of Committee Bay, the southern end of the Gulf of Boothia, drawn in 1847. This confirmed the Inuit map drawn for Ross (MAP 129, page 77) and, by connecting to Ross's map at Lord Mayor Bay, determined once and for all that Boothia Felix was a mainland peninsula, the Boothia Peninsula.

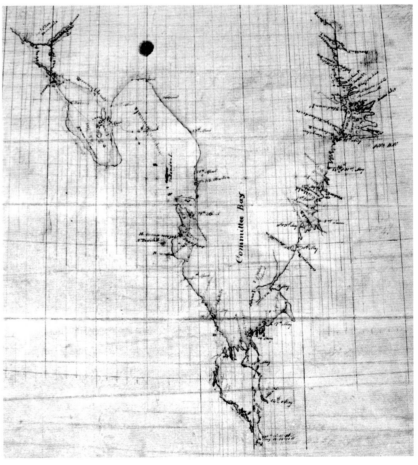

Alexander was explored. They found that there was a narrow strait between the mainland and King William Land (Island) to the north, widening to the east, although they still did not realize that it was an island, for their map (MAP 135, left) allows for a posible connection to the mainland on its eastern side. The strait was named Dease and Simpson Strait; today Dease has been dropped and it is just Simpson Strait. By 16 August the pair reached Montreal Island, deep in Chantrey Inlet, and the following day were at Cape Britannia, on the

81

inlet's eastern shore. From here the unstoppable Simpson made yet another foray just a little farther, reaching a river he named the Castor and Pollux River, after their boats, at 96° W.

On their return, the southern coasts of both King William Island and Victoria Island were surveyed. Simpson and Dease had boldly completed the mapping of the northern coast of North America; only a small part remained to be explored. The Hudson's Bay Company had their licence renewed.

It was at this stage of geographical knowledge that the fateful voyage of Franklin was planned. His disappearance would give birth to a flurry of activity that would define much of the remaining unknown Canadian Arctic. This is the subject of the next section (see page 84).

In the meantime, one further expedition was planned, again by the Hudson's Bay Company, to fill in that last small gap in the mapped coastline of northern North America.

In June 1846 John Rae sailed north from York Factory to Repulse Bay and then crossed what is now the Rae Isthmus, the southern neck of the Melville Peninsula, but did not proceed farther that year due to the lateness of the season. Instead he returned to Repulse Bay and constructed a stone house he called Fort Hope, in which the winter was spent. The following year he set out in April, crossed the isthmus again, and traveled north along the western shore of the southern end of the Gulf of Boothia, which he named Committee Bay, reaching territory previously explored by John Ross at Lord Mayor Bay (MAP 131, page 78). Later that season he also explored the eastern shore of the gulf, coming within 30 km of the western end of Fury and Hecla Strait, previously mapped by Edward Parry (MAP 116, page 69). MAP 138, *below*, shows how Rae's survey connects with those of Ross and Parry.

In August 1847 Rae sailed back to York Factory. His map (MAP 137, *previous page*) showed conclusively that Ross's Boothia Felix was a peninsula of mainland North America, and thus there were no connections here with any southern Northwest Passage. Now the only unmapped part of the northern coast of North America (except for the 30 km gap just south of Fury and Hecla Strait) was a small stretch of coast on the southwestern end of the Boothia Peninsula. To all intents and purposes, the North American coast was defined.

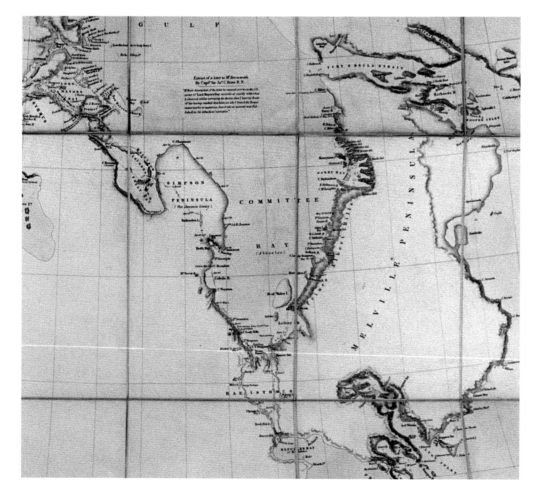

MAP 138 (*left*).
This is part of a map published by British mapmaker John Arrowsmith in 1848 to show the various coasts revealed by the different explorers. This portion shows John Rae's explorations in 1847, in red, connecting at Lord Mayor Bay with those of John Ross (yellow) and at Fury and Hecla Strait with those of Edward Parry (blue). The Rae Isthmus is the narrow neck of the Melville Peninsula, which Rae crossed several times from his base at Repulse Bay.

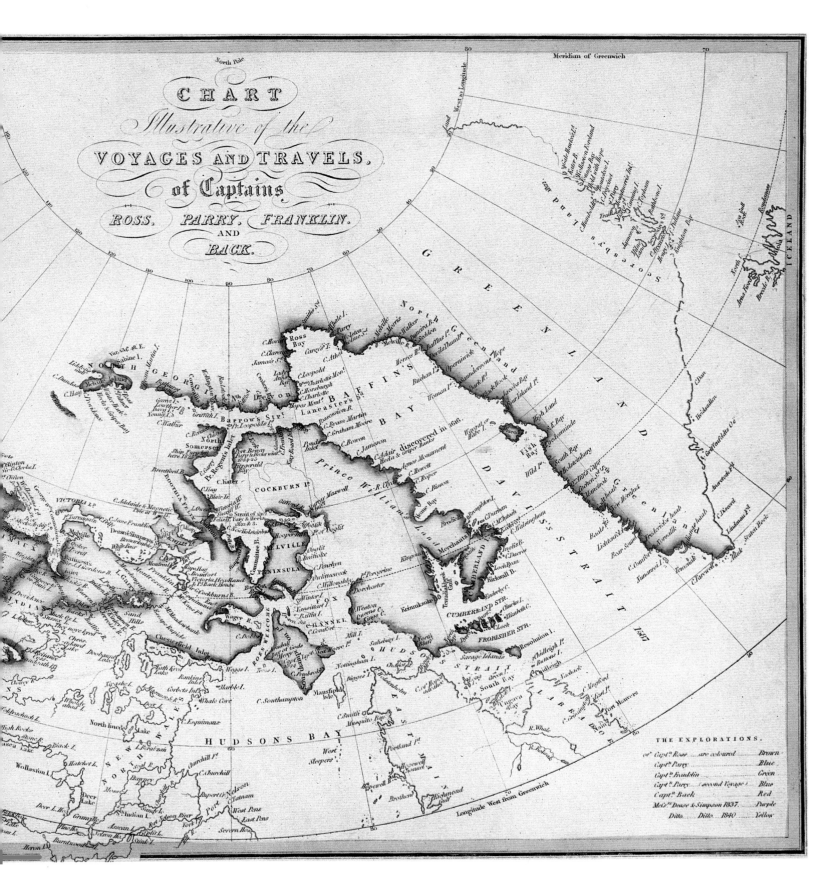

MAP 139.
Another map by John Arrowsmith. Although this map shows discoveries up to 1847, it has completely ignored John Rae. His Committee Bay is there, which he found in 1847, finally proving that Prince Regent Inlet and the Gulf of Boothia did not connect westwards with any other water body. (It does connect through the tiny Bellot Strait, but this was not discovered until 1852.) But Rae's Committee Bay is shown on this map with its western half as discovered by Ross (brown) and its eastern half as by Parry (blue), when in fact neither reached that far. Frederick Beechey's contribution in Alaska is also ignored. Dease and Simpson's connection in 1837 of the coast surveyed by Beechey with that by Franklin is, however, shown (in purple). Grey-green is from Franklin's maps and yellow from Dease and Simpson in 1838–39. George Back's 1834 descent of the Back (*Great Fish*) River is shown (purple), but not accurately, for he reached the shores of the Arctic Ocean at Chantrey Inlet. In an attempt to reconcile the map of George Back with that of John Ross, Arrowsmith has shown King William Island, *K. Willᵏ. Lᵈ*, connected to the Boothia Peninsula.

For Franklin—and Glory

On his return from his second land expedition in 1827, John Franklin remarried (his first wife having died from tuberculosis while he was away). Jane Griffin was a vivacious and intellectual woman who, as Lady Jane Franklin, was destined to play a significant part in future exploration of the Arctic, despite never actually visiting the region herself. For it would be her driving will that would be responsible for many searches for her husband.

Franklin was knighted in 1829, spent several years in the Mediterranean, and from 1836 to 1843 was lieutenant-governor of Van Diemen's Land—Tasmania. It was here in 1840 that he hosted James Clark Ross, who stopped for magnetic observations on his way to the Antarctic in HMS *Erebus* and *Terror*.

On Franklin's return to London he was appointed to lead another expedition to find the Northwest Passage, a new proposal put forward by a now eighty-year-old John Barrow a month before he resigned. Barrow thought that Banks Land (seen by Parry in 1819 and 1820), Wollaston Land (seen by Richardson in 1826), and Victoria Land (seen by Simpson and Dease in 1838) were merely small islands, when in fact they are parts of two very large islands.

Franklin was therefore instructed to sail from the last known point to the south of Barrow Strait (Cape Walker, on a small island named Russell Island at the north end of Prince of Wales Island, west of Somerset Island) south southwest to find the now mapped coast of North America and follow it to Bering Strait. The second part of these instructions might have worked, but the first part would have had Franklin sailing through land. An alternative route was given to Franklin, should he be unable to sail south, and this was to sail northwards up Wellington Channel, out into a presumed open polar sea, and then west to Bering Strait. It was as though nothing had been learned. This alternate instruction is important because it explains why so many of the subsequent search expeditions probed to the northwards.

The two ships were to be *Erebus*, commanded by Franklin, and *Terror*, commanded by Francis Rawdon Moira Crozier. They had just returned from the Antarctic under James Clark Ross and were now fitted with auxiliary steam power in the form of two railway engines, with screw propellors that could be raised when they were not in use. The engines were intended to be used only for short periods when required to push through ice. The ships were provisioned for three years—and supplied with maps of the Pacific Ocean.

Franklin sailed in May 1845. He was seen by some whaling ships in northern Baffin Bay as he headed for Lancaster Sound. Then he disappeared.

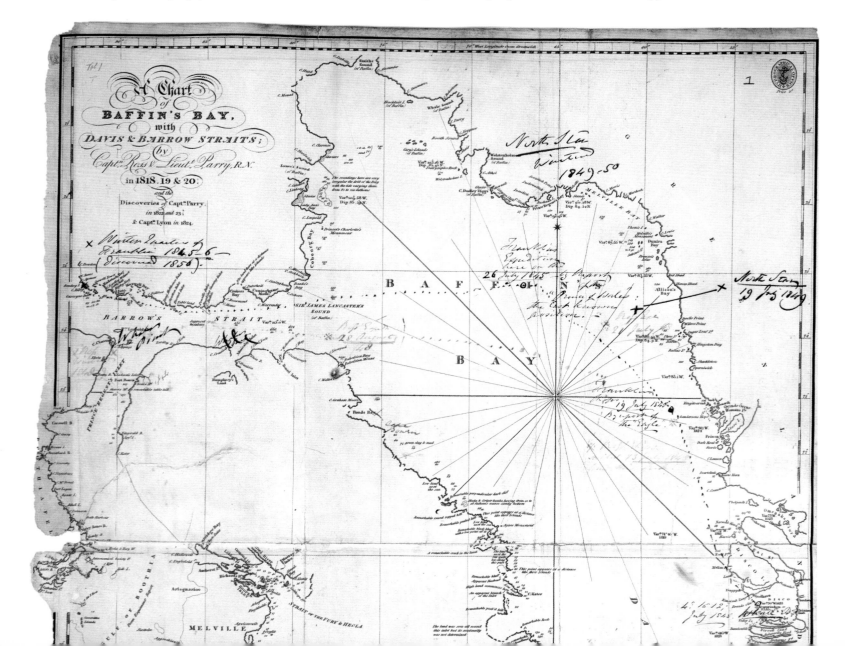

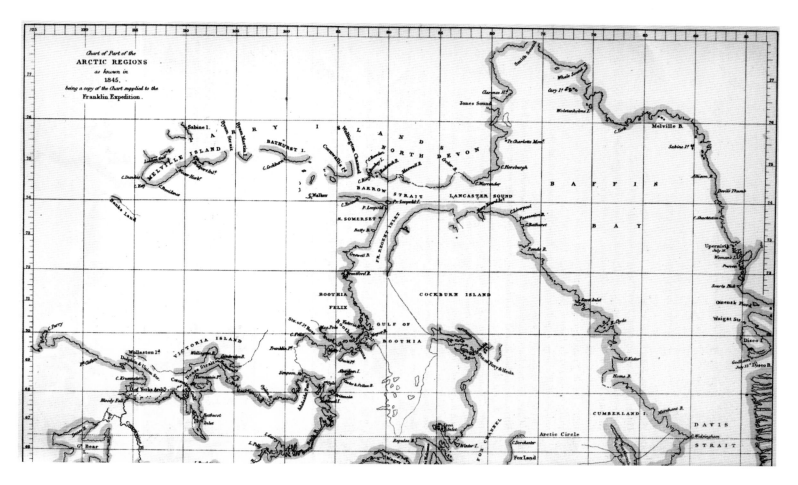

MAP 140 (*left*).
An Admiralty Chart used in 1850 to plot Franklin's known and probable tracks. Written on it are notes showing the positions of James Clark Ross's ships and a supply ship, the *North Star*, sent to resupply them but which became trapped in ice in 1849. The finding of Franklin's wintering place, at Beechey Island, off the southwest tip of Devon Island, is also noted.

MAP 141 (*above*).
This copy of the map supplied to John Franklin in 1845 shows the extent of what he knew when he sailed. All he had to do, it seemed, was to find a way to connect the known pathway to the north—Lancaster Sound westwards—with the now largely known northern coast of mainland North America. Could this be so difficult?

Franklin did sail north into Wellington Channel that year, presumably because the ice conditions would not let him go south, and he reached 77° N. He returned to Beechey Island, to the southwest of Devon Island, to winter. The following year he sailed south through Peel Sound and Franklin Strait into what is now Larsen Sound, the right direction—were it not for the ice. The multi-year ice here surges from the main pack southeast down M'Clintock Channel. On 12 September 1846 Franklin became irreversibly trapped in the ice. They were at 70° 05´ N, 98° 23´ W, just off the northern tip of King William Island.

The ice refused to release the ships the following summer, and the ships remained beset. Men were dying, and on 11 June 1847 Franklin also died. The following April Crozier, leading 105 survivors, abandoned the ships and headed south, landing at Victory Point on King William Island, where records were left. Crozier then led his men south, intending to make for Back's Great Fish River (the Back River). The last survivors died at Starvation Cove, on the Adelaide Peninsula, immediately south of King William Island.

Franklin left no maps, but the massive search for him that was to follow would end up mapping much of the Canadian Arctic Archipelago.

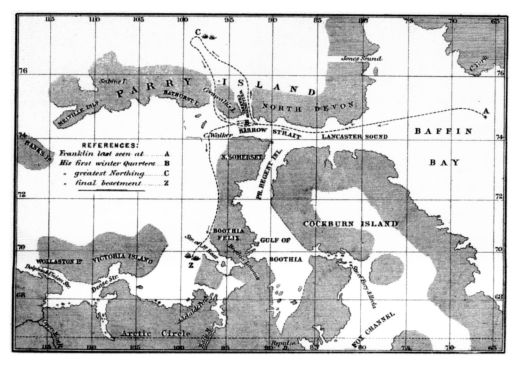

MAP 142.
The supposed track of Franklin deduced during the search for his missing expedition and drawn by Leopold M'Clintock in 1859. The track is superimposed on a map of the Arctic as known to Franklin.

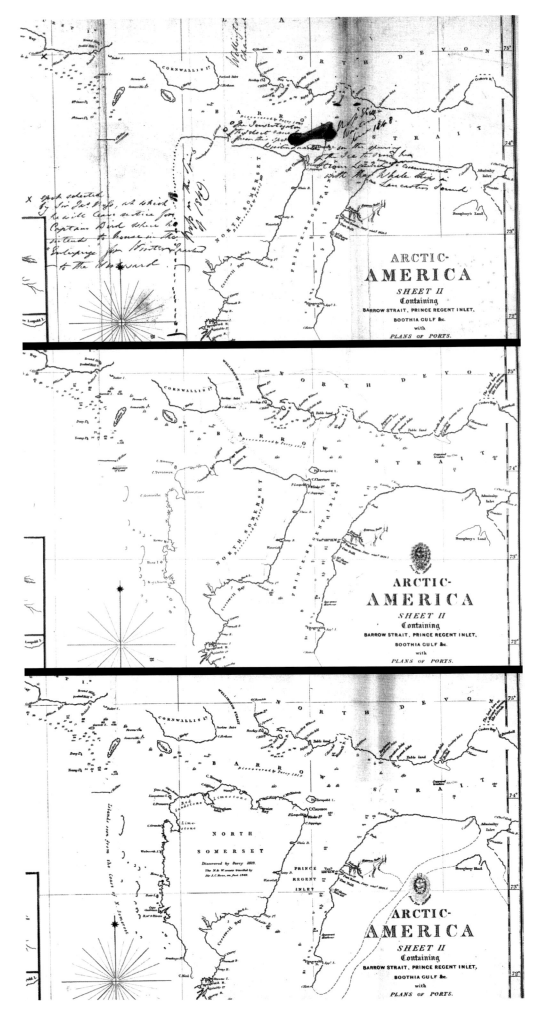

Map 143 (*top*), Map 144 (*middle*), and Map 145 (*bottom*). This sequence of maps shows how the process of updating geographical knowledge was applied. All the maps are part of the *Arctic America* sheets issued by the Admiralty. Map 143 (*top*) was that used by James Clark Ross in the winter of 1848. X marks the spot selected for Edward Bird to overwinter, a plan that was abandoned due to heavy ice. The winter of 1848–49 was spent at Port Leopold, at the northeast tip of *North Somerset* (Somerset Island). In the spring of 1849 sledge parties led by Ross and Francis Leopold M'Clintock surveyed and mapped the northwest and west coasts of Somerset Island, producing the survey shown as Map 144 (*middle*). The final map, Map 145 (*bottom*), shows the results of this survey incorporated into the printed chart after Ross returned to England late in 1849.

By 1847 concern was beginning to be felt, and pemmican was manufactured in England and sent to York Factory. John Richardson at the age of sixty volunteered to lead an overland search. Accompanied by John Rae, he searched the coast between the Mackenzie Delta and the Coppermine River in 1848. The following year Richardson returned home, but Rae tried again in 1849, traveling to the mouth of the Coppermine and searching to the northwest. Ice prevented him from getting very far that year, but in 1851 he was back, sent by the Hudson's Bay Company, and this year he searched, and explored and mapped for the first time, over 1 000 km of the southwest and southeast coasts of Victoria Island (Map 152, page 90).

Richardson and Rae's 1848 search was actually part of a three-pronged effort the Admiralty orchestrated that year.

The second search was that of James Clark Ross, who with HMS *Enterprise* and Edward Bird on HMS *Investigator* was to approach from the east, following Franklin's instructed route. The ice, however, was unco-operative, and he only got as far as the northeastern tip of Somerset Island, although sledge parties did search Barrow Strait and Prince Regent Inlet. Ross himself, with Leopold M'Clintock, searched and mapped the west coast of Somerset Island (Map 144, *left*, *middle*).

Thirdly, Thomas Moore in HMS *Plover* and Henry Kellett in HMS *Herald*, the latter diverted from surveying duties in the Pacific, were to search through Bering Strait as far as the Mackenzie Delta. William Pullen and William Hooper actually carried out the search, starting in July 1849 to the east of Wainwright Inlet, where *Plover* was stationed. To just beyond Point Barrow they were accompanied by a pleasure yacht, *Nancy Dawson*, that had been on a round-the-world voyage until its owner, Robert Shedden, had decided to assist in the Franklin search. The

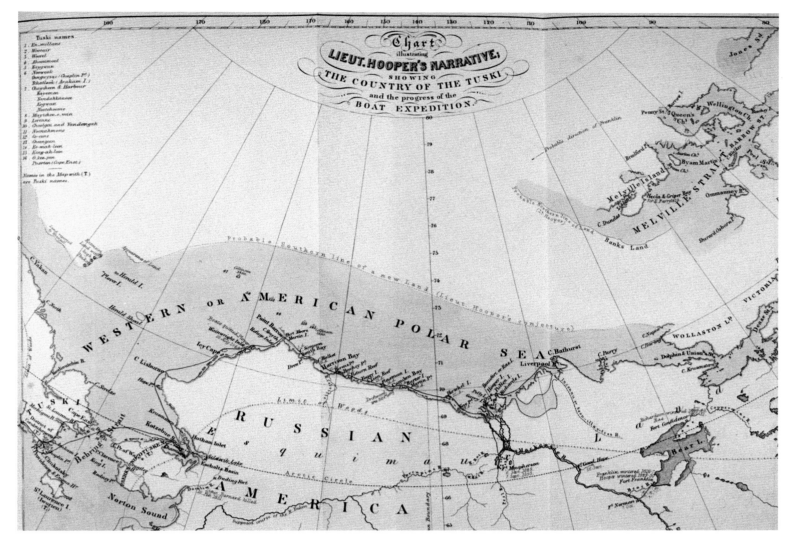

MAP 146 (*above*).
The map from a book by William Hooper, showing his search with William Pullen of the north coast of Alaska from Wainwright Inlet and east to Cape Bathurst in 1849–51. Also shown is Kellett's *Herald I.* and *Plover I.*[slands], found by Henry Kellett in 1849, with *Land reported by Wrangel Land, Extensive Land with high Peaks*, and *Appearance of Land* marked on a vast northern land which would have, had it existed, formed the northern shore of the Northwest Passage. Farther east it is marked *Probable Southern line of a new land (Lieut. Hooper's conjecture)*. At least he was honest.

MAP 147.
This is part of a world map from a book by Berthold Seeman, a naturalist with Kellett on the *Herald*. Although small-scale, it does shown the *Herald*'s circuitous track, and also *Herald I.*[sland]. Again a vast northern land is shown, marked *Probable Line of Land*. Obviously the Arctic mirages were convincing that year.

Nancy Dawson turned back because its increasingly nervous crew began to mutiny.

For two years Pullen and Hooper searched the coast as far east as Cape Bathurst in two small boats, retiring up the Mackenzie to Fort Simpson to winter twice. This search was a not inconsiderable feat, although, because Franklin himself had explored this coast before, no new geographical discoveries were made except for two small islands in the Mackenzie Delta, which were named Pullen and Hooper Islands (MAP 148).

In 1849 Henry Kellett searched north of Siberia, thinking Franklin could have perhaps drifted in that direction. Here he discovered a small island, which he named Herald Island (Ostrov Geral'd). He saw another island in the distance which he named Plover Island; this may have been a mirage or it could have been a misplaced Wrangel Island (Ostrov Vrangelya; see page 116). Whatever islands he may have found, he did not find Franklin.

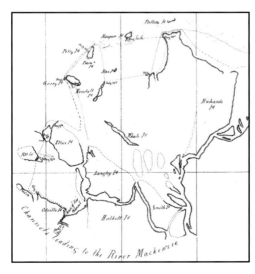

MAP 148.
This map by William Pullen shows part of the Mackenzie Delta with the only new discoveries he and Hooper made, the small islands they named *Pullen I*[slan]*d* and *Hooper I*[slan]*d*. Mackenzie's *Whale I*[slan]*d* persists, as it would right up to the 1940s, when aerial survey finally showed that it did not exist. Garry Island was Mackenzie's Whale Island.

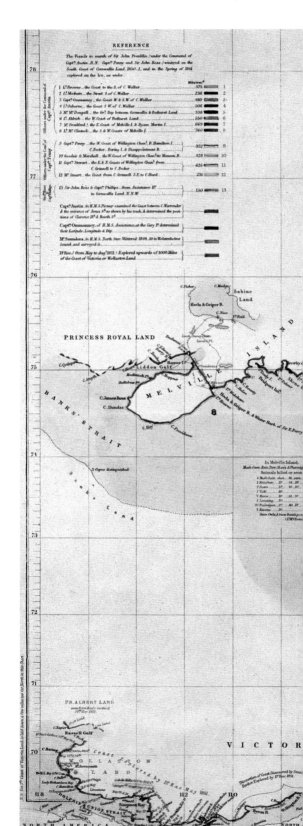

MAP 149 (left).
A map used by Charles Forsyth in 1850, showing the track of his ship *Prince Albert*, together with notes derived from information from other ships. After Erasmus Ommanney found Franklin's first wintering place at Beechey Island, Forsyth thought the information so important he sailed back to England with it.

In 1850 came the first of two massive efforts by the Admiralty. Four ships were sent to search from the east as a fleet, under the command of Horatio Austin in HMS *Resolute*. The others were Erasmus Ommanney in HMS *Assistance*, John Bertie Cator in the steamer HMS *Intrepid*, and Sherard Osborn in the steamer HMS *Pioneer*. Two others, initially privately organized by Jane Franklin, but then taken under the Admiralty's wing, also joined the effort: William Penny in HMS *Lady Franklin* and Alexander Stewart in HMS *Sophia*. These were whaling ships that had been temporarily converted into naval vessels.

This small armada was to try a new method of searching. From the ships, still trapped in the ice in the spring, men would fan out, pulling sledges in all directions, covering vast amounts of territory at a time in the season when the ships could not move. This arduous technique proved successful, covering far more territory and revealing much more of the geography of the Arctic than would otherwise have been possible.

It was at this time, and as part of the Royal Navy's search plans, that two other ships were sent out to sail through Bering Strait and search eastwards. HMS *Enterprise*, under Richard Collinson, and HMS *Investigator*, under Robert M'Clure, disappeared from public view for five and four years, respectively, and we will consider them when they reappear (page 97).

In addition that year the effort was joined by John Ross in a venture financed by the Hudson's Bay Company and public subscription, including a large donation from Felix Booth, no doubt still pleased at having his name festooned across the Arctic. He joined in with a schooner, *Felix*, towing *Mary*, his own small yacht. As if this was not enough, Jane Franklin financed the *Prince Albert*, commanded by Charles Codrington Forsyth.

Even the Americans pitched in, sending two ships, the *Advance*, under overall commander Edwin Jesse De Haven, and the appropriately named *Rescue*, under Samuel

MAP 150 (below).
A detailed summary map of the expeditions of 1850–51, drawn by London mapmaker John Arrowsmith in April 1852. Although Franklin was not found, long stretches of Arctic coastline were mapped; the key gives details. Captain Phillips was John Ross's second-in-command; James Saunders in the supply ship HMS *North Star* had been sent out in 1849 to resupply James Clark Ross's expedition of 1848–49, but was caught in the ice of Wolstenholme Fjord in northwestern Greenland; the details of parties under each captain refer to sledge parties sent out in the spring of 1851. This technique, it was thought, would cover vastly more territory than would be possible by ships alone. But the British didn't want to use dogs to pull the sledges, and insisted instead on using men, as shown in the painting at right, from the journal of Walter William May, mate of the *Resolute*.

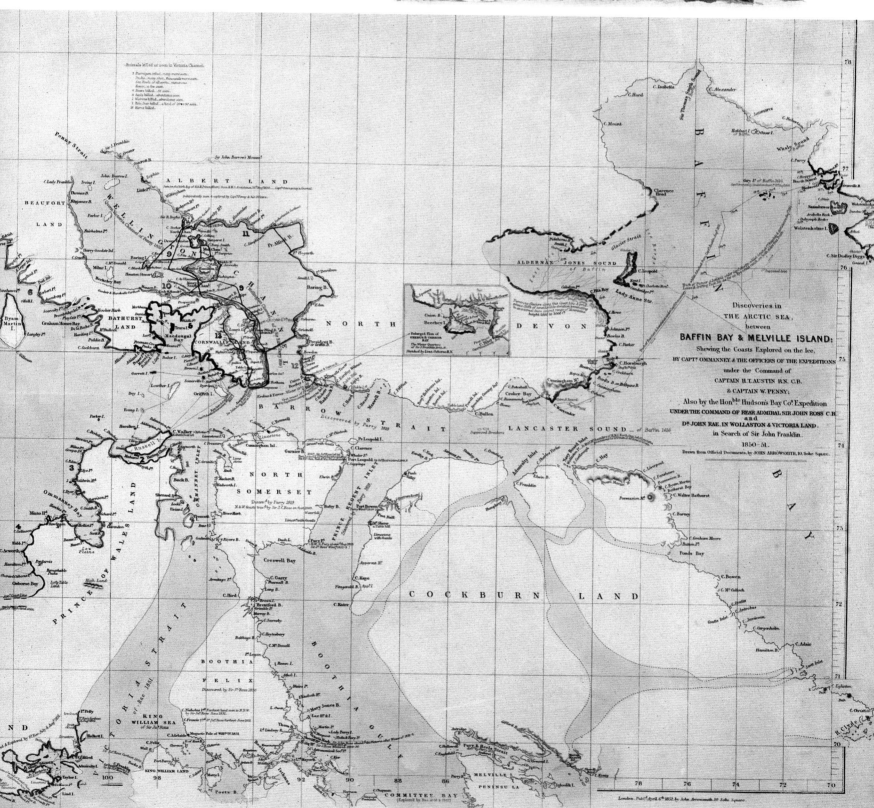

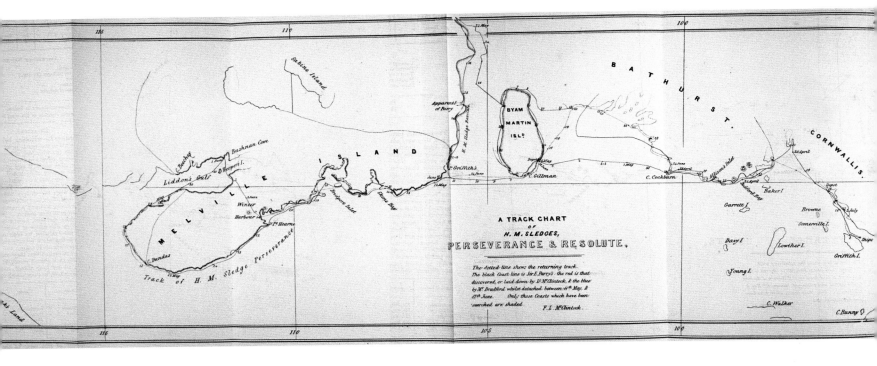

MAP 151 (above).
Many of the maps of the individual sledge parties sent out in the spring of 1851 were published in parliamentary reports the following year. This one shows the track of Leopold M'Clintock, who was later considered one of the most expert sledgers, around Byam Martin Island and the southern part of Melville Island. The sledges, just like the ships, were all given names; M'Clintock's was HM Sledge *Perseverance*. His sledge trip was made between 15 April and 4 July 1851. Abraham Bradford, the surgeon from the ship *Resolute*, now in command of the sledge *Resolute*, accompanied M'Clintock for much of the time, but searched the eastern part of Melville Island individually, as shown.

MAP 152 (below).
John Rae's map of roughly 1 000 km of new coastline he discovered and mapped on Victoria Island, his *Wollaston Land* (the Wollaston Peninsula) and *Victoria Land*.

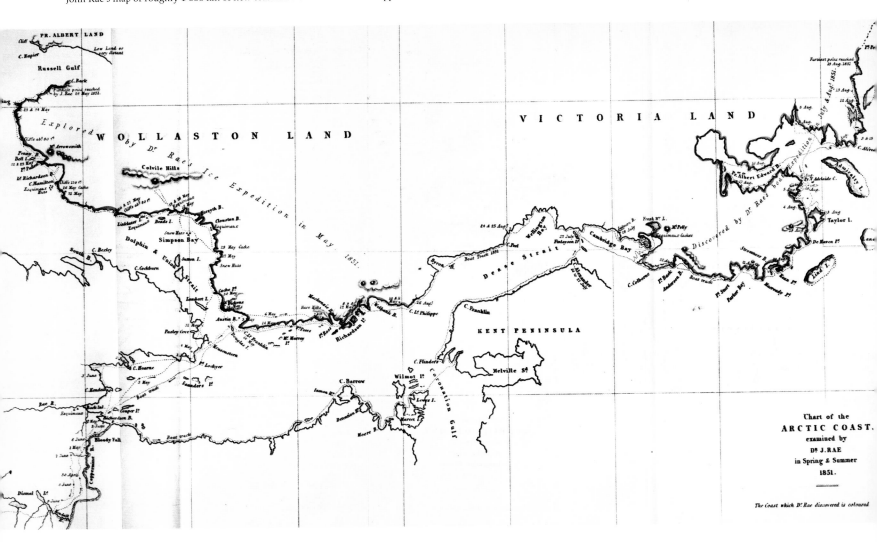

Griffin. This expedition was sponsored by philanthropist Henry Grinnell. De Haven managed to get his ships caught in the ice of Wellington Channel, which then drifted north to 75° 24´ N, from where undiscovered land, the Grinnell Peninsula of Devon Island, could be seen.

With this huge concentration of resources, a lot of coastline was mapped, but Franklin was not found. Evidence of where he had been, however, was. In August 1850 Erasmus Ommanney discovered Franklin's wintering place in 1845–46 at Beechey Island, off the southwest tip of Devon Island (see MAP 166, page 102). But there were no messages, and no indication as to where Franklin had headed in the 1846 season. Elisha Kent Kane, surgeon on the *Advance*, later wrote that he thought this "an incomprehensible omission." The graves of three of Franklin's crew were found here in 1984, with the bodies well preserved in the permafrost. They had been given a

MAP 153 (*right*).
In 1852 German geographer Augustus Petermann suggested that a search for Franklin should be made northwards through "the wide opening" between Spitsbergen and Novaya Zemlya, which, he thought, offered the "easiest and most advantageous entrance into the open, navigable Polar Sea." It was never tried. This map shows his *Probable position of Franklin*, within the red dashed line north of Bering Strait. Ironically, Petermann, at the time an advocate of the existence of an open polar sea, later produced a map that suggested Greenland was connected to Wrangel Island (Ostrov Vrangelya), which was completely at odds with his earlier theory.

MAP 154 (*below*).
Edwin Jesse De Haven's map of his discoveries during the Grinnell expedition, the first American attempt to find Franklin. The track of his two ships, *Advance* and *Rescue*, is shown; the wavy part of this is their drift north in Wellington Channel after they were trapped in the ice in September 1850 until they were released again the following June. Maury Channel, north of Cornwallis Island, was named by De Haven after Matthew Fontaine Maury, an oceanographer who advocated the open polar sea concept; De Haven thought this channel must lead to such a sea.

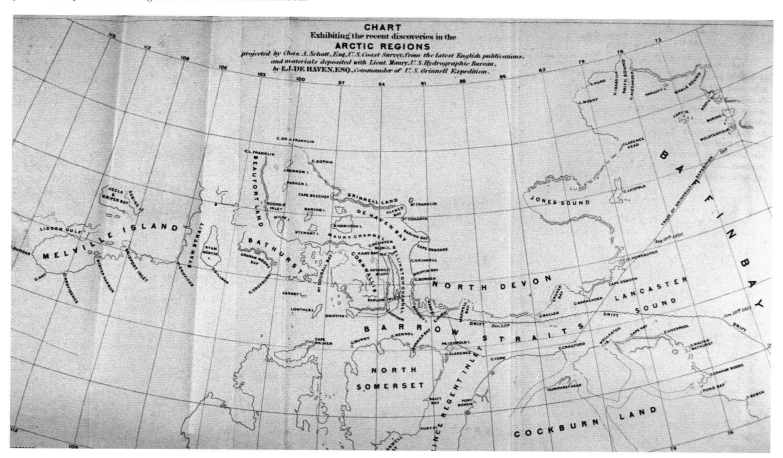

proper burial in coffins, perhaps a better fate than that which befell the rest of the men.

In 1851, while Austin and Penny's teams continued with their search, another expedition was dispatched, organized again by the determined Jane Franklin and financed both by her and with public subscription. It was led by William Kennedy in the *Prince Albert*, with a French naval lieutenant, Joseph-René Bellot, as second-in-command. Volunteers, they had each suggested that the other be the commander.

Kennedy attempted to enter Port Leopold on Somerset Island to winter, and he made it onto the shore, only to find the ship drifting away without him. A clear-headed Bellot took the ship to Batty Bay, to the south, and quickly mounted an overland rescue mission for Kennedy. Sometimes the rescuers needed rescuing.

In 1852, Kennedy and Bellot set out using dogsleds to search the nearby coasts for traces of Franklin. After traveling south along the east coast of Somerset Island they discovered a narrow channel cutting what until then they had thought was Boothia Felix into two, with Somerset Island to the north and the Boothia Peninsula to the south. Consequently they had discovered the northernmost point of the North American continent. Kennedy named the channel Bellot Strait.

The pair passed through their newfound strait to Peel Sound, crossing also Prince of Wales Island to the recently explored Ommanney Bay; they then sledged north to Cape Walker, crossing again to Somerset Island, this time covering some of the same ground as James Clark Ross in 1849. Their 1 800 km trek, shown on Kennedy's map (MAP 155, *below*), was one of the longest sledge journeys made during the search for Franklin, but they, sensibly, used dogs; most of the expeditions, and certainly all the naval expeditions, refused to do so. Right through this period and into the 1870s, the British navy would never admit to the superiority of dogs, pinning its hopes instead on poor British seamen.

There is a sad footnote to this story. In 1853 the intrepid volunteer Bellot was back again, this time on a naval supply vessel, HMS *Phoenix*, under Edward Inglefield. On a mission to carry dispatches to the main British fleet he was carried away on an ice floe and disappeared.

A final massive effort to find Franklin was made in 1852, under the general command of the surprisingly inept Edward Belcher, a hard-working but tyrannical taskmaster selected for his navigating and surveying skills rather than his command of men. Belcher in HMS *Assistance*, Sherard Osborn in the steamer HMS *Pioneer*, Henry Kellett in HMS *Resolute*, Leopold M'Clintock in the steamer HMS *Intrepid*, and William Pullen in HMS *North Star*, the latter a supply ship used as a base at Beechey Island, again mounted a huge sledge search.

Belcher's instructions were to search northwards up Wellington Channel and westwards to Melville Island. Apparently no one at the Admiralty thought it worthwhile to search to the south or southwest, the primary direction that Franklin had been ordered to investigate.

Belcher and Osborn searched north in Wellington Channel by sledge in the spring of 1853, mapping considerable new lengths of coastline (MAP 157 and MAP 158). Belcher explored the north coast of Devon Island as far west as Jones Sound. Belcher began to move his ships southwards in mid-July, but they became trapped, and he was forced to winter another year near Cape Osborn, in Wellington Channel. It was August 1854 before the ice released its grip on the ships, but they could not force a channel enough to

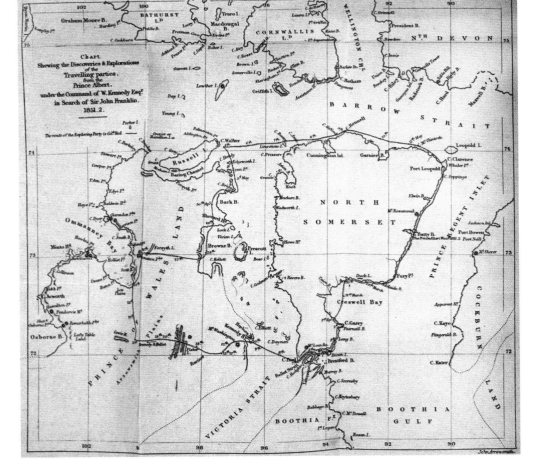

MAP 155.
William Kennedy's map showing his search path with Joseph-René Bellot in 1852. This map, drawn by John Arrowsmith for Kennedy's 1853 book, is the first depiction of Bellot Strait, the channel he named after his friend and rescuer. It is easy to see why John Ross and others had missed this narrow channel before.

MAP 156.
Bellot Strait (D.[etroit] de Bellot) shown on the map in Joseph-René Bellot's book, published posthumously in 1854.

MAP 157 (*right*).
A map of the northeast coast of Bathurst Island made by Sherard Osborn—with his sledge *John Barrow*—during his search in June and July 1853. Queens Channel, to the north of Cornwallis Island and Wellington Channel, lies between the Grinnell Peninsula of Devon Island, at top right, and the eastern shore of Bathurst Island, to the left. This map is typical of many made by the various commanders during their sledge journeys.

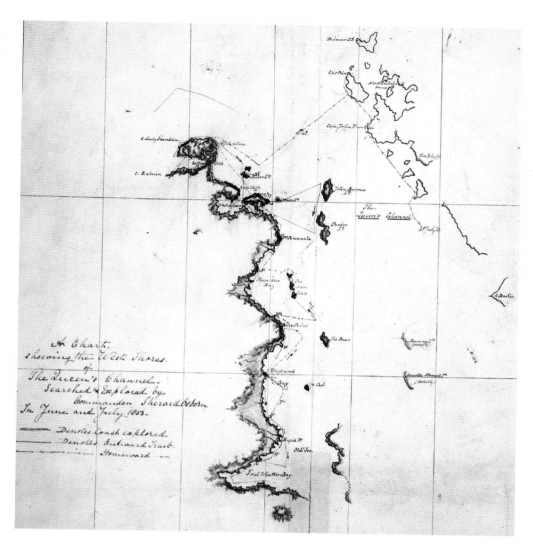

escape. Belcher ordered the ships abandoned, and the men sledged to the *North Star* at Beechey Island.

The second part of the Belcher fleet was the *Resolute,* under Henry Kellett, and the steamer *Intrepid,* under Leopold M'Clintock. They were detailed to carry out the search to the west. They were to search for Franklin, certainly, but also for Collinson and the *Enterprise* and M'Clure and the *Investigator,* who had sailed to Bering Strait to search for Franklin from the west in 1850 and had disappeared themselves.

The *Resolute* and the *Intrepid* left the other ships in mid-August 1852 and sailed and steamed west to the south coast of Melville Island, where they found a safe haven at Dealy Island. Sledge parties fanned out to lay depots for the search the following spring. One of these, led by George Frederick Mecham with the sledge *Discovery* and George Nares with the sledge *Fearless,* stopped at Winter Harbour—Parry's wintering place on his pioneering voyage in 1819—where they found a note left by Robert M'Clure of the *Investigator* detailing his ship's entrapment in ice at Mercy Bay, on the north coast of Banks Island. It was too late in the season to do anything right away, but it was this note that led to M'Clure's rescue the following year by the first sledge to set off in the spring (see page 99).

The sledge parties searching for Franklin set out on 4 April 1853. Mecham sledged through Liddon Gulf to the western end of Melville Island. He then crossed Kellett Strait to Eglinton Island and then Crozier Channel to the southern part of Prince Patrick Island. He returned via the northern end of Eglinton Island and was back at his ship on 6 July. Mecham sledged over 1 800 km, one of the longest journeys made during the Franklin search, and he also produced one of the finest of the "sledge maps"; it is reproduced overleaf in its original manuscript form (MAP 160).

Another similarly long trek was made by Leopold M'Clintock with his sledges *Star of the North* and *Satellite.* He crossed Melville Island to Hecla and Griper Bay and

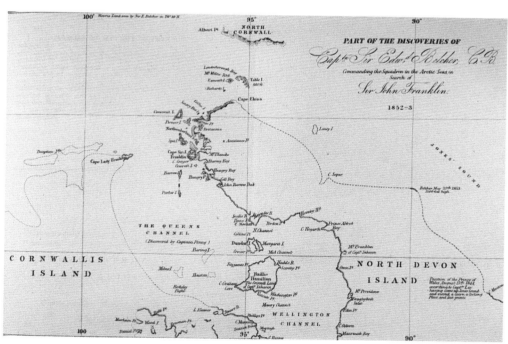

MAP 158 (*above*).
Edward Belcher's map of some of his discoveries in 1852–53 from his book, optimistically entitled *The Last of the Arctic Voyages.* In May and June 1853 Belcher, with sledge *Londesborough*, traveled eastwards along the north coast of the Grinnell Peninsula (the main body of land in the center of the map) of Devon Island, as far as Cardigan Strait, which leads to Jones Sound. The strait between the Grinnell Peninsula and Cornwall Island, here *North Cornwall*, is today named Belcher Channel. There is an interesting notation at the top of this map: *Victoria Land seen by Sir E. Belcher in 78° 10´ N.* This is the correct position of the hills at the southern end of Ellef Ringnes Island, ostensibly undiscovered until Otto Sverdrup found it in 1901 (see page 156).

MAP 159 (*below*).
An enlarged part of the printed version of George Frederick Mecham's map of his explorations of Melville, Eglinton, and Prince Patrick Islands, showing Dealy Island and the position of *Resolute* in the winter of 1852–53. The steamer *Intrepid* was also at Dealy Island, but Mecham, as lieutenant on the *Resolute,* has chosen to show only his ship. A little drawing of a ship and the words *HM Ship Resolute* are on the original version at right. The two ships wintering at Dealy Island were equipped with the latest technology: an electric telegraph ran between the two ships, a wonderful convenience when communication was necessary at fifty below.

MAP 160 (*right*).
Perhaps the finest of the maps of sledge journeys made during the various searches for Franklin is this one of the southern part of Melville Island, Eglinton Island, and the western and southern parts of Prince Patrick Island drawn by Frederick Mecham between April and June 1853. The latter two islands had not been mapped before. Not only is the map carefully and skillfully drawn, but four views are also given at top, plus a sketch of the sledge *Discovery*, complete with a sail. The poor seamen are doing the actual pulling while the officer, Mecham, scouts out the route using his telescope. On the map, the red line is Mecham's track from his ship, the *Resolute,* at Dealy Island (bottom right), to the west coast of Prince Patrick Island and his return. Each camp is marked with the date. Just west of Dealy Island is Parry's Winter Harbour, where Mecham found the note from Robert M'Clure late the previous season. The key at far right reveals that depots, laid down in 1852, are shown as a red rectangle enclosing a dot. Mecham's round trip shown on this map totalled about 1 800 km.

In May 1854 Mecham embarked on another epic sledge journey, which was to prove to be the longest of any during the Franklin search. He was to search this time to the south for Richard Collinson and the *Enterprise*, which, like M'Clure and the *Investigator*, had sailed into Bering Strait to search for Franklin from the west. He found records from *Enterprise* in a cairn on the Princess Royal Islands, and more records at the southern end of the strait. He then returned to his ships, only to find that they had been abandoned on Belcher's orders. Mecham and his crew had to sledge east to the *North Star* at Beechey Island, where they arrived on 12 June. This was the longest man-hauled sledge journey ever, a distance of 2 470 km.

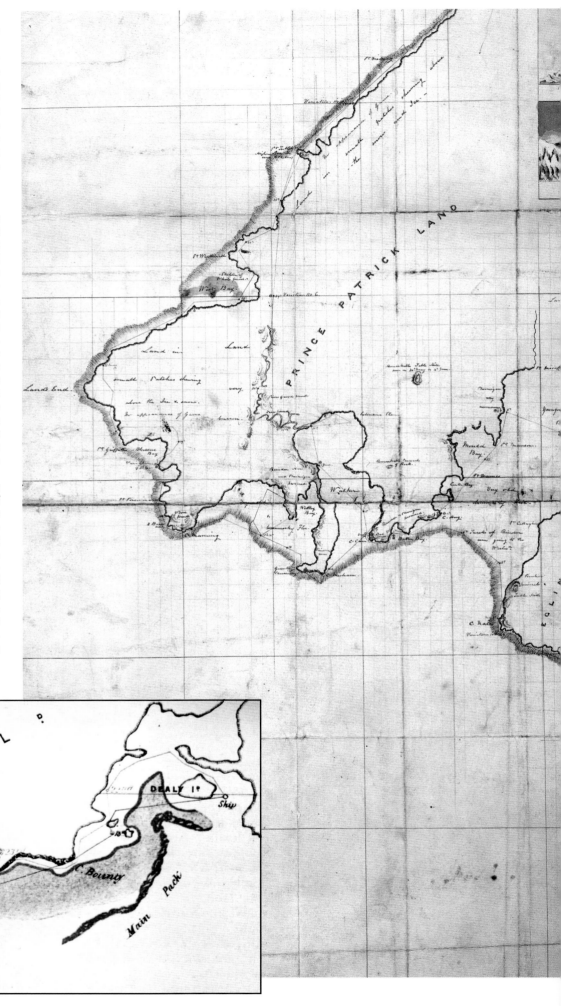

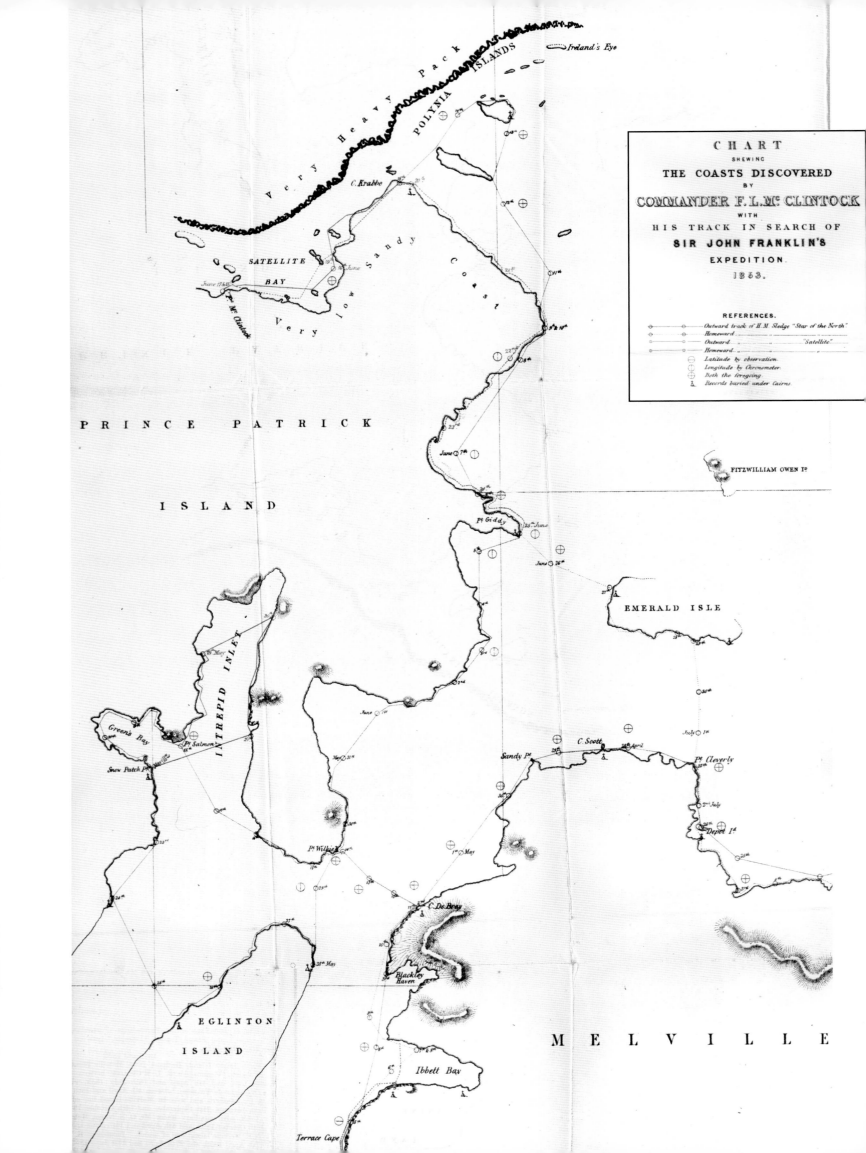

MAP 161 (left).

Leopold M'Clintock's map of his epic sledge journey in 1853, searching for Franklin—and incidentally discovering unknown coasts—on the northern part of Prince Patrick Island, one of the outermost islands of the Canadian Arctic Archipelago. The multi-year ice invades the western shore of this island virtually all the time; it is truly at the edge of the Polar Sea. This map was a printed version of one originally drawn, like Mecham's (MAP 160), in the field. It was published, as they all were, in *Parliamentary Papers* in 1855.

searched to the island's northernmost point. He then found Prince Patrick Island's northern end and followed its coast northwards and westwards to Satellite Bay, on the edge of the multi-year ice pack. M'Clintock's journey, which was, like Mecham's, one of the longest of the Franklin search and also about 1 800 km, is shown here (MAP 161).

Another significant sledge journey was made by Bedford Pim and William Domville, a lieutenant and surgeon on *Resolute*, respectively. They were detailed to sledge to Mercy Bay on Banks Island and find Robert M'Clure and the *Investigator*. To relate how M'Clure got to Mercy Bay we must backtrack to 1850, when, as part of the first concentrated effort to find Franklin, M'Clure, and Richard Collinson in the *Enterprise*, were dispatched to Bering Strait to search from the west eastwards. Both ships disappeared, hence the instructions given to Kellett and M'Clintock to search for them as well as for Franklin.

In August 1850, *Investigator* passed Point Barrow and sailed east along the north coast of Alaska, becoming the first ship to navigate in the Beaufort Sea.

On 7 September M'Clure discovered the southern coast of Banks Island but, not realizing that it was part of the same island Parry had sighted in 1819 when he looked southwards from his position off Melville Island—in what is now M'Clure Strait—he gave it a new name, Baring Island. M'Clure then sailed north into Prince of Wales Strait, between Banks Island and Victoria Island, making it to within 55 km of Parry's Viscount Melville Sound on 17 September. But it was too late in the season; the ship became surrounded by ice and they were forced to retreat, finding a wintering place in the tiny Princess Royal Islands. During October, M'Clure sledged north to Russell Point, at the northwestern tip of Banks Island, and was able to visually confirm not only where he was, but that a Northwest Passage existed, which must have been a quite momentous revelation given the long history of the search for such a passage.

The following April, sledge parties set out to explore and search the surrounding area. The northwest coast of Banks Island was explored

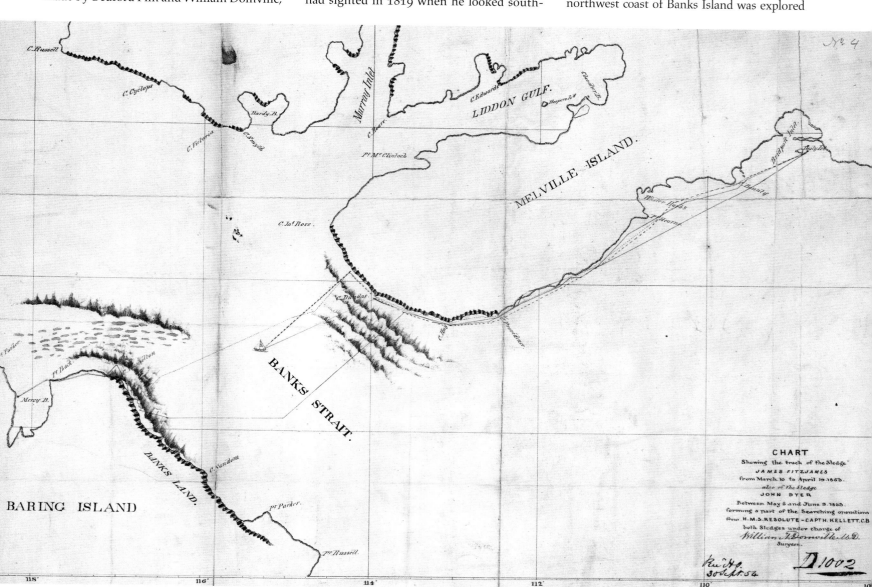

MAP 162.

Map drawn by William Domville, *Resolute*'s surgeon, showing his sledge journey to Mercy Bay in company with Bedford Pim's sledges to rescue Robert M'Clure. *Resolute*'s position at *Dealy Is*[lan]*d* is at top right, *Mercy B.*[ay] at left. Domville has used both Parry's original name of *Banks Land*, applying it just to the coast, and M'Clure's name, *Baring Island*. Today it is Banks Island.

by Samuel Gurney Cresswell, and the north coast of Victoria Island by Robert Wyniatt east to Wyniatt Bay. The east coast of Victoria Island was explored by William Haswell as far south as Prince Albert Sound. Quite by coincidence, this was only 75 km from the northernmost point reached by John Rae only ten days later (see MAP 152, page 90). M'Clure, taking his interpreter Johann Miertsching, sledged south to talk to a group of Inuit at the southern end of Prince of Wales Strait.

M'Clure attempted to sail north once more in July, but by 15 August it was clear that the ice would not let him pass. It must have been very frustrating for him. He knew the passage existed, he knew what direction to take to get through it—yet the ice would not let him. So he took the *Investigator* south once more and began an attempt to reach Viscount Melville Sound via the west and north coasts of Banks Island. What M'Clure did not understand, as no one did at this time, was that the multi-year ice pack pressed eastwards on the outer islands of the archipelago and was unlikely to allow him to pass.

The ship made a fast run, reaching Cape Kellett, at the southwest tip of Banks Island, three days later. The timing must have been just right, for another day saw them almost at the northwest tip of the island, but then the ice closed in again and they were trapped. Luckily for M'Clure and his crew, in mid-September the ice allowed them to work their way around to the north coast, where a safe haven was found at what M'Clure later named the Bay of Mercy (Mercy Bay). But as it turned out, the ship was too safe, for it would never leave this harbor.

The following year, 1852, M'Clure sledged across the strait to Winter Harbour, the bay Edward Parry wintered in during his pioneering first penetration west in 1819–20. Here he left the note describing his position that would be found later that year by Frederick Mecham. The *Investigator* could not be moved that summer. By the following spring, M'Clure was getting ready to abandon ship, sledging towards the Coppermine River or Greenland, Franklin-like.

Thus it was that in March 1853, knowing now where M'Clure was, Kellett sent Bedford Pim and William Domville to rescue him. On 6 April 1853, Pim located the *Investigator*. He described what happened:

5.0 PM, arrived within 100 yds. without being observed, then however two persons taking exercise on the ice discovered that I did not belong to their ship, upon beckoning they quickly approached and proved to be Capt. McClure and Lt. Haswell, their surprise and I may add delight at the unexpected appearance of a stranger (who seemed as it were to drop from the clouds) it is needless to describe. One of the men at work near them conveyed the news on board and in an incredible short time the deck was crowded, every one that could crawl making his appearance, to see the stranger and hear the news. The scene which then presented itself can never be effaced from my memory nor can I express any idea of the joy and gladness with which my arrival was hailed.

A bustle of sledging activity ensued between *Investigator* and *Resolute*. M'Clure was still hopeful of completing the Northwest Passage and was reluctant to abandon his ship, but he was overruled by the more senior Henry Kellett, and six weeks later M'Clure transferred all his men to the *Resolute* and *Intrepid*.

Samuel Gurney Cresswell, taking some of the invalids, left *Resolute* with dispatches and headed to Beechey Island. Here he was able to board HMS *Phoenix*, a supply ship, and arrived back in England later in 1853 with the news of M'Clure's salvation. By so doing, Cresswell and his small group of men became the first to complete a transit of a Northwest Passage.

The *Resolute* and *Intrepid*, meanwhile, became trapped by ice and had to spend another winter in the Arctic. In the spring they received orders from Belcher to abandon the ships, despite Kellett's protestations. The crews had to sledge to Beechey Island and returned to England on the *North Star* and two supply ships which had just arrived, *Phoenix*, returning again, and *Talbot*.

Thus did Robert M'Clure lay claim to being the first through the Northwest Passage. His journey was from west to east, his ship had been lost, and the transit was only possible because of the assistance of Kellett and his men. But as commander of the first men to travel from Pacific to Atlantic, he had a legitimate claim.

Others thought so too. Sherard Osborn wrote to John Barrow (John Barrow senior's son, who had inherited his position at the Admiralty) from *Pioneer* on 20 July 1853, "I only wish I could see your face when you hear the N.W. Passage has been discovered, and accomplished in a way, by that marvellous man McClure." The junior Barrow no doubt would have liked his father to have heard this news.

Belcher's decision to abandon four of the five ships of his squadron has never been satisfactorily explained. He was court-martialled, but because he had been given the authority to abandon ships if necessary, he was acquitted. The Admiralty never dreamed he would abandon four, all without the concurrence of his senior officers.

Kellett's ship *Resolute* was not about to be abandoned. In what was really an amazing indictment of Belcher's decision, the lonely

This dramatic painting of the *Investigator* nipped by two large ice floes off the coast of Banks Island in September 1851 was painted by Lieutenant Samuel Gurney Cresswell.

MAP 163 (*left*).
Robert M'Clure's map of the coasts of Banks and Victoria Islands discovered by him between September 1850 and October 1851, likely drawn towards the end of 1851. The position he reached in September 1850 and his meeting with the ice is shown at the north end of Prince of Wales Strait. His wintering place at the Princess Royal Islands in the middle of the strait is marked *Ship wintered 1850–51*. The track of the *Investigator* from there around the west coast of Banks Island (Baring Island, M'Clure's name, on the map) is shown, as is the bay on the north coast which was to prove the ship's final resting place, where the winters of 1851–52 and 1852–53 were spent by M'Clure and his crew. This was the harbor M'Clure called the Bay of Mercy (Mercy Bay). The bay is not named on this map but marked *Ship wintered 1851–52*. M'Clure named the bay after his deliverance.

MAP 164 (*overleaf*).
A magnificent summary map of the Franklin search expeditions and the new lands discovered by them. It was published by London mapmaker John Arrowsmith in November 1853, rushing to press the geographical discoveries of the Belcher armada and the considerable new information from M'Clure's transit of the Northwest Passage. The information was all carried back to Britain by Samuel Gurney Cresswell, ahead of the return of either Belcher or M'Clure. This map was an updated version of Arrowsmith's previous map (MAP 150, pages 88–89) published only the preceding April (1853), and it is instructive to compare the two.

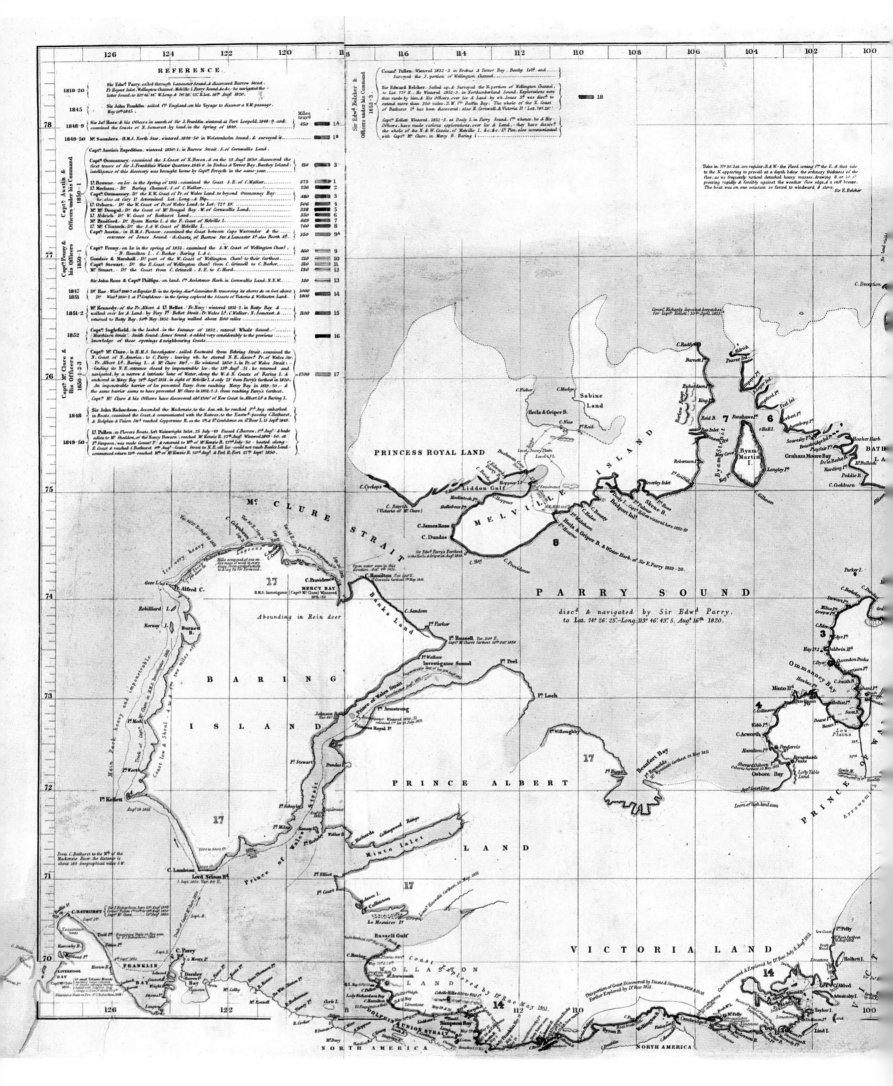

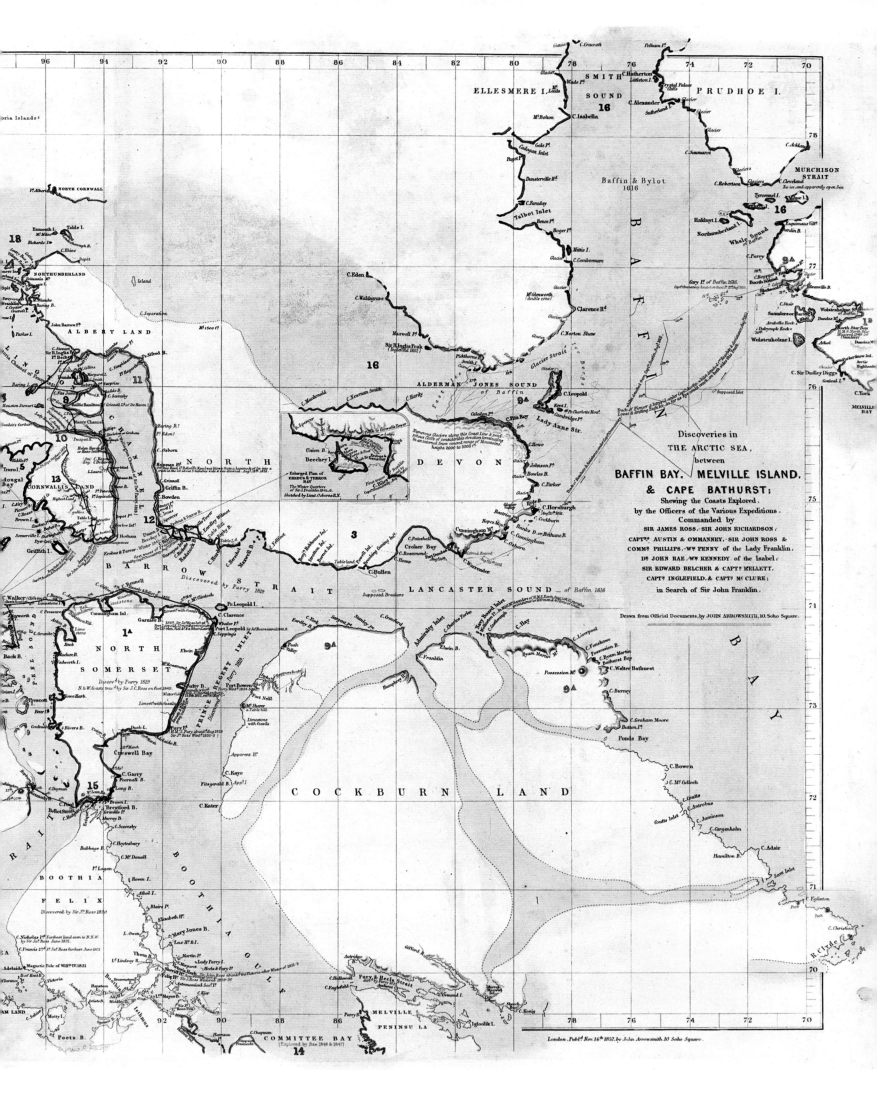

ship drifted to the west, through Lancaster Sound, and south to Davis Strait, where it was found by the whaler *George Henry*. The *Resolute* had drifted about 2 500 km in sixteen months. She was purchased by the United States government, restored, and presented as a gift to Queen Victoria.

With the return of Belcher, the British government now essentially gave up the search for Franklin, reasoning that if he could not be found with such a massive infusion of resources, he was probably never to be found, and was certainly not likely to be alive anyway. Future searches would be left to private individuals.

The ship that had initially accompanied *Investigator* for the search from the west in 1850 was that of Richard Collinson, the *Enterprise*. His turned out to be the longest of all the voyages associated with the Franklin search; Collinson left England in January 1850 and did not return until May 1855.

Collinson could not get farther than Point Barrow in 1850, and so he wintered in Hong Kong. The following year he tried again, and this time he found himself following the track of M'Clure but never quite catching up. He too sailed north in Prince of Wales Strait hoping to make it into Viscount Melville Sound, but he too was stopped by ice. He could not follow M'Clure around the west coast of Banks Island, however, for the fickle multi-year ice had let M'Clure pass only by luck of timing. Collinson wintered in 1851–52 in Walker Bay, on the southwest coast of Victoria Island.

In the spring of 1852 sledging parties fanned out, searching a wide area. In the summer Collinson first tried to sail east through Prince Albert Sound, but found it to be an inlet. He then tried farther south, passing round the Wollaston Peninsula and through Dolphin and Union Strait, Coronation Gulf, and Dease Strait, north of the Kent Peninsula, finding a wintering place at Cambridge Bay, on the south coast of Victoria Island.

In the spring of 1853 Collinson led sledge parties along the east coast of Victoria Island, only to find, near the end of his explorations, a note left by John Rae in 1851. It seemed that everwhere Collinson went, someone had been there before.

Collinson decided to take the *Enterprise* west again, but he could not escape that season. He was forced to winter once more, at Camden Bay, on the north coast of Alaska. The following August, at Point Barrow, he met up with *Plover*, the supply ship stationed so many

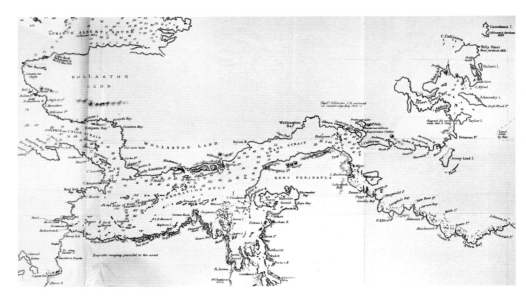

MAP 165 (*above*).
Map illustrating the voyage of Richard Collinson printed in the British *Parliamentary Papers* for 1855. Collinson's wintering at Cambridge Bay in 1852–53 is noted. The map also incorporates the discoveries of John Rae; Collinson covered much of the same ground to the east of the island. Only at Gateshead Island (at right), named by Collinson, did he progress beyond Rae's farthest.

years near Bering Strait, and both ships sailed south. Collinson's route pioneered much of that taken by Amundsen in 1906–09, the route that is today considered the most practical Northwest Passage, but because others had been to various parts of it before him he has received relatively little credit for what was a tremendous feat in a sailing vessel.

It fell to John Rae to find the first evidence of Franklin's demise. In 1853 the Hudson's Bay Company dispatched Rae to complete the map of the northern coast of North America by surveying the southwestern coast of the Boothia Peninsula. This he did in 1854. From Inuit he learned they had seen forty white men north of King William Island, walking south, and later they had seen thirty corpses west of the Great Fish (Back) River. But most definitive of all, Rae purchased from the Inuit a silver plate engraved with Franklin's name and some cutlery, finding, incredibly, that even as they fought for their very survival, Franklin's men were lugging the comforts of home across the ice.

In September, boarding a company supply ship, Rae himself took the news to Britain.

Essentially the final major voyage of the Franklin saga was the voyage of Francis Leopold M'Clintock in *Fox*, from 1857 to 1859. This was a private venture financed by a persistent Lady Franklin and a sympathetic public to try to find more evidence of what had happened to Franklin. From the eastern end of Bellot Strait, in April to June 1859, using dog teams, M'Clintock made a sledge journey right around King William

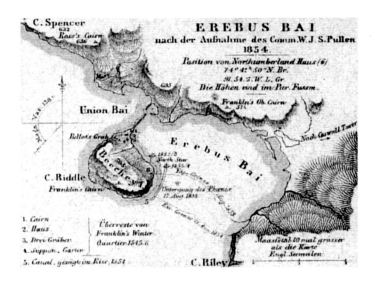

MAP 166.
An inset from a German atlas map of the period showing Beechey Island and the positions in which the *North Star*, the supply ship used as a base, was anchored. The island is connected by a narrow bar to the mainland of the much larger Devon Island. The map was derived from a survey done by William Pullen.

MAP 167 (right).
This map was drawn by Rochfort Maguire, who in July 1852 took over command of the depot ship *Plover*, stationed in Bering Strait, from Thomas Moore. During the winter of 1852–53 *Plover* wintered at Elson Lagoon (Elson's Bay) at Point Barrow (the northernmost point on the map). During March and April Maguire sledged east along the coast of Alaska as far as Franklin's Return Reef (at right). Maguire has also marked known positions of *Investigator* and *Enterprise*, such as HMS *Investigator Last seen Aug/50* at right.

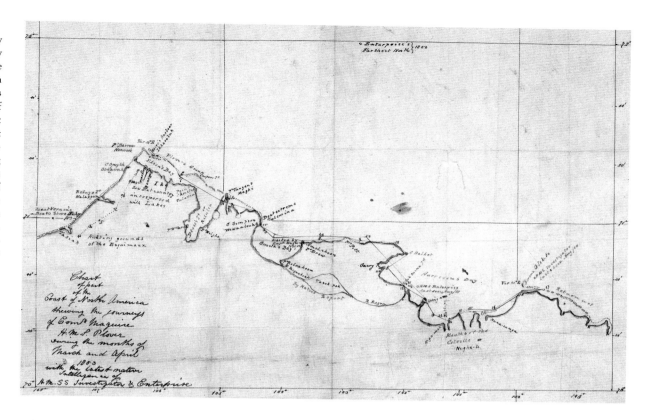

MAP 168 (below).
A map of John Rae's journey in 1854, published by the Royal Geographical Society. The map includes the southwestern coast of the Boothia Peninsula. This last piece in the puzzle of the northern coastline of North America, between the coasts mapped by John Ross and those mapped by Thomas Simpson, was filled in by Rae that year. Items belonging to the Franklin expedition were purchased from Inuit here.

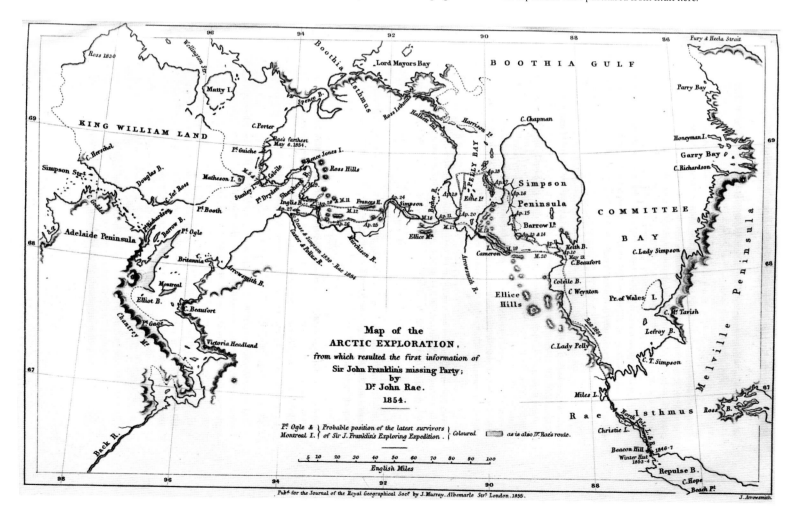

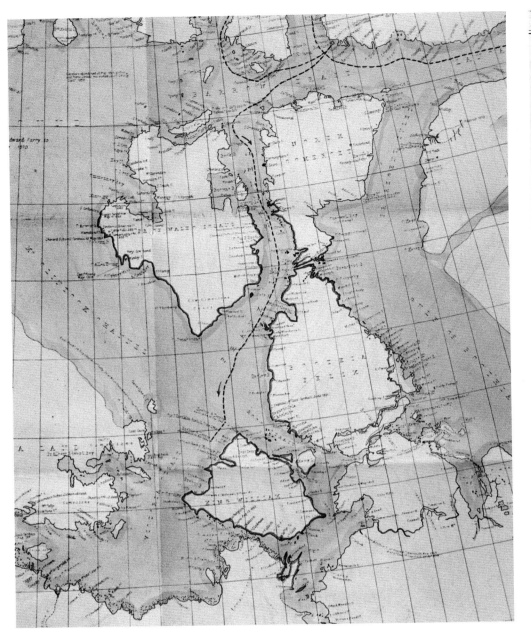
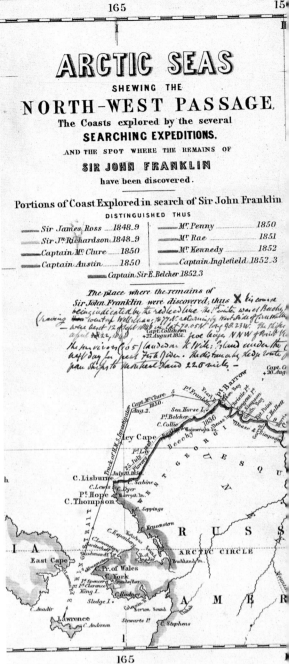

MAP 169.
By the time Leopold M'Clintock returned he had figured out quite accurately what had happened to Franklin. This map, prepared by John Arrowsmith for M'Clintock's book, published late in 1859, shows his postulated track of Franklin's expedition, ending abruptly north of King William Island. The coasts traveled by M'Clintock and his men are shown in red. The text on this map is indistinct. It is a proof map found in the sleeve of a dummy copy of M'Clintock's book, an advance copy of a book that repeats sections of the same pages to give an idea of its finished size and feel. Blank versions of such books are routinely made by printers for publishers today. This particular book, however, used to belong to the Scott Polar Research Institute in Cambridge, England, and was discarded presumably because it was a dummy book, but perhaps without the realization that this map was in the pocket, where it was later found by an astute antiquarian book dealer.

Island, finding skeletons and relics. He also visited Montreal Island on Chantrey Inlet and purchased relics from Inuit.

M'Clintock's second-in-command, William Hobson, sledged with dogs to the west coast of King William Island, where he found a boat left by Franklin's men and the only written evidence of the fate of the expedition, a note saying that all was well, signed by Franklin and dated 28 May 1847. Around the margins of this note was another scribbling added on 28 April 1848, signed by James Fitzjames and Francis Crozier. The note contained many of the details we now know about the fate of the Franklin expedition, such as the date of Franklin's death (11 June 1847), and ended with "and start tomorrow 26th for Backs Fish River."

Second officer Allen Young explored the coasts of Prince of Wales Island, discovering and naming M'Clintock Channel. This is the main channel that delivers the multi-year ice to the vicinity of King William Island, the very ice that trapped the hapless Franklin.

The exploit gained M'Clintock a knighthood and promotions for all, and Parliament later voted £5,000 to the officers and men of the *Fox*. All Britain was relieved to know what had happened to poor Franklin.

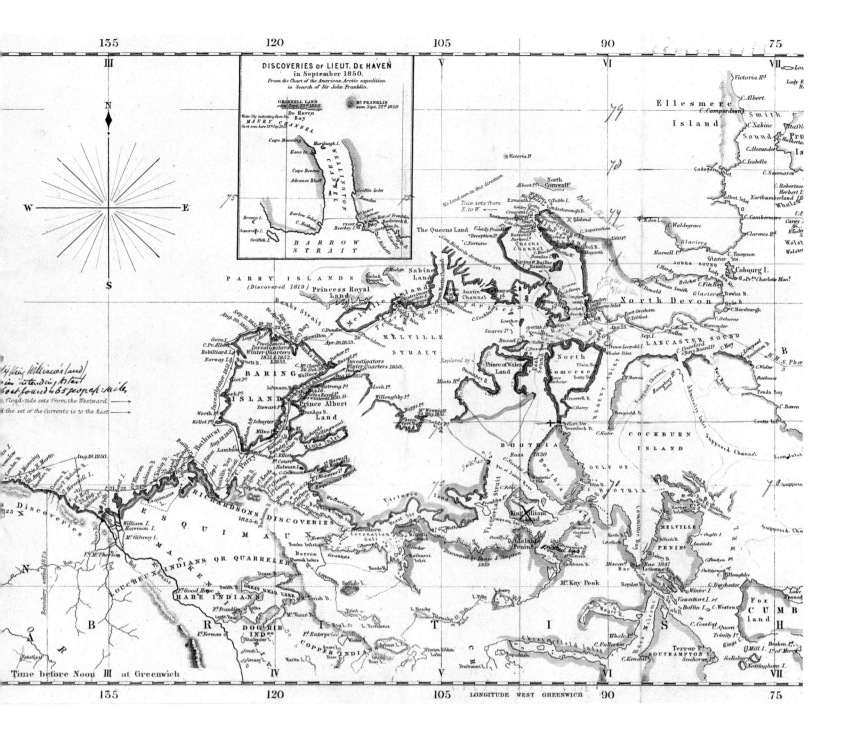

MAP 170 (above).
This summary map of searches for Franklin up to 1854 was published by Edinburgh mapmaker W. and A.K. Johnston that year. The written notation about the location of Franklin's demise has been added much later, after the return of M'Clintock's expedition, because it mentions the boat found by Hobson. A red *X* on King William Island indicates the presumed general location of the Franklin tragedy. Franklin's presumed route is marked, now much faded. The coasts explored by each of the expeditions are shown in different colors. One, that of the 1852 expedition of Edward Inglefield into Smith Sound, is detailed on page 125, for it turned into a push northwards rather than a search for Franklin. Inglefield was the captain of the supply ship *Phoenix* in 1853, when it returned to England with Samuel Gurney Cresswell from the *Investigator*. Elisha Kent Kane's expedition in 1853–55 was also technically a search for Franklin but in reality an attempt to explore northwards (see page 126).

MAP 171 (overleaf).
A rather spectacularly illustrated general map of the Canadian Arctic Archipelago published in 1856 by commercial mapmakers Fullarton & Co. It is in fact the top part of a map that included the polar projection map shown on this book's title pages (pages 2–3).

Just before the return of M'Clintock, one of the most different and—initially, at any rate—unlikely Arctic explorers began his career, one that would end in 1871 at a new "farthest north" (see page 132). Charles Francis Hall was a Cincinnati newspaper proprietor who suddenly decided in 1860 that he was destined to become the discoverer of the fate of Franklin. Inspired by stories of Elisha Kent Kane (see page 126), Hall jettisoned his job and to all intents and purposes his family too, and took off on an expedition, financed by the redoubtable Henry Grinnell and public subscriptions. It was not enough to afford a ship, but this did not stop Hall. He obtained a

H.M.S. TERROR, THROWN UP BY THE ICE IN FROZEN STRAIT.

"PRINCE ALBERT" SURROUNDED
(W. Par

ESQUIMAUX MAN & BOY.

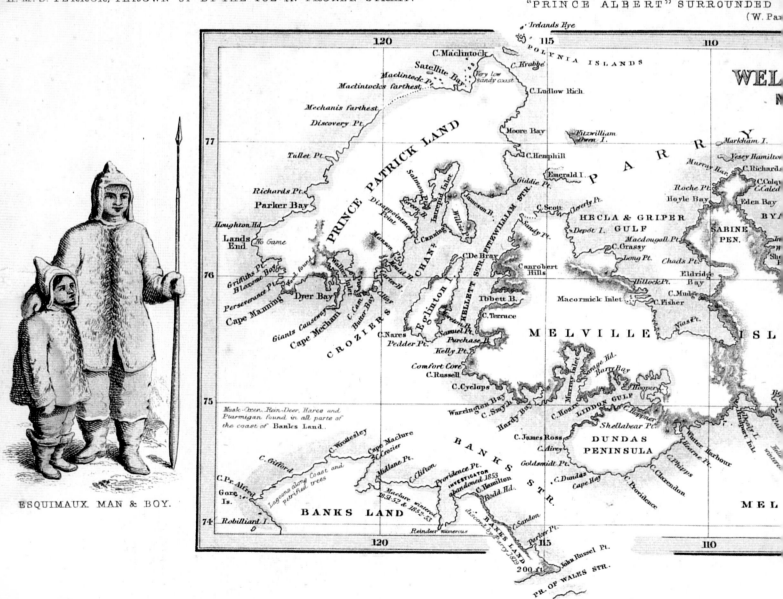

ERGS IN MELVILLE BAY.

CAPE HOTHAM, WITH H.M.S. PIONEER AND RESOLUTE.

ESQUIMAUX FEMALE.

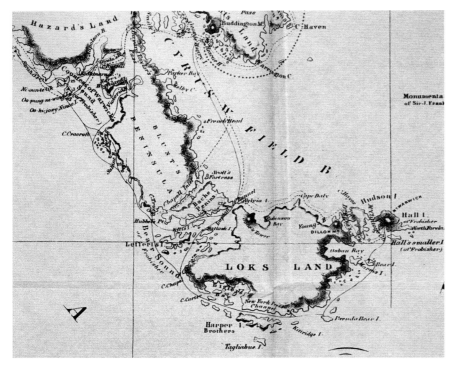

Map 172.
Charles Hall's map of part of the north shore of Frobisher Bay, compiled from his surveys in 1861–62. Note *Kodlunarn Island* (Frobisher's Countess of Warwick's Island) in *Countess of Warwick's Sound. Hall I*[sland], at the western end of *Loks Land,* was named by Frobisher for his sailing master Christopher Hall, not, as it might seem, by Charles Hall after himself.

free passage on a whaler going to Baffin Bay. He was to be dropped off at Frobisher Bay, which was still assumed to be a strait, so that he could travel in a smaller boat through the strait towards King William Island, where, he was convinced, men from Franklin's expedition might still be alive, living with the Inuit.

Hall soon learned that his Frobisher Strait was a bay, first from the Inuit and later from his own experience. It was here that Hall befriended an Inuit couple who were to prove his loyal and constant companions on all three of his Arctic expeditions, Tookoolio and Ebierbing, called Hannah and Joe. And it was here that Hall made a truly monumental discovery, for he found remnants of the three expeditions of Martin Frobisher almost three centuries before. Up to this time, no one was sure where Frobisher had landed, and even his "strait" had been misplaced in Greenland for a century. In 1861 Hall was told of white men coming in two ships, then three, and then fifteen, many generations before, and realized that this information seemed to be describing Frobisher. Later in 1861 and again in 1862 Hall not only proved Frobisher's inlet to be a bay, but discovered numerous relics that were without a doubt from Frobisher's voyages—sea coal, red bricks, iron, glass, pottery, and evidence of mining. He had set out to find Franklin and instead found Frobisher.

Hall surveyed the Frobisher Bay area, and his map shows the locales from the Frobisher expedition, including Kodlunarn Island ("White Man's Island") in the Countess of Warwick's Sound, where most of the expedition's mining activity had occurred (Map 172).

In the late summer of 1862 Hall returned on a whaler to the United States—and the Civil War. He brought with him Hannah and Joe and their baby, who died while there.

Hall didn't want anything to do with the war, but it made organizing expeditions much harder. Nevertheless, such was the continued interest in

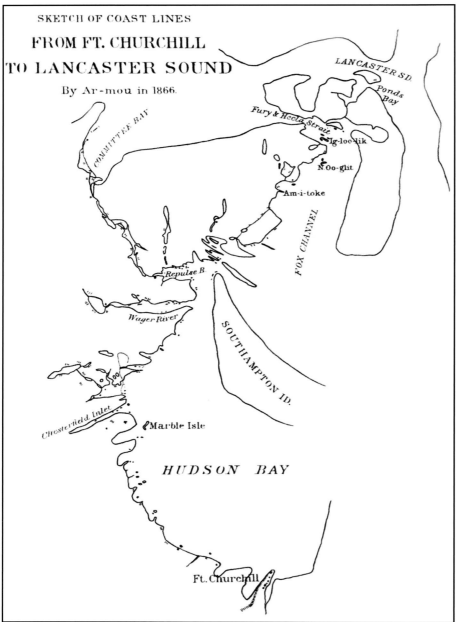

Map 173 (*left*).
Map of the west coast of Hudson Bay north to Lancaster Sound, as drawn for Hall in March 1866 by an Inuit man named Ar-mou (the Wolf). It was a map of the waters and lands he had voyaged over during his lifetime. Hall noted that the distance covered was about 960 nautical miles (1,100 statute miles, or 1 775 km), and six times that if all the indentations of the coastline are taken into account. Apart from the depiction of the eastern ends of Baffin and Southampton Islands, which were beyond Ar-mou's range, the map is quite accurate. It is, however, possible that Ar-mou had at some time seen maps belonging to the fur traders of the Hudson's Bay Company.

Franklin, he managed to raise enough money by public subscription and lectures, which included Hannah and Joe. This time Hall intended to go to Repulse Bay and trek overland to King William Island.

Hall had learned to like living among the Inuit on his previous expedition, and this he was to do again. His endeavors were to take him much longer than he realized; the travels begun in 1864 were to last until 1869. With Hannah and Joe, he was landed, by mistake, 60 km south of Wager Bay, and this cost him a whole year. In 1865 he reached Repulse Bay and in 1866 set out for King William Island, but turned back when he encountered hostile Inuit. The following year he was unable to obtain dogs, essential for travel, or men, due to competition from whalers. In 1868 Hall sledged to Fury and Hecla Strait, having heard reports of two white men there that he thought likely to be Franklin's men; he found none. The loneliness of the Arctic was beginning to tell on Hall by this time; in July he shot and killed one of the contracted whaling men, Patrick Coleman. The incident has never been fully explained because Hall's account is the only surviving one.

Hall finally made it to King William Island in 1869 and was rewarded for his efforts, for he encountered Inuit with tales of struggling white men that could only be Franklin's. On 12 May, Hall reached Todd Island, a small island at the southern tip of King William Island, where he found bones of humans and was told of an encounter with Europeans desperate for food. The Inuit had not been able to support such a number of men without endangering themselves and had departed, as was their way, leaving Franklin's men to their fate. Here Hall gave up the idea that there might yet be Franklin survivors.

On the shores of King William Island itself Hall encountered Inuit with all manner of Franklin relics—cutlery bearing Franklin's crest, pieces of a mahogany desk, and more. And, he was told, when the remains of the white men were found, "arms, legs, &c., were found cut off to be eaten, and the cut of the bone had always showed this to have been done by a saw." Franklin's men, it seems, had resorted to cannibalism in their desperate last days.

In August 1869 Hall took passage on a whaler back to New York. He felt he had not achieved much, but he had in truth achieved what he set out to do; this time he had set out to find Franklin, and he had.

His search for Franklin was over, but Hall's Arctic adventures were not. Two years later he would set out not to find Franklin but to find the Pole (see page 132).

In 1878 reports apparently derived from Inuit sources that journals or other written records of the Franklin expedition might still exist in a cairn on King William Island led the American Geographical Society to sponsor an expedition to try to find them. An expedition of five men, led by Frederick Schwatka, a lieutenant in the U.S. army, traveled to Hudson Bay on a whaling ship in the summer of 1878 and wintered near Daly Bay, just north of Chesterfield Inlet.

The reports of records proved false, but Schwatka decided to search anyway. The following April, with fourteen Inuit including

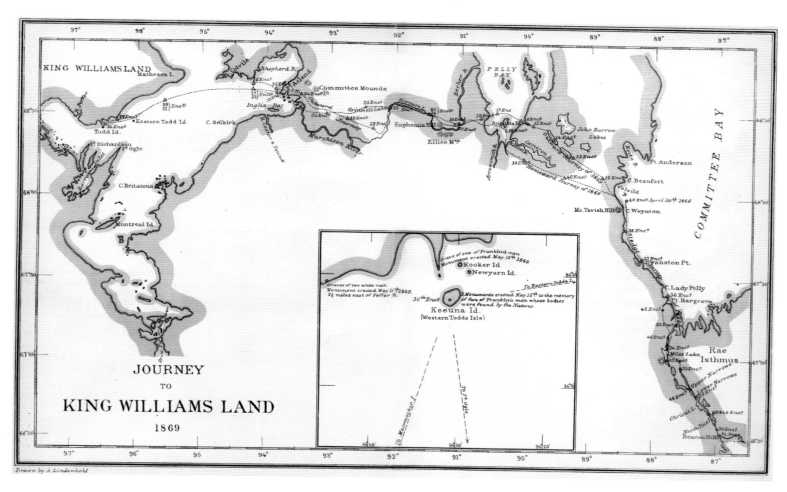

MAP 174.
Map of Charles Hall's travels to King William Island (from Committee Bay, at the southern end of the Gulf of Boothia) in 1869 in search of Franklin survivors. *Keeuna Id.* (Todd Island), where bones of Franklin's men were found, is shown on the inset map, together with the locations of other Franklin graves on King William Island.

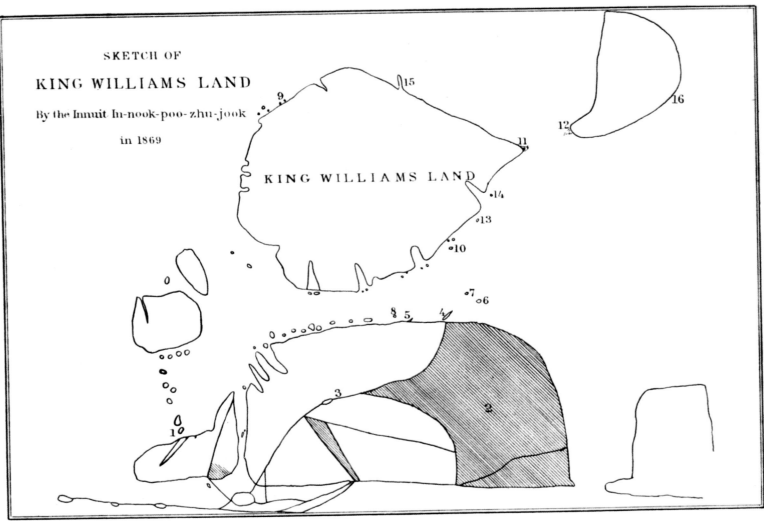

MAP 175 (above).
Map of King William Island drawn in 1869 by In-nook-poo-zhee-jook (*In-nook-poo-zhu-jook* on the map), one of the Inuit who showed Hall a silver spoon belonging to Franklin. The numbers assigned by Hall generally refer to native names for places, but 3 is the Great Fish or Back River; 9 is Too-noo-nee, where In-nook-poo-zhee-jook said he found the two boats from the Franklin expedition; 10 is Kee-u-na (Todd Island, off the southern coast of King William Island), where he found "the remains of five white men," one of which was believed by the Inuit at that time to be the remains of John Ross. A can of meat—unopened—was found in the grave. His body was unmutilated, whereas the other four showed signs of cannibalism; limbs had been severed and flesh taken. Reference was also made by In-nook-poo-zhee-jook's map to an unnamed islet in an unnamed inlet to the west of Richardson Point on the Adelaide Peninsula, where a boat and the remains of "a great many whites" were found.

MAP 176 (right).
This 1885 map published by the German geographer Augustus Petermann shows the routes taken and coasts explored by Charles Hall in 1868 and 1869 in red, and by Frederick Schwatka in 1879–80 in green.

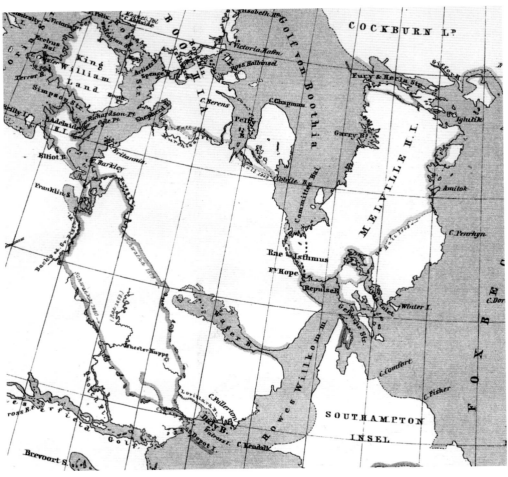

Ebierbing, Hall's Joe, he sledged with dogs to King William Island. That summer, when the land was largely bare, Schwatka combed the island. More relics and skeletons were found, but nothing written. Inuit reports were recorded, however, which led Schwatka to believe that the last of the Franklin survivors had died on the Todd Islands immediately south of King William Island, rather than on Montreal Island, in Chantrey Inlet. Other Inuit, it appeared, had found a box of records in a boat at a place Schwatka called Starvation Cove, just to the west of Richardson Point, on the Adelaide Peninsula, but being unintelligible and of no obvious use, they had been lost.

Exploring the Adelaide Peninsula late in the season Schwatka discovered Sherman Inlet and Sherman Basin.

The return journey was begun on 5 December and they reached Hudson Bay three months later. The whaling ship that was supposed to pick them up was not there, and so they continued south, eventually finding another ship at Marble Island.

Schwatka sledged over 5 000 km during the twelve-month odyssey, a record at the time and rarely surpassed since.

To this day people continue to search for Franklin relics, fascinated by the tale of hardship. In 1993 a scientific expedition found the remains of another boat as well as bones. When subjected to laboratory tests, the bones showed high levels of lead, supporting the idea that Franklin's demise may have been at least partly due to lead poisoning from the welds in tin cans. Evidence of cannibalism first described by Hall was confirmed, adding to the dramatic aura that has come to surround the ill-fated expedition of Captain Sir John Franklin.

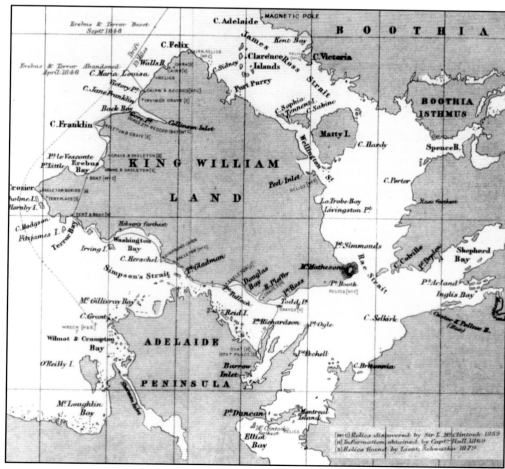

MAP 177 (*right, above*).
This "provisional map" of Schwatka's expedition was published by the German magazine *Petermann's Geographische Mittheilungen* in 1880. The magazine continued after the suicide in 1878 of its founder, Augustus Petermann, the geographer whose theory of an open polar sea was becoming less and less able to be believed by that time. *PGM* tried to publish maps and articles of cutting-edge geographical exploration, and thus featured much on the Arctic during its early days. This map shows Frederick Schwatka's route, carefully covering the entire west coast of King William Island, where Franklin evidence was most likely to be found.

MAP 178 (*right*).
This map of King William Island and vicinity was published by Britain's Royal Geographical Society in 1880 and was the first summary of all the evidence found by M'Clintock, Hall, and Schwatka pertaining to the fate of Franklin. The map shows the probable position of *Erebus & Terror Beset Septr. 1846*, the drift of the ships in the ice, and *Erebus & Terror Abandoned April 1848*, together with the likely track of survivors south (the solid red line) and the probable drift of the abandoned ships (dashed red line). The Back River, at the time called the Great Fish River, flows into Chantrey Inlet, in which is Montreal Island, where the southernmost Franklin relics were found; it was this river that the Franklin survivors hoped to reach and follow south.

Discoveries in the Russian Arctic

In 1820, the Russian Admiralty dispatched two of its officers to survey the northeastern coast of Siberia and to search for the land that, persistently, was still reported off the coast. This was the mysterious "Sannikov Land" reported by Pyotr Pshenitsyn and Yakov Sannikov in 1811 (see page 43). No empire likes to have land off its coasts that it knows nothing about.

The officers were Baron Ferdinand Petrovich Vrangel' (often anglicized as Wrangel) and Petr Fedorovich Anzhu (or Anjou). Although dispatched together, they were to explore different sections of the coast and to all intents and purposes their expeditions were separate.

Vrangel' made sledge journeys northwards across the ice north of the Kolyma River, reaching 71° 43´ N, about 230 km from the mainland, in spring 1822, before being stopped by impassable ice. The following year he reached 72° 02´ N, about 262 km north of the mainland, before giving up due to hummocky ice. In 1823 he set out to find land that had been reported as visible from Mys Yakan on the mainland when the weather was clear. This time he reached 70° 51´ N, 150 km from the coast, when he was stopped by open water. Had he managed to proceed about

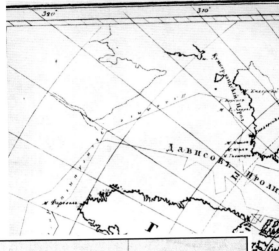

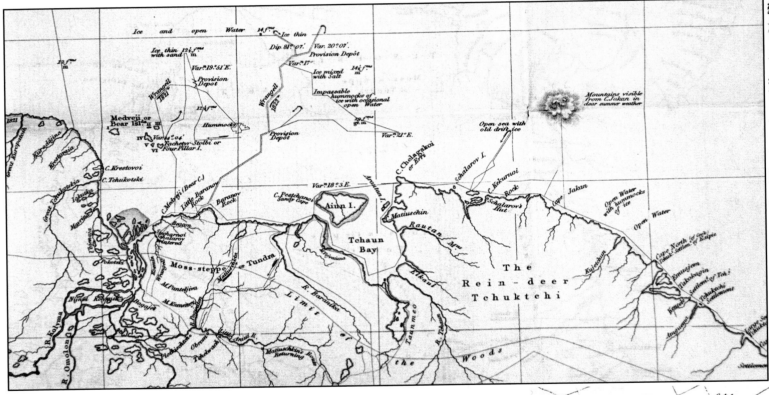

Map 179 (above).
Ferdinand Vrangel' 's map of his explorations in 1820–24, from the English translation of his book, not published until 1840. Note the *Mountains visible from C. Jakan* [Mys Yakan] *in clear summer weather*. This is the approximate location of Ostrov Vrangelya (Wrangel Land) that Vrangel' tried to reach in 1823. Thomas Long clearly had this map with him in 1867 when he finally found the island, for it was he who named it Wrangel Land. Vrangel' 's tracks in 1821 and 1822, farther north than in 1823, are also shown. The track north from the coast at far left is that of Anzhu.

Map 180 (right).
This Russian map of the polar regions was drawn about 1824. The illusory "Sannikov Land," east of Novosibirskiye Ostrova (the New Siberian Islands) is shown as a detailed dotted line suggesting a vast land extending northwards. Apart from some of the offshore islands, the coastline of northern Asia is quite well defined by this time, in considerable contrast to the northern coast of North America. Here details are limited to the discoveries of Hearne and Mackenzie, and the latest additions are the discoveries of Parry in 1819–20.

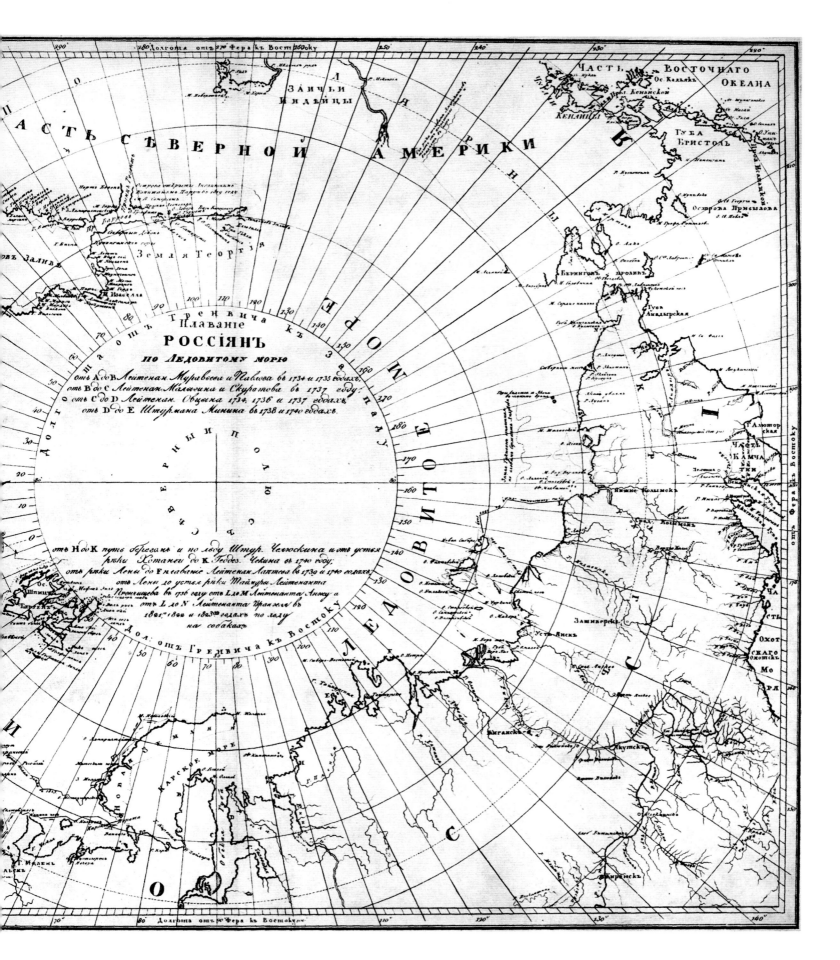

50 km farther, he would likely have discovered the island that now bears his name. The name was given to the island by Thomas Long in 1867, convinced that Vrangel' had in fact seen the island before him.

Petr Anzhu was trying to find Sannikov Land somewhat farther west, searching to the east, west, and north of Novosibirskiye Ostrova (the New Siberian Islands). He reached farther north than Matvey Gedenshtrom in 1808 (see page 42) but was twice stopped, by open water one year and thin ice the next.

Although Ostrov Vrangelya remained undiscovered for another forty years, the surveys of Vrangel' and Anzhu accurately defined the northeastern coast of Siberia. The contrast in known geography between it and the northern coast of North America at this time (MAP 180) was considerable.

The region was visited once more, this time by ship, in 1855. The seas north of Bering Strait were becoming known to American whaling ships at this time and consequently more frequented. In 1885, John Rodgers, commanding the USS *Vincennes*, extended a survey of the Pacific coast of Asia with a probe through Bering Strait (MAP 184, *overleaf*). The *Vincennes* was part of a squadron of ships of the United States North Pacific Exploring Expedition.

Rodgers reached Henry Kellett's Herald Island (Ostrov Geral'd; see page 87) on 13 August, then sailed to the area in which Kellett had reported seeing his "Plover Island," in much the same position where Vrangel' had placed land on his map thirty years before. But he could find no land. The *Vincennes* was probably slightly too far south to have seen Ostrov Vrangelya—Wrangel Land—but as a result of Rodgers' voyage, the

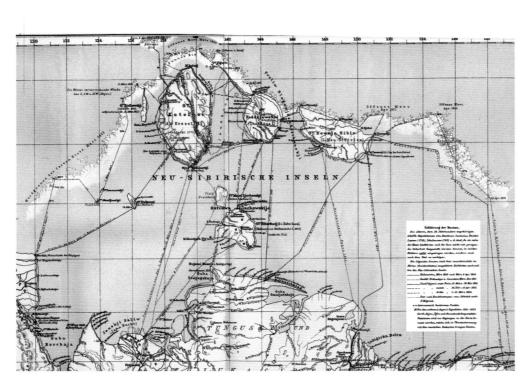

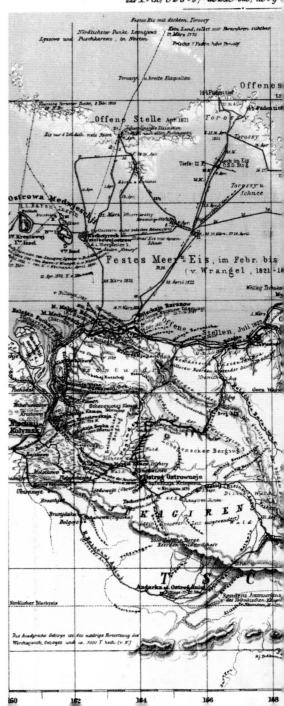

MAP 181 (*above*).
A map published by *Petermann's Geographische Mittheilungen* in 1879 shows Petr Anzhu's travels over the ice around Novosibirskiye Ostrova (the New Siberian Islands) in search of new lands in 1821, 1822, and 1823. The tracks of earlier explorers are also shown: Matvey Gedenshtrom (*Hedenström* in the key) with Yakov Sannikov in 1808–11, and Pyotr Pshenitsyn (*Pschenizyn* in the key) in 1811. The very existence of Novosibirskiye Ostrova seemed to suggest that other islands might also be close by, but Anzhu and Vrangel' were searching for something much bigger, a vast land stretching northwards. Later in the century, with this land still undiscovered, some mapmakers, convinced of its existence, drew it on their maps anyway—as witness Petermann's map, MAP 182, *right*. Petermann, long an advocate of an open polar sea, thought that Wrangel Land was connected to Greenland (see MAP 225, page 146).

MAP 182 (*right*).
This interesting summary map of Vrangel' 's expedition and others in the region north of eastern Siberia and Bering Strait, also published in *Petermann's Geographische Mittheilungen* in 1879, but composed by Petermann before his death. The tracks of Vrangel' are shown in red; the tracks of two of his men, Prokopy Kosmin and Fedor Matyushkin (*Matiuschkin* on the key), are shown in yellow and blue, respectively. The tracks of earlier explorers Billings and Sarychev are also shown, in green. Also shown are ships' tracks: John Rodgers in the *Vincennes* in 1855 and Thomas Long, discoverer of Wrangel Land, in the *Nile* in 1867 (not 1876 as in the key; see page 116). Wrangel Land has been shown as the southeastern part of a vast Arctic land which includes Kellett Land, from the land Henry Kellett, discoverer of Herald Island, saw in the distance in 1849, his "Plover Island."

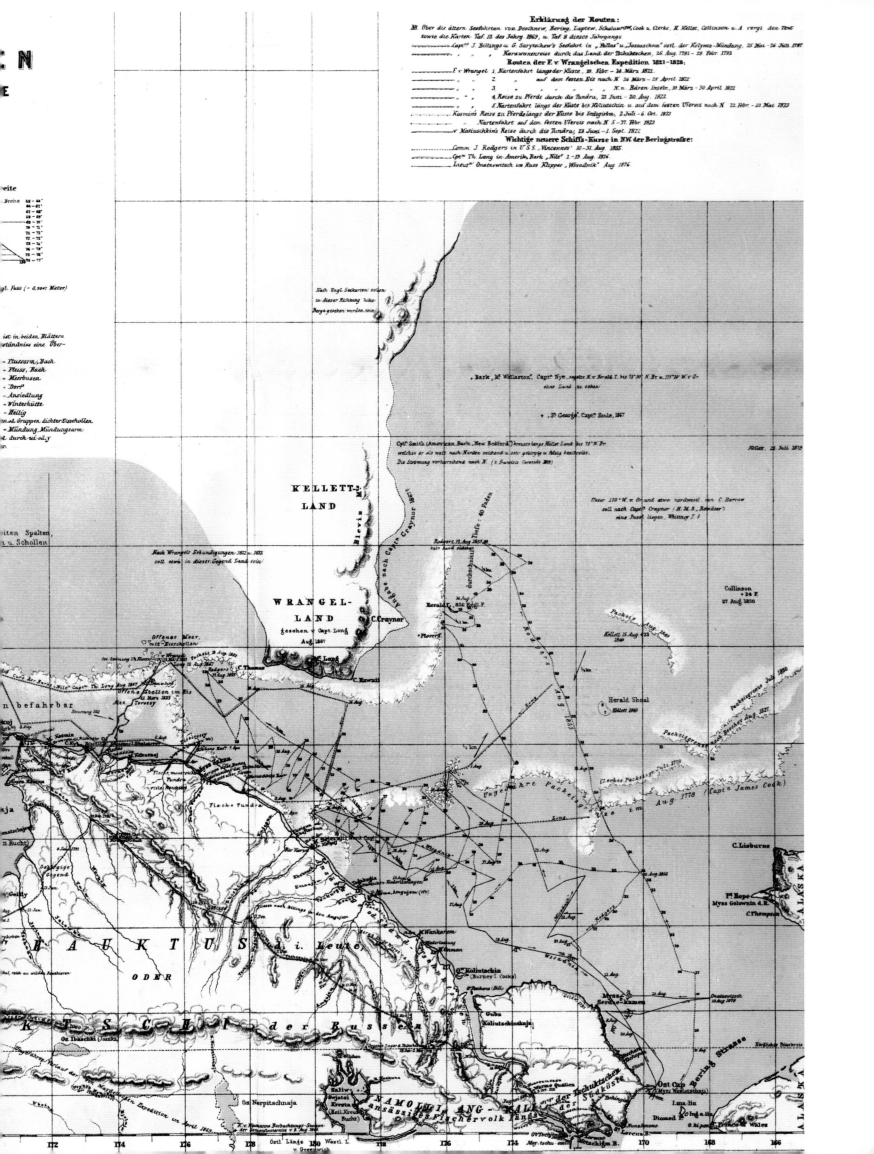

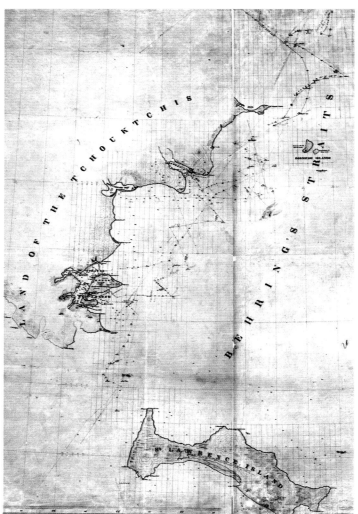

existence of both Plover Island and Wrangel Island remained in doubt until the latter island was found by Thomas Long in 1867.

Thomas Long was captain of the *Nile,* a whaling ship. The Chukchi Sea north of Bering Strait was proving to be a profitable area for whalers like Long, and he was taking advantage of good ice conditions to explore farther west than he had been before. But he was looking for whales, not land. He sailed west along the coast of Siberia and returned at a higher latitude when, on 14 August, he found the land that so many had searched for. He named it Wrangel's Land, knowing that Vrangel' had claimed to have seen land in the vicinity. Other ships had been following the whales that season, for no less than five of them later claimed to have seen the land, but the priority of claim is Long's. The strait between Ostrov Vrangelya and the mainland is today named Proliv Longa (Long Strait) after him. Long's map is shown here (MAP 186).

Long's discovery was soon turned into much more than it really was by the armchair geographers. Augustus Petermann, in particular, was to show it on his maps extending far to the north (MAP 182, *previous page*), and in 1868 produced a map which showed Long's Wrangel's Land as the the tip on the Asiatic side of a vast continent extending across the Pole and joining with the northern part of Greenland, which at this time had not been explored. This astonishing leap is shown on MAP 225, page 146.

MAP 183 (*above*).
A general map of Bering Strait and the region to the north, with soundings, from the United States North Pacific Exploring Expedition. It was drawn about 1855 and shows information gathered from other sources, such as *Track of Baron Wrangel in 1823.*

MAP 184 (*left*).
A survey of the coast of the Chukotskiy Poluostrov (Chukchi Peninsula), which forms the Asiatic side of Bering Strait. East Cape, the westernmost point of Asia, is at the top. The survey was carried out in August 1855 by John Brooke, a master surveyor and oceanographer and a lieutenant on the *Vincennes.* The ship sailed into Arctic seas without him, picking him up on its return. Had it not returned, it would have been Brooke's responsibility to organize a relief expedition.

MAP 185 (*right*).
John Rodgers' map of Herald Island in 1855. The notes refer to his search for Kellett's Plover Island and his inability to find it. The view of Herald Island has been moved from its original position on the map. Today it is Ostrov Geral'd.

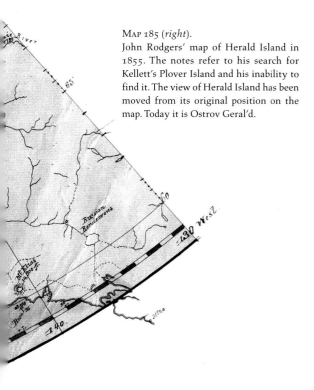

MAP 186 (*below*).
Thomas Long's map of his discovery of Wrangel Land (Ostrov Vrangelya) in 1867, in his ship the *Nile*.

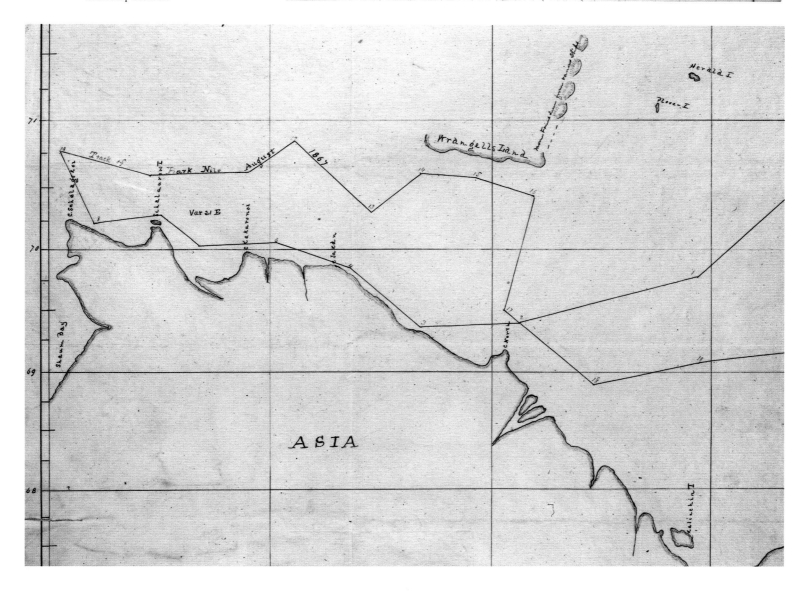

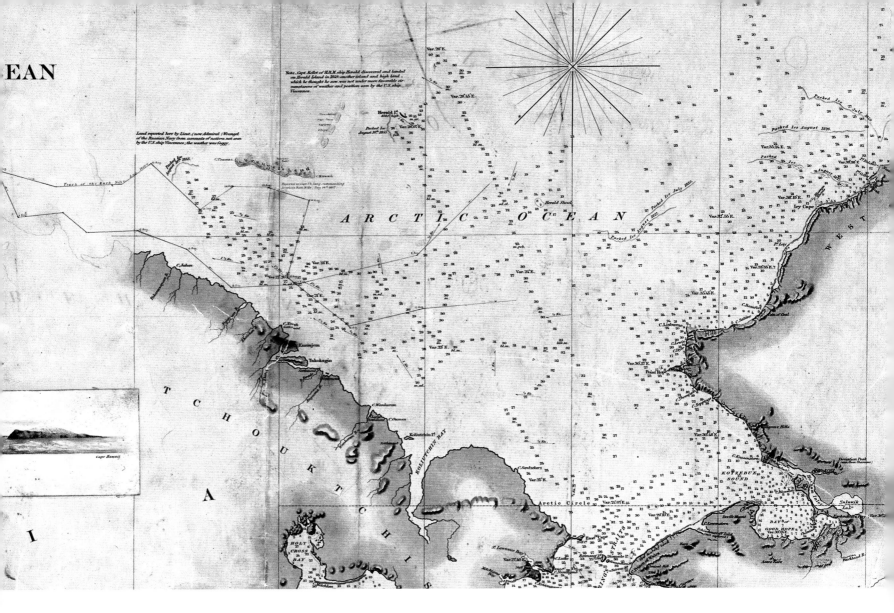

Map 187.
This printed map incorporates information from John Rodgers' voyage through Bering Strait in the *Vincennes* in 1855. Written on it, in red ink, is information gained from Thomas Long's discovery of his *Wrangel's Land* in 1867. There is land to the north of it, reported *by various ships*. It was this sort of map that theorists like Augustus Petermann used to "fill in the gaps" and produce maps such as Map 182, pages 114–15.

Soon after Petermann's death—he committed suicide in 1878—maps started getting more realistic. This was not to say that he was the only one responsible for wild theories; the progression to reality was largely the effect of further exploration.

Starting in 1879 a series of voyages was begun by the United States to patrol the now busy whaling grounds north and south of Bering Strait, the most famous being the voyages of the American Revenue Marine Service ship *Thomas Corwin*, commanded by Calvin Leighton Hooper and later by Michael Healy. In 1879 the ill-fated polar expedition under George Washington De Long, attempting to reach the North Pole through Bering Strait, was crushed by ice east of Novosibirskiye Ostrova, but not before discovering some new small islands and determining that Wrangel Land was in fact an island. This was the voyage of the *Jeannette* (see pages 135–38).

In 1881 Robert Mallory Berry, commanding the USS *Rodgers*, which had been sent to search for De Long, landed on Ostrov Vrangelya, Wrangel Island, and trekked into the interior, where, from a mountain, he could see the entire island (see Map 213, page 137).

A footnote is in order here. In 1921 Canadian explorer Vilhjalmur Stefansson (see page 166) made an abortive attempt to settle on the island, claiming it for Canada and embarrassing his government in the process. The American press rumored that the United States was going to claim Wrangel. Russia, meanwhile, was becoming annoyed, for it saw the island as clearly Russian. In 1924 the island was seized by the armed icebreaker *Krasny Oktobr* (*Red October*) claiming it forcefully for the U.S.S.R.

The other major discovery of new land in the north Asian Arctic during the nineteenth century was that of the large archipelago located between Spitsbergen and Novosibirskiye Ostrova, now called Zemlya Frantsa-Iosifa—Franz Josef Land—after the emperor of the Austro-Hungarian Empire. The new land was found by two of his subjects in 1873.

And it was our friend Augustus Petermann who had much to do with getting the ball rolling. He had organized Karl Koldeway's expedition north to try to sail across the North Pole in 1868 and again in 1869–70 (see pages 129–31), but the scientific community was still not quite sure that Petermann's idea of an open polar sea was wrong. A group of Russian scientists proposed a major expedition to the seas northeast of Novaya Zemlya to settle the matter. A financing request backed by the Russian Geographic Society was being considered when Petermann himself, unable to raise support in Germany because the country was now at war with France, looked for a "Germanic" nation to continue his quest for an open polar sea to the east of where Koldeway had tried. He obtained the backing of Austrian Graf Johann von Wilczek—who would be rewarded with his name on two islands in the Franz Josef group. Although there was a hope that Bering Strait might be reached across Petermann's open polar sea, the specifics were left to circumstances.

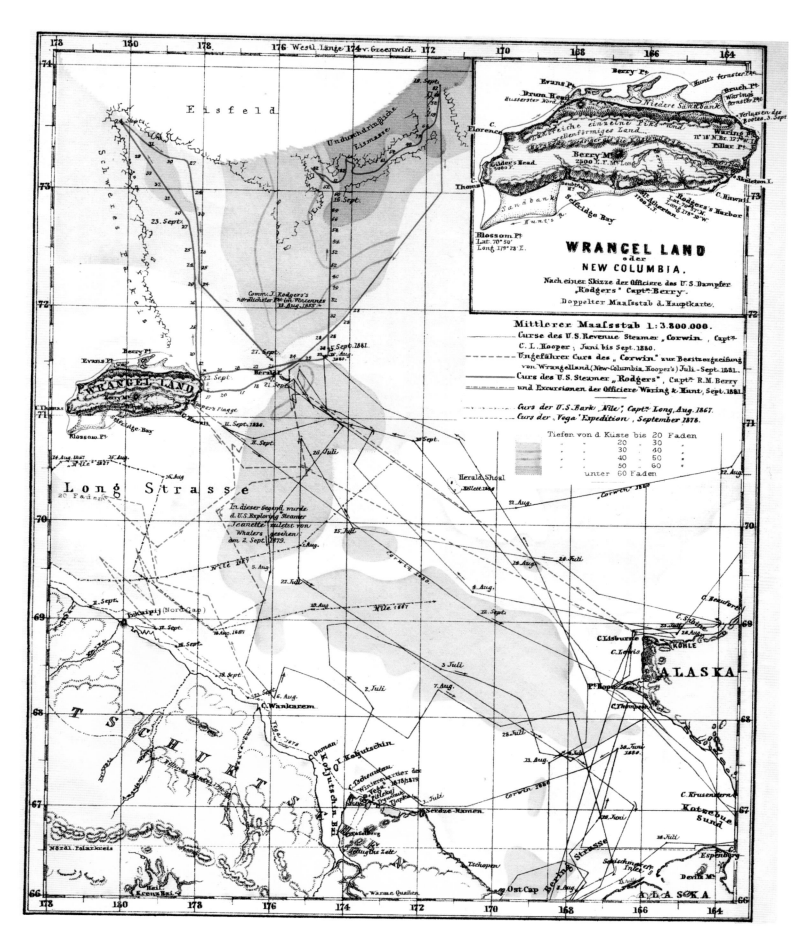

MAP 188.
This map from *Petermann's Geographische Mittheilungen* in 1882 has suddenly dispensed with Petermann's vast northern continent shown so graphically only three years before, when Petermann's influence was felt, as in MAP 182, pages 114–15. *Wrangel Land* (Island) is shown as such, and the land to its north has gone, replaced with Robert Berry's *Eisfeld* (ice pack) and *Schweres packeis* (heavy pack ice). From Petermann's death onwards, the maps in the journal he founded tend to be a much more factual statement of geography as reported, not theorized. Berry's track in the *Rodgers* in 1881 is the solid red line; the solid black line is the track of Calvin Hooper in the *Thomas Corwin* in 1880, the dashed red line in 1881.

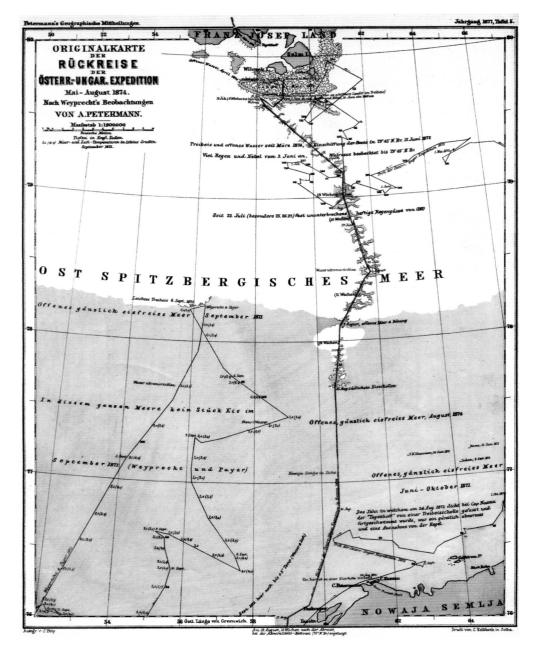

MAP 189 (above).
The map published in *Petermann's Geographische Mittheilungen* in 1877 showing part of the erratic drift in 1872 of the *Tegetthoff* north from Novaya Zemlya (at bottom of map) to discover Franz Josef Land (at top), and all of the return route south in 1874 after the ship was abandoned (the relatively straight dotted and blue-colored line), first over the ice, then in an open sea. The thin black line at left is the track of Weyprecht and Payer's preliminary expedition in 1871. The title states that the map is a copy of the original map of the return of the Austro-Hungarian Expedition, May–August 1874. The edge of the pack is shown prominently.

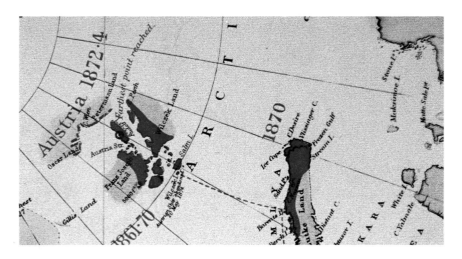

MAP 190 (left).
Part of a British map published in 1874 showing the Arctic discoveries of various nations. The Austro-Hungarian Empire has been added to the roster of Arctic exploring nations. The northern end of Novaya Zemlya, which had been incorrectly elongated eastwards on maps up to this time, was redrawn as a result of the voyage of a Norwegian walrus hunter, Edvard Holm Johannesen, in 1870. Unlike most hunters, Johannesen surveyed the coast he visited, and quite accurately too. He circumnavigated Novaya Zemlya that year.

Karl Weyprecht, a naval lieutenant, was selected to command the expedition while at sea, and Julius Payer, an army lieutenant, commanded on land; whether ice was to count as sea or land was not clear.

After a preliminary voyage in 1871 the pair sailed in *Tegetthoff* from Bremerhaven in June 1872. They followed open water northwards along the west coast of Novaya Zemlya much as Willem Barentsz had done nearly two hundred years before. On 20 August they met the ice and tried to push through it; the ship became trapped as a result. It would never be released. In fact, Weyprecht and Payer made all their discoveries while drifting at the mercy of the ice.

The *Tegetthoff* drifted northwards for just over a year—no doubt prompting some concerns—when on 30 August 1873 new land was sighted at 79° 43´ N, which was promptly named in honor of their emperor. Drifting with the ice, they did not come near enough to put ashore until 1 November.

The following March to May, Payer made sledge journeys to explore the new lands, naming two islands in honor of his heroes, Hall Insel (now Ostrov Gallya) and Insel-MacClintock (Ostrov Mak-Klintoka). One large island was named Wilczek Land (Zemlya Vil'cheka) after his backer. And, perhaps inevitably, Payer thought he saw land even farther north, and this was shown on his map as—what else?—Petermann Land.

Although they seemed to be discovering lots of land, clearly they were not in a very good position as far as getting home again was concerned. On 20 May the still-icebound *Tegetthoff* was abandoned and boats were hauled laboriously over the ice. It took until 14 August before they found open sea about 200 km to the south. They were rescued off the coast of Novaya Zemlya.

It would have been doubtless to Augustus Petermann's chagrin that now only one other *major* land still lay undiscovered to the north of Asia, the archipelago of Severnaya Zemlya, which would remain hidden until 1913 (see page 168). The myth of land in the high Arctic would not finally be put to rest until the advent of the airship and airplane.

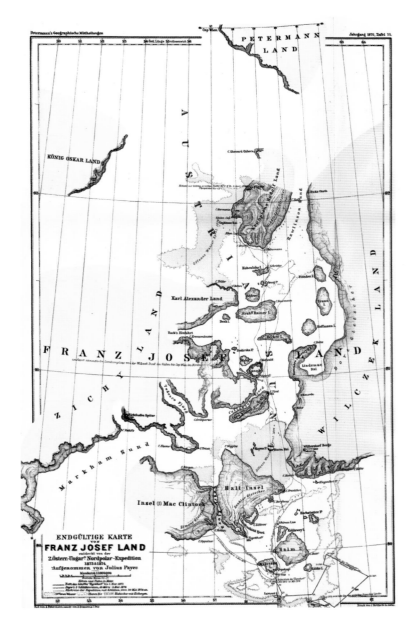

MAP 191 (*above*).
Julius Payer's map, again published by Petermann, showing details of the southern part of Payer's Franz Josef Land, from his explorations in 1874. Note the number of geographical features named after Arctic explorers: Leopold M'Clintock; Charles Hall; Albert Markham (who achieved a farthest north in 1876, before this map was published; see page 133); Adolf Erik Nordenskiöld (see page 140); Karl Koldeway (see page 131); Isaac Hayes (see page 128); and other explorers such as Fedor Litke (Lütke). To the far north is *Petermann Land*.

MAP 193 (*above*).
Augustus Petermann, of course, had to explain how Weyprecht and Payer's discovery of their Franz Josef Land fit into his theories. Petermann must have given up much hope of his open polar sea by this point, because he published this map which showed how the landmass of northeastern Greenland now veered off to the east to incorporate the new archipelago and the vast land that Payer saw beyond it to the north—*Petermann Land* and another he called *König Oscar Land* (King Oscar Land). These lands did not exist, and were likely mirages or ice. To the west of Payer's discoveries Petermann has shown another small land, *Gilles Land*, which he neatly fits into his northen landmass. *Gilles Land* was said to have been seen in 1707 by Dutch whaling ship captain Cornelius Giles but again was just a mirage or a misplaced Nordaustlandet, the easternmost large island of the Svalbard (Spitsbergen) group.

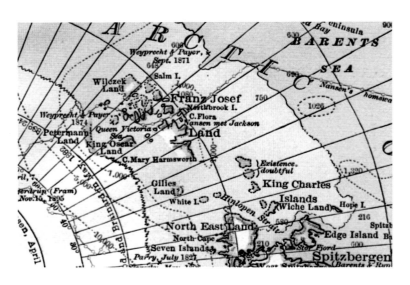

MAP 192 (*left*).
In 1893–95, Norwegian explorer Fridtjof Nansen drifted across the Arctic Ocean in his specially constructed ship, the *Fram* (see page 148). In 1895 he left the ship and attempted to reach the North Pole, and on his return south, by sledge and boat, he passed right through the position that Julius Payer had placed Petermann Land. That was the end of poor Augustus Petermann's theory, though others would claim there was land somewhere in the northern ocean for many years yet.

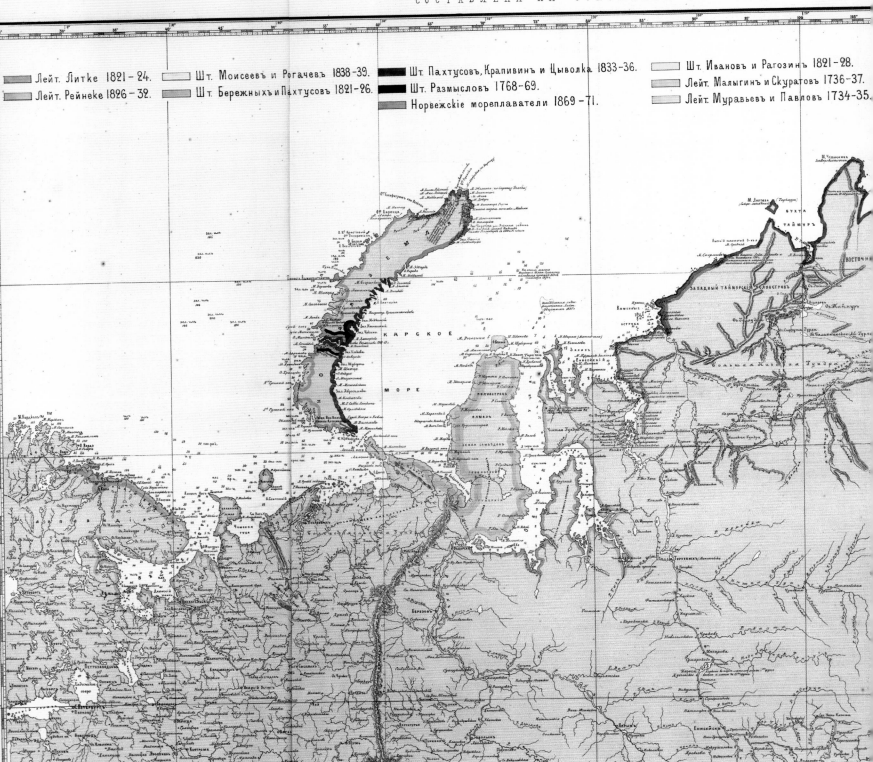

Map 194.
A Russian summary map of the explorations of the Arctic coast of Asia from 1734 to 1871. The map was published in 1874. Not all explorers are shown, and some are shown that have proved untraceable. A translation of the key follows; dates have been retained for recognition purposes. The names are in column order, left to right. (?) before a name means that the first name is not known, while ? after the name means the spelling in English is uncertain. GNE denotes part of the Great Northern Expedition (see page 38).

Fedor Litke 1821–24
(?) Reinecke? 1826–32

Stepan Moiseyev and (?) Rogachev? 1836–39
Il'ya Berezhnykh and Petr Pakhtusov 1821–26

Petr Pakhtusov, August Tsivol'ka and (?) Krapivin?
    1833–36

Fedor Rozmyslov 1768–69
Norwegian explorers 1869–71

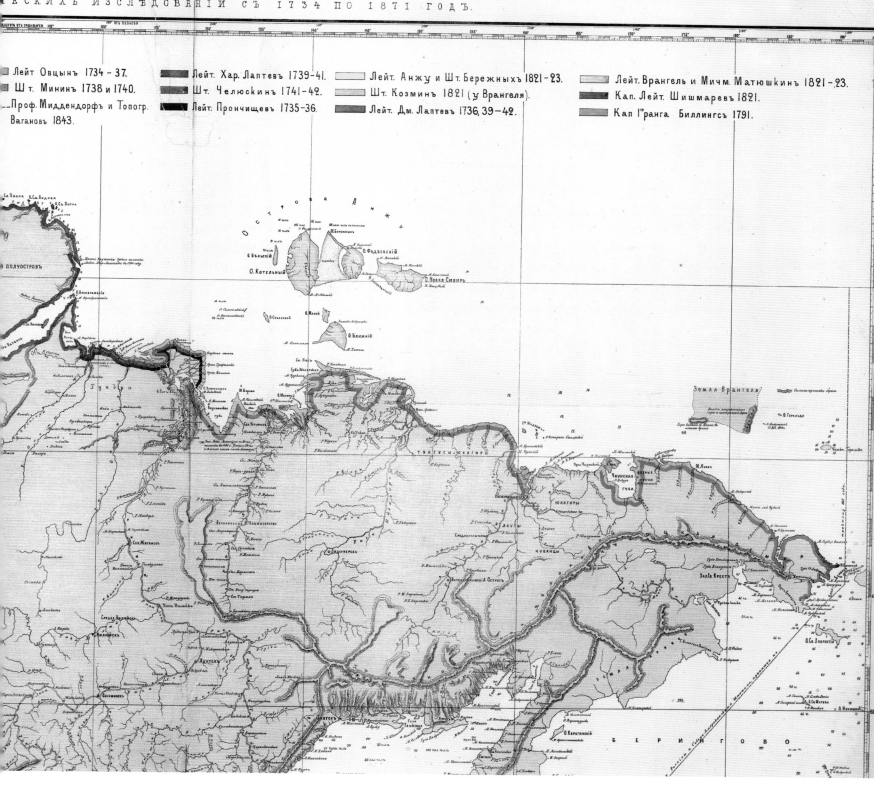

Ivan Ivanov and (?) Ragozin? 1821–28
Stepan Malygin and Aleksey Skuratov (GNE) 1736–37
Stepan Murav'yev and Mikhail Pavlov (GNE) 1734–35

Dmitriy Ovtsyn (GNE) 1734–37
Fedor Minin (GNE) 1738–1740
Aleksandr Middendorf and (?) Vaganov? 1843

Khariton Laptev (GNE) 1739–41
Semen Chelyuskin (GNE) 1741–42
Vasiliy Pronchishchev (GNE) 1735–36

Petr Anzhu and (?) Berejnukh 1821–23
Prokopiy? Koz'min 1821 (Vrangel)
Dmitri Laptev (GNE) 1736, 1739–42

Ferdinand Vrangel (Wrangel) and (?) Matyuskin 1821–23
Nikolay Shishmarev 1821
Joseph Billings and Gavriil Sarychev 1791

The Path to the Pole

The British navy sent expeditions north to try to sail to the North Pole in 1773 and 1818; both, of course, failed (see pages 55–57). Despite continuing expectations of finding an open polar sea, it was realized that, at the very least, there was a barrier of ice at lower latitudes that would have to be overcome first. Just perhaps, there might be ice or even land nearer the Pole. Whether over sea, ice, or land, a properly equipped expedition should be able to reach the Pole, it was thought (indeed correctly; but ideas as to what "properly equipped" meant would not match the needs of reality for some time). Thus began a long series of attempts to reach the Pole over the ice in which the principal players were British and later American.

The first over-ice attempt was in 1827, by the British navy, who selected none other than the by now highly experienced veteran of the Arctic, Edward Parry.

In his old ship *Hecla* he accompanied the English whaling fleet to Spitsbergen, from where, on 21 June, Parry and his men set out towards the Pole in two small boats they named *Enterprise* and *Endeavour*, the first commanded by Parry and the other by his second-in-command, James Clark Ross. The boats had been fitted out with steel runners, so that they could be hauled across ice. And hauled they were, for Parry spent forty-eight days on the ice. But it was gruelling work, as the boats, laden with supplies, were very heavy. The Admiralty's John Barrow thought it an "unusual and unseaman-like labour," but it was a method the British navy would stick to for many years. Parry had obtained eight reindeer in Hammerfest intending that they should pull the boats, but it was evident that this would not work well in an environment of slushy and hummocky ice and water leads.

Parry set off northwards from Sorgfjorden, at the northeast tip of the main island of Spitsbergen, on 21 June. The going was slow, and although he was reluctant to believe it, Parry soon became aware that they were drifting back southwards towards Spitsbergen, making forward progress even slower. Eventually he realized that their trek was hopeless and returned, reaching the ship again on 21 August. They had attained, Parry reckoned, 82° 45´ N, at the time the farthest north reached by man, a record that would stand for nearly fifty years.

One of the main pathways to the Pole was to be the water between Greenland and Ellesmere Island. The first venture into this region was not by an expedition intent on the

MAP 195.
The map accompanying Edward Parry's book about his 1827 attempt to reach an assumed open polar sea. Although Parry reached 82° 45´ N, he did not consider it a significant enough achievement to show it on his map; the track of the northward expedition goes off the top of the map.

(*top*) Men of the Nares expedition (page 133) man-haul their heavy sledge north over hummocky ice.

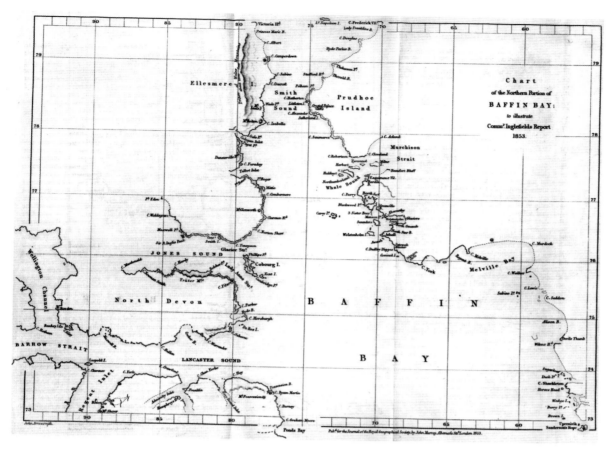

MAP 196 (*left*).
Edward Inglefield's map of his 1852 voyage, reproduced by the Royal Geographical Society in 1853. He found and named Cape Sabine at the narrow point of Smith Sound on the Ellesmere Island side. This would become better known later due to the misfortunes of the Adolphus Greely expedition of 1881–84 (see page 138). Inglefield's *Murchison Strait*, not a strait, is today Inglefield Bredning (Widening).

MAP 197 (*below*).
This map from Elisha Kent Kane's book curiously emphasizes a large landmass reaching from northern Greenland to the approximate location where Thomas Long would find his Wrangel Land in 1867. The open polar sea that Kane believed in is limited to two sections on the American and Asiatic sides of this landmass. But *Dr. Kane's Line of Descent* through Smith Sound leaves some room for doubt as to how far water might reach towards the Pole. Note the *Induced Area of Franklin Position* to the north of Penny Strait, itself north of Wellington Channel, where the *Advance* had been searching in 1850–51.

Pole, however, but by one of the ships searching for John Franklin, in 1852. Edward Inglefield was appointed to command a private venture financed by Jane Franklin and public subscription, using a steam yacht, *Isobel*. Although his expedition was originally planned as a supply voyage to Beechey Island for Edward Belcher's squadron, Inglefield, with his eye on the Pole, persuaded Jane Franklin that he should instead search—he meant explore—north of Baffin Bay.

Inglefield penetrated Smith Sound and pushed through the ice to 78° 28′ N, farther than any recorded voyage in this region, although it seems likely that a few whalers had gone that far before. The geographical results of Inglefield's voyage are shown in his map (MAP 196) and in the comparative sequence of maps on pages 128–29.

Perhaps the biggest single jump in knowledge of the Smith Sound route to the Pole was made in 1853–55 by another expedition nominally searching for Franklin, but in fact seeking to prove the existence of an open polar sea. This was the expedition of the American Elisha Kent Kane.

Kane, a doctor, had accompanied Edwin De Haven as a surgeon on the *Advance* in 1850–51, during the Grinnell expedition, the first American effort to search for Franklin. If there ever was an image of a popular heroic Arctic explorer, Kane was its epitome.

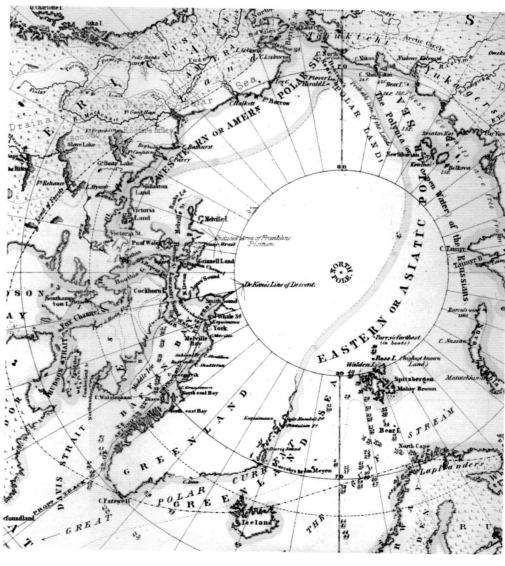

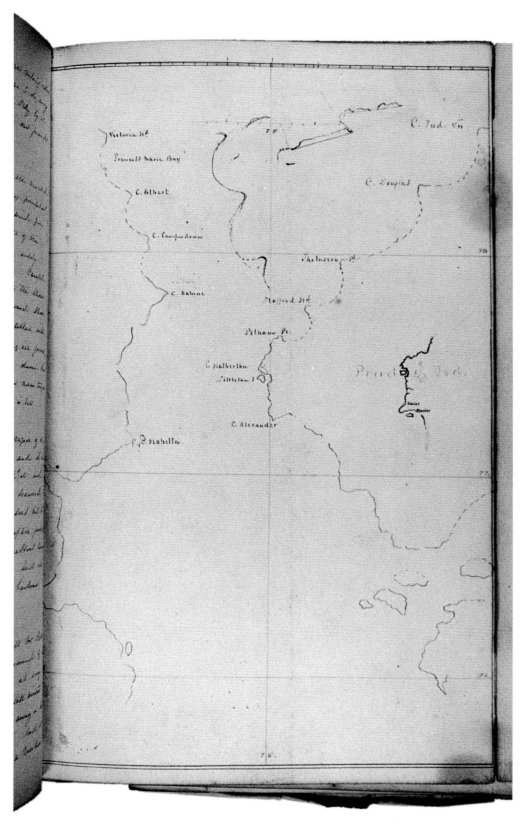

MAP 198.

This map from Elisha Kent Kane's original journal is his copy of the map of Edward Inglefield drawn in 1852 and published in 1853. This map is opposite the entry in the journal for 13 August 1853, as Kane approached Smith Sound. C[ape] *Isabella* on the Ellesmere Island side of the entrance to Smith Sound is at center left, while C[ape] *Alexander*, on the Greenland side, is at center right. Another attempt at a coastline is drawn in Kane Basin, and another small map showing the coastline of his Prudhoe Land (marked on MAP 199) is drawn at right. Compare this map with MAP 196 (*previous page*).

On his return to the United States in 1851 Kane immediately proposed his new expedition and managed to persuade Henry Grinnell to finance it. Kane was going to find the open polar sea. This theory had been revived by Matthew Fontaine Maury, the famous American naval officer widely regarded as the father of the science of oceanography. But although Maury was good at synthesizing the "big picture," unfortunately many of his ideas were wrong—including his theory of an open polar sea, a vast polynya around the North Pole saved from freezing by the twenty-four-hour daylight in summer, upwelling of warmer water, and the displacement of warmed waters northwards via the Gulf Stream and other currents.

Kane sailed from New York in May 1853 in the *Advance*. He met relatively good ice conditions, and he was able to follow the west coast of Greenland north farther than anyone before him, finding a harbor for the ship at Rensselaer Bugt (Bay) at 79° 37´ N. From there boat and sledge parties explored the Greenland coast to nearly 80° N. This despite the fact that all their dogs died of some mysterious illness.

The following season one such foray north, by the steward William Morton and an Inuit guide, Hans Hendrik, reached 80° 34´ N. From a massive headland they called Cape Independence (now Kap Constitution) they looked north and saw—or thought they saw—the open polar sea. It was again likely a mirage, creating a wavelike effect on the surface of the ice, but Morton believed it was open sea, and so did Kane. "From an elevation of five hundred and eighty feet, this water was still without a limit," he wrote, "moved by a heavy swell, free of ice, and dashing in surf against a rock-bound shore." This observation, from the expedition's most northern point, in Kane's mind fully justified the trials to come.

Kane's surgeon, Isaac Israel Hayes, led another party north along the Ellesmere Island coast to about 80° N, and others reached the massive Humboldt Glacier, where the Greenland ice sheet spills out into the Kane Basin.

Getting the ship so far north so relatively easily in 1853 had its downside; the ice decided not to release the ship in 1854. Even blasting could not free the ship. Kane wrote: "It is horrible—yes that is the word—to look forward to another year of disease and darkness to be met without fresh food and without fuel." Confinement in close quarters with Kane's argumentative crew through even one winter was bad enough, without going through another. An attempt was made with five men to reach Edward Belcher's ships, but this failed. In what was really a mutiny, since Kane ordered them to stay, Hayes took half the men and headed south, intending to make for Upernavik on the west Greenland coast farther south, but he too failed and the group struggled back to the ship in December, wondering if Kane would permit them to come back; he did. Kane and the rest of the men,

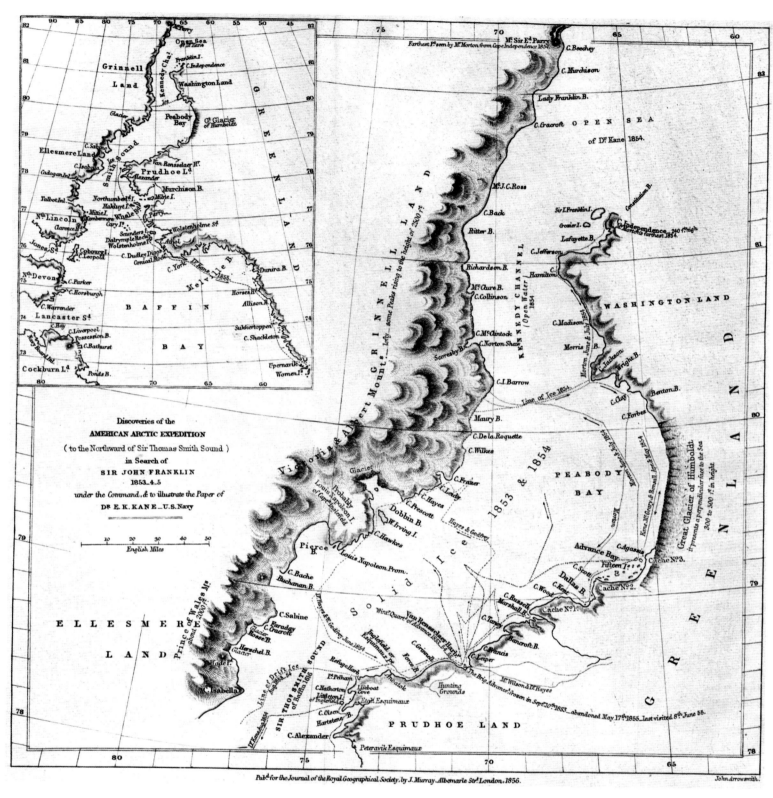

MAP 199.
Elisha Kent Kane's map of the results of his 1853–55 expedition in a version published by Britain's Royal Geographical Society. Kane's *Open Sea* is marked at top, in the position where the Hall Basin, off Kennedy Channel, is today. Kane wanted to name his open sea after Maury, who had predicted its existence. Maury deferred, saying it should be named after Kane; perhaps it was appropriate that there was never a "Maury Sea," since there is no open polar sea. Kennedy Channel was named by Kane for a friend. Kane's expedition, for all its drama, added significantly to geographical knowledge in this region.

meanwhile, survived only because they were able to obtain food from Inuit.

On 17 May 1855 *Advance* was abandoned by all and an epic journey south by boat and on foot ensued; in early August they reached Upernavik, and continued south to Godhavn (Qeqertarsuaq). Here they were found by a relief ship that had been sent north to find them. The appropriately named relief ship *Release* had reached Cape Inglefield (Kap Inglefield), only a short distance south of where the *Advance* was trapped.

Kane's exploits led to his being revered as an American icon, and his well-written and well-illustrated book sold 65,000 copies and made him rich. It was said that only the bible was seen on more American tables. Kane suffered from rheumatism and had heart problems, all of which served to increase the drama, as did a strange liaison with a fraudulent psychic who claimed to tap out messages from the deceased. When Kane died at thirty-seven he was given a state funeral.

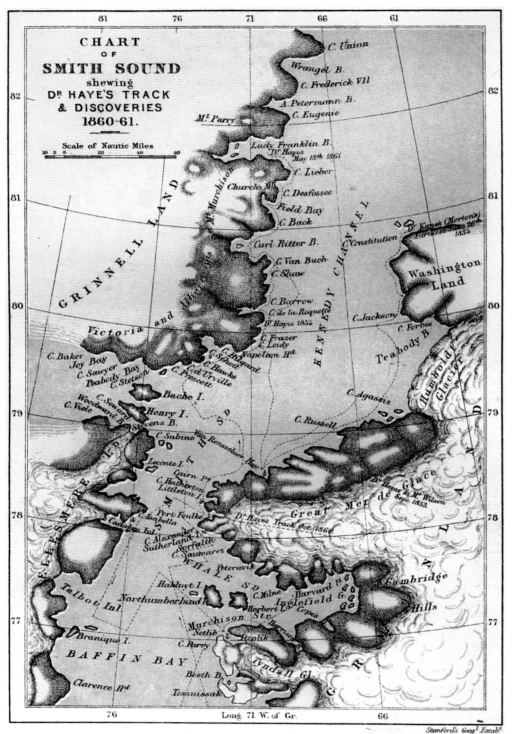

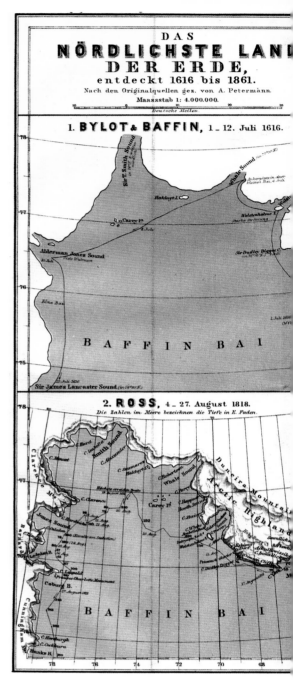

MAP 200.
Isaac Hayes's map from his book entitled—what else?—*The Open Polar Sea*, which was not published until 1867, due to the American Civil War. Hayes was undaunted by not finding what he expected to see; he was convinced it was there anyway. He claimed to have reached 81° 35´ N, but he probably reached no farther than 80° 14´ N. His descriptions of the terrain at the higher latitude have been called inacccurate and even imaginary. But then his open polar sea was imaginary too.

MAP 201 (*right, above*).
This informative series of maps was published by Augustus Petermann in 1867. It illustrates the march of geographical knowledge successively farther north from the time of Baffin in 1616 to that of Hayes in 1861.

Isaac Israel Hayes, Kane's surgeon, was an even more avid believer in the concept of the open polar sea than was Kane, and he determined to try once more to break through what he thought was merely a "belt" of ice surrounding the Pole and enclosing an otherwise ice-free sea. Encouraged by the fact Morton swore that he had seen this open polar sea (shown on Kane's map, *previous page*) when he stood at Kap Constitution, Hayes resolved to reach it. It was but a short distance from where Kane's ship lay, but even short distances have a way of becoming insurmountable barriers in the Arctic.

Hayes sailed from Boston in May 1860 in a small schooner named the *United States*, in which he thought he would have a better chance of moving through the ice. But 1860 was not 1853, and ice prevented him from reaching farther than 78° 20´ N, near Smith Sound's narrowest point.

From here, in the spring of 1861, Hayes set off to follow the coast of Ellesmere Island north to the open polar sea. He claimed to have reached 81° 35´ N on 18 May, farther than Morton, but this is now disputed. The land in the distance veered off to the west, or so he thought, and the ice he saw would soon be clear. "All the evidences showed that I stood upon shores of the Polar Basin," he wrote. Wishful thinking seems to have translated into the required perception. Hayes decided to return to Boston because he felt his ship unseaworthy to continue northward. It was

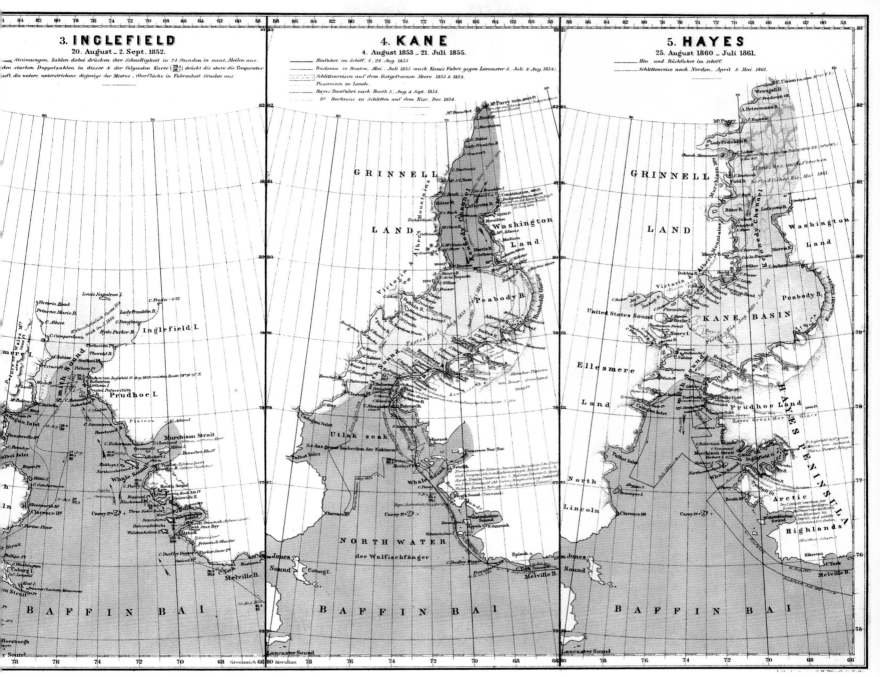

his intention to try again the following year; this plan, however, was frustrated by the outbreak of the American Civil War.

At this time, another possible route to the open polar sea was thought possible north along the *eastern* coast of Greenland. This was the favorite choice of Augustus Petermann, who in 1868 raised funds to dispatch an expedition, intended as a preliminary to another the following year which was to sail to the North Pole through the "hole" in the barrier of ice created, or so Petermann thought, by the warm waters of the Gulf Stream. On his 1868 map (Map 225, page 146) Petermann had conveniently left the Pole covered by water

Map 202 (*right*).
A explanation of the reasons for an open polar sea by Silas Bent, 1872. The positions of the warm currents did not agree with those of Petermann, but the presumed effect was the same. This rare map is from *An Address Delivered Before the St. Louis Mercantile Library Association January 6th 1872 Upon the Thermal Paths to the Pole*.

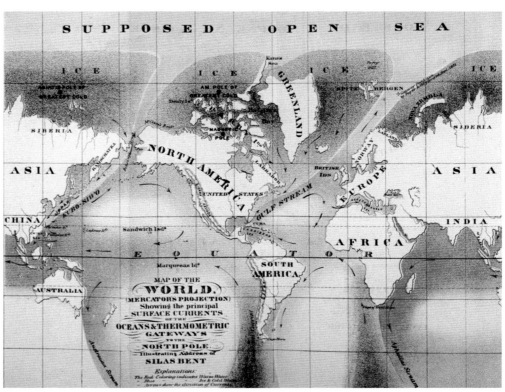

129

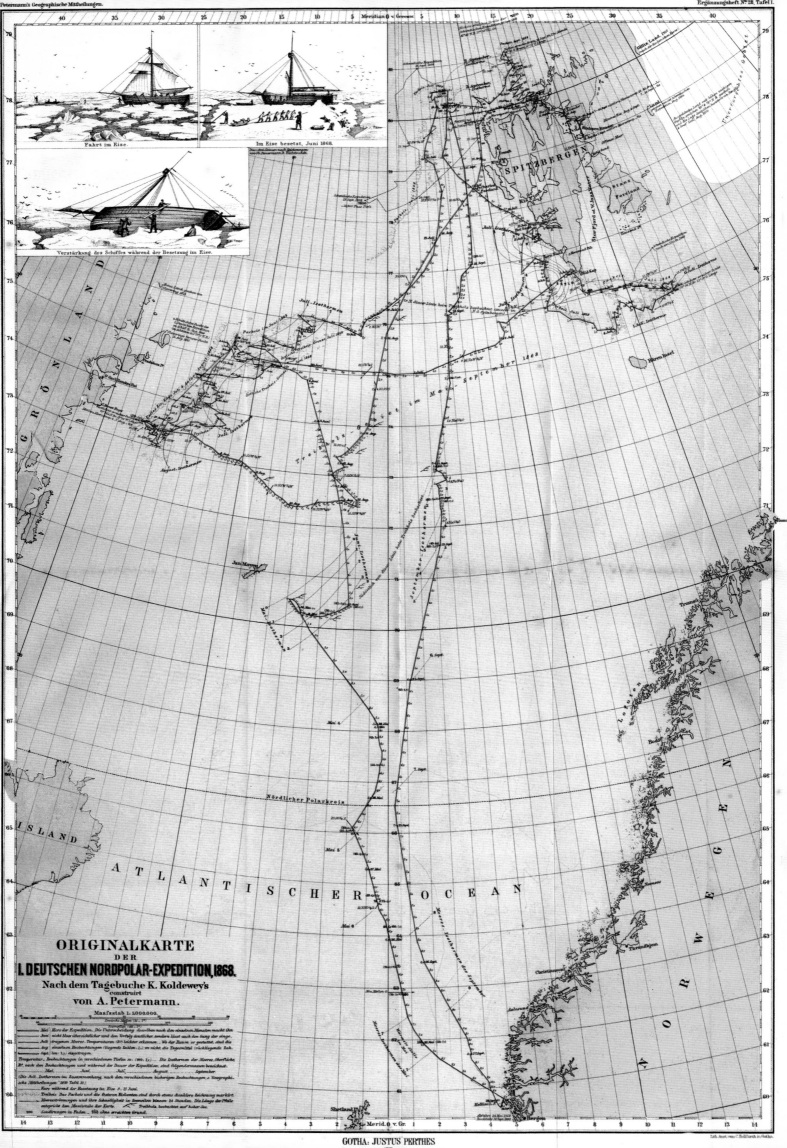

rather than the vast land he had also shown so that although he believed in a land stretching north from Greenland, it would also be possible to sail to the North Pole. As it turned out, the first expedition got farther north than the second. Both were led by Karl Koldeway.

In 1868 Koldeway tried to sail north along the eastern coast of Greenland but met the ice that clogs the northeastern part. He therefore headed east along the edge of the ice pack, searching for open water. He diverted briefly to search for "Gilles Land" to the east of Spitsbergen and, not finding it, tried to sail north from the west coast of Spitsbergen. He reached 80° 30´ N before being stopped by ice.

The "main" expedition the following year had two ships; one was wrecked off the Greenland coast and the other reached only 75° 30´ N before meeting the ice barrier. Petermann's theory was not disproved, however, because he believed that there was a ring of ice around the polar regions through which a path simply had to be found, which Koldeway had failed to do that year.

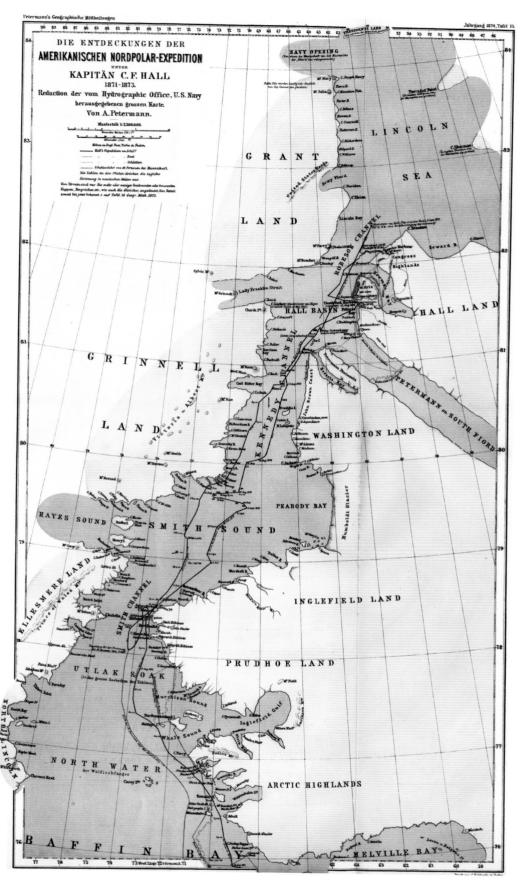

Petermann switched his sights eastwards. Since Germany was now at war with France, he found a supporter in Graf Johann von Wilczek in Austro-Hungary; together they organized the expedition of Karl Weyprecht and Julius Payer, which found Franz Josef Land (see pages 118–21).

MAP 204 (above, left).
This map was sent to Petermann in 1874 by whaling captain David Gray, whose voyage to the Greenland Sea that year found a huge swath of open water, as shown on the map. Gray, like Petermann, thought that this was the best route north.

MAP 205 (above).
Petermann's map of the discoveries of Charles Hall's expedition of 1871–73, extending knowledge of the long strait between Greenland and Ellesmere Island to Robeson Channel and the very edge of the Arctic Ocean.

MAP 203 (left).
Petermann's illustrated map of the voyage of Karl Koldeway in 1868. It was supposed to be a preliminary voyage for one the following year, but Koldeway got farther north in 1868 than he did in 1869.

On the west coast of Greenland again, the next attempt on the Pole was by Charles Hall. He was by now an experienced polar explorer, having lived among the Inuit for several years (see pages 108–10). In 1870 Hall managed to persuade the American government to support a new attempt on the Pole. The government provided a ship, USS *Periwinkle,* which after considerable refitting was given a more appropriate name, USS *Polaris.* Hall left in July 1871. It was a good ice year, and *Polaris* progressed rapidly through Smith Sound, Kane Basin, and Kennedy Channel to a new widening, later called Hall Basin. Able to continue even farther north, they steamed through the northernmost channel before the Arctic Ocean. This channel Hall named Robeson Channel, after the U.S. secretary of the navy. The ship finally reached 82° 11′ N.

It was a spectacular achievement, and a new record for a ship. Retreating to a harbor he called Thank God, on a bay he named after his ship (MAP 205), Hall was now superbly positioned for a real attempt on the Pole the following year.

But it was not to be. On his return from a sledge reconnaisance on 24 October, Hall drank a cup of coffee and became ill, and on 8 November he died. The day before his death, Hall had composed a prayer to be used on his arrival at the North Pole.

The reason for Hall's sudden demise remains a mystery to this day, although we now know the immediate cause. In 1968 Chauncey Loomis, who was writing a biography of Hall, exhumed Hall's body. It was quite well preserved, having been frozen, and an autopsy and analysis of fingernail samples showed he had been poisoned by arsenic. How this happened is still unknown. Arsenic was a component of medical chests at the time and Hall could have overdosed. But it seems likely he was murdered, perhaps by the expedition's doctor, Emil Bessels. Certainly he and Hall were at odds from time to time, but this seems insufficient explanation for a murder. We will likely never know.

The expedition rapidly unraveled after Hall's death. Steaming south in August 1872 *Polaris* became stuck in the ice and drifted south. On 15 October the ship was nipped by the ice and preparations were made to abandon ship. While nineteen people were on the ice unloading supplies, it broke up and they were stranded on a large floe. The story of their six-month odyssey became one of the most dramatic Arctic tales of near death and survival. The floe drifted slowly south, becoming smaller and smaller, until on 1 April they were forced to take to a small boat. Mainly because of the hunting skills of two Inuit, Hans Hendrik and Hall's loyal companion Joe, and the leadership of the first mate, George Tyson, the group survived. They were picked up by a Newfoundland sealer off the Labrador coast on 30 April 1873.

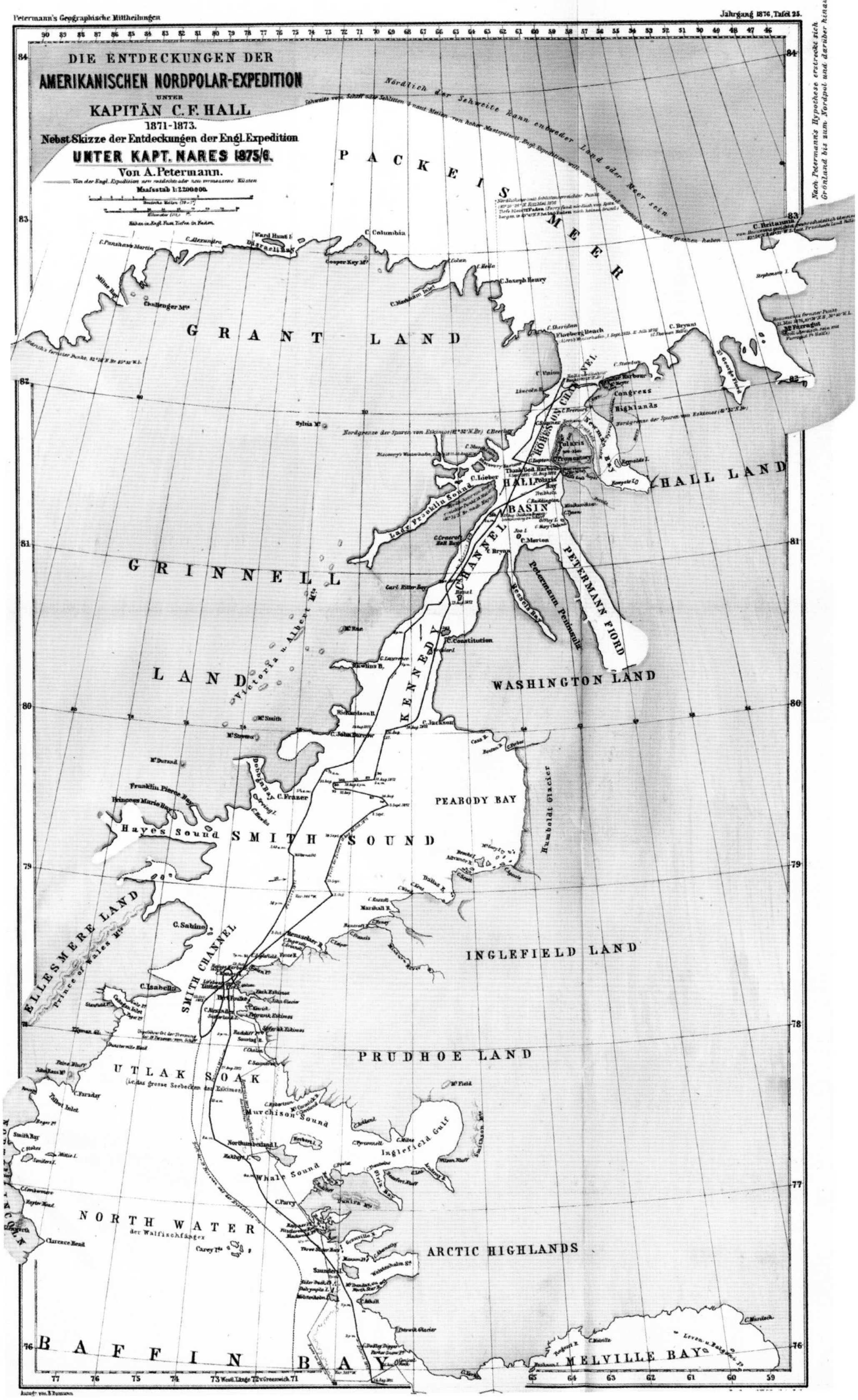

MAP 206.
A map showing the discoveries of the Nares expedition in 1875–76, again by Augustus Petermann. Markham and Parr's track northward across the ice is shown; Aldrich's farthest point west on the north coast of Ellesmere Island is marked at top left, as is Beaumont's on the northern Greenland coast, at top right. This being a Petermann map, he has suggested the Arctic Ocean might be open sea beyond the ice he thought was attached to the land.

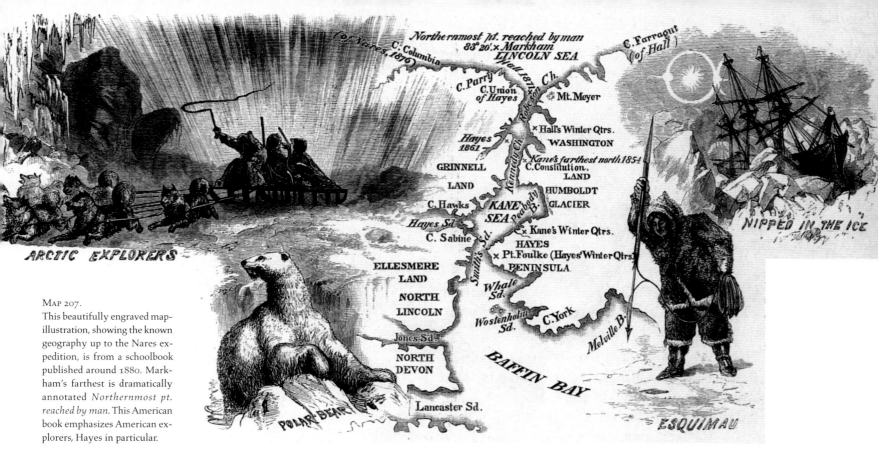

MAP 207.
This beautifully engraved map-illustration, showing the known geography up to the Nares expedition, is from a schoolbook published around 1880. Markham's farthest is dramatically annotated *Northernmost pt. reached by man*. This American book emphasizes American explorers, Hayes in particular.

The fourteen men still on *Polaris* had, by comparison, an easy time. On 3 June 1873 they abandoned the ship and headed south in whaleboats. They were rescued by a whaler in Melville Bugt (Bay) twenty days later.

In 1875 the British Royal Navy jumped back into the Arctic arena with an attempt on the Pole itself. Since 1865, Sherard Osborn, who had been the captain of the *Pioneer* with Edward Belcher in 1852–54 (see page 92), had been in a verbal jousting match with Petermann over the open polar sea concept and the best way to the Pole. Osborn thought that there was a good probability of ice all the way from the Greenland–Ellesmeres Island region to the North Pole. He was, in fact, right. And, asked Osborn, what nation was the most experienced at sledging expeditions? The British thought Britain, where it was still considered heroic for men to pull heavy loads over enormous distances and unsporting to use dogs.

A well-financed expedition with two steamships, HMS *Alert* and HMS *Discovery*, was fitted out, and a veteran of Arctic sledging, George Strong Nares, was transferred from his position as captain of HMS *Challenger*, a pioneering oceanographic ship, to lead it. His instructions were to reach the North Pole through Smith Sound, and explore northern Ellesmere Island and northern Greenland. Nares was to find he could not carry out the first instruction, but he could the others.

MAP 208 (right).
The track of Markham and Parr northwards out over the ice in 1876 is shown on this map from Nares's book.

The *Discovery* found winter quarters at Discovery Harbour, on the north side of Lady Franklin Bay, on northern Ellesmere Island, while the *Alert* continued north through Robeson Channel to Floeberg Beach, near Cape Sheridan, which at 82° 28´ N was the highest latitude to date for any ship, and for any wintering. This alone was a major achievement by Nares.

In April 1876, three sledging parties set out. Albert Hastings Markham, assisted by Alfred Parr, was assigned the Pole. With fifteen men hauling two sledges, the *Marco Polo* and the *Victoria*, they followed the coast to Cape Joseph Henry and then, on 7 April, left the land and headed north across the ice. The going was hard right from the start, with broken, hummocky ice, through which a path had

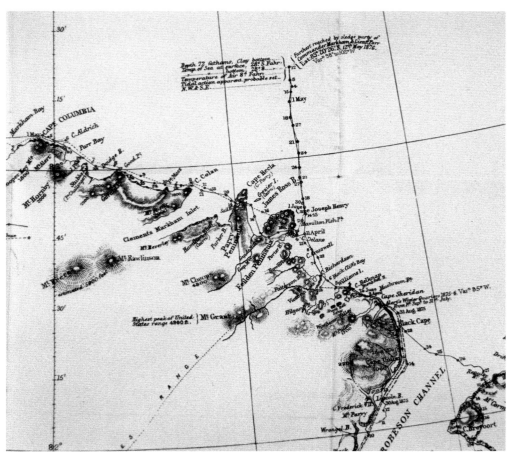

133

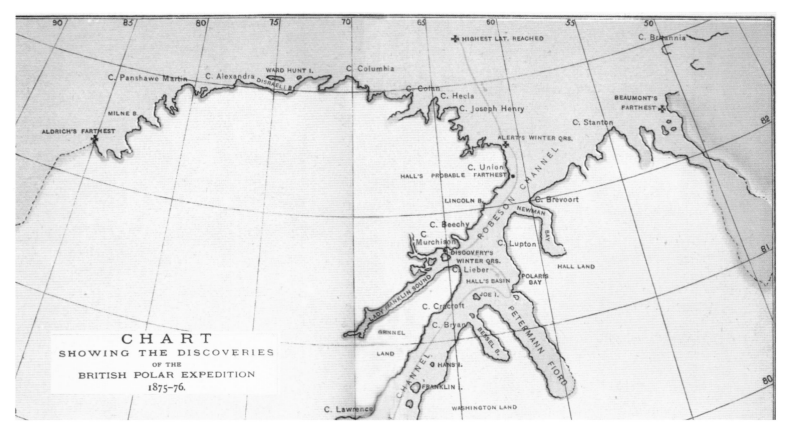

MAP 209.
A map of the discoveries of the Nares expedition from an 1877 book. Markham's track is not very accurately shown; he sledged first to Cape Joseph Henry before trekking north. The choice of this cape as the jumping-off point for the trek northwards across the ice meant that the latitude reached, 83° 20′ 26″ N, was only 14′ or about 26 km north of the northernmost point of land at Cape Columbia, farther west along the coast. Once the ice conditions between Ellesmere Island and the Pole were appreciated, Cape Columbia became the spot chosen to launch out over the ice by several later expeditions, including that of Robert Peary in 1909 (see page 160). Note that Cape Fanshaw Martin, named by Aldrich, is misnamed on this map.

to be cut, making for slow progress. It would have been difficult even if the terrain had been smoother for they were lugging two large boats, in case they met with water. Each day they would have to haul for a much longer distance than they "made good," due to having to go back and forth to haul what was far too much of a load. As it turned out, they would travel only 117 km from the ship but would have to trek more than 800 km to do it.

On 12 May, after thirty-nine days of arduous labor with a steadily weakening crew, Markham decided that they should go no farther. More than half of their food was gone, and his men were starting to suffer from scurvy. They planted a flag, drank a magnum of Scotch given to them to be drunk at their highest latitude (but which, along with many other unnecessary things, had had to be hauled over the ice), and sang their national anthem. They had reached 83° 20′ 26″ N, a new record for farthest north.

The scurvy worsened on the return trip, and they were only saved by a rescue party which came to meet them and was able to transport the sick back to the ship, which was reached on 14 June. Of Markham's original fifteen men, one was dead and only three were still able to haul a sledge.

Meanwhile, two sledges were exploring westwards along the northern edge of Ellesmere Island—the Arctic Ocean shore.

MAP 210.
Part of a map of the polar regions published in 1877, after the return of the Nares expedition. On the northern coast of Greenland, C[ape] Britannia was the farthest point seen by Lewis Beaumont, and is about 75 km farther to the northwest than Beaumont actually reached, on Sherard Osborn Fjord. Cape Britannia is in fact at the southwest tip of a small island, John Murray Ø, and not the next headland, as the map suggests.

They were led by Pelham Aldrich, with the sledge *Challenger*. They had accompanied Markham's party to Cape Joseph Henry and then struck out westwards.

It was just as well they were traveling by sledge, for the coastline here is much indented by fjords and ringed by high mountains. Aldrich discovered and named a string of headlands and fjords, including Cape Columbia and Cape Aldrich, at the most northerly land (83° 06′ N) in North America (except for Greenland). They reached Alert Point,

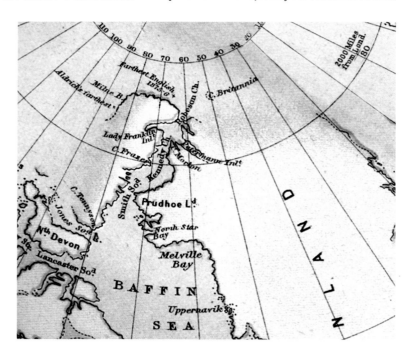

at the northwest tip of Ellesmere Island, having mapped some 400 km of previously unknown coastline. Like Markham, Aldrich's party was affected by scurvy on their return trip and were aided by a relief party sent out from the ship.

The third sledge party was sent from *Discovery* on 6 April to explore eastwards along the northern coast of Greenland. It was led by Lewis Beaumont, with the sledge *Sir Edward Parry* and twenty-five men. Scurvy was soon a problem and many of the sick men were sent back to the ship. Nevertheless Beaumont struggled on, and managed to explore and map the coast as far east as Sherard Osborn Fjord, turning back on 22 May. Returning towards the ship, all the men were stricken with scurvy. And they could not cross Robeson Channel, as the ice had melted. They remained on the Greenland shore opposite Discovery Bay from 25 June until August, when a relief party found them. They arrived back at the ship on 15 August.

The widespread problems with scurvy on both ships led Nares to decide not to risk another winter in the Arctic, and so, on 20 August, they sailed for England.

The Nares expedition was wildly regarded as a failure in Britain, simply because the Pole had not been reached. Public opinion, as had been the case for centuries, expected far too much of Arctic exploration and did not appreciate the incredible difficulties and hardships faced. One newspaper, which had boosted the expedition on its departure, headlined a story "The Polar Failure," saying the expedition had "gone out like a rocket and come back like the stick." Such criticism was grossly unfair, for Nares had achieved a great deal with what we now know to be highly inappropriate resources for Arctic travel. And the Arctic shore had been mapped, the cap of a continent defined.

The expedition achieved one more thing, for they found the grave of poor Charles Hall. On 13 May 1876, in the presence of twenty-four officers and men, the captain of the *Discovery*, Henry Stephenson, hoisted the American flag over Hall's grave and placed a brass tablet to commemorate one "who sacrificed his Life in the advancement of Science, November 8th, 1871."

One of the officers on an expedition that had searched for Charles Hall in 1873 was George Washington De Long. Enthralled by his Arctic adventure, De Long approached James Gordon Bennett of the *New York Herald*—the man who had sent Henry Morton Stanley to Africa to search for David Livingstone—with the idea of an expedition to the North Pole. This was at the time rejected, but after Bennett met Augustus Petermann in 1877 he changed his mind, for he was convinced of Petermann's latest modification to his theory, that the Pole could be reached via an open polar sea accessed through Bering Strait. Here, Petermann now maintained, the warm Kuro Siwo current would keep a pathway through the encircling ice "belt" clear.

Although financed by Bennett, De Long's expedition was outfitted and instructed by the U.S. navy. De Long sailed in the steamship *Jeannette* from San Francisco on 8 July 1879. By September he was in the Arctic Ocean, sighting Herald Island (Ostrov Geral'd) early in the month. But although De Long intended to winter there, the ice closed in and on 6 September the ship was locked in an embrace with the pack that would never be relinquished. Nansen was later to write, "Scarcely anywhere have polar travellers been so hopelessly blocked by ice in comparatively low latitudes."

Now began an erratic drift which took the ship slowly and inexorably to the northwest. "This drift of ours is in no sense uniform or capable of being foreseen," wrote De Long. By the end of the year they were north of Ostrov Vrangelya (Wrangel Island), which he was able to confirm as an island. Yet by November the following year the *Jeannette* was back at almost the same spot as it had been in April.

But then a distinct drift to the northwest set in. On 16 May 1881 two islands were found, which De Long named Jeannette Island (Ostrov Zhannetty) and Henrietta Island (Ostrov Genriyetta), the latter after Bennett's mother. On 2 June a sledge party was landed on Henrietta Island to take formal possession for the United States. "Our voyage, thank God, is not a perfect blank, for here we have discovered something, however small it may be," wrote De Long in his journal.

Then—disaster. The ice, turning ugly, closed in and started to crush the ship. De Long ordered supplies, boats, sledges, dogs, and all crew out onto the ice, and at four in the morning on 13 June, the *Jeannette* vanished beneath the cruel ice.

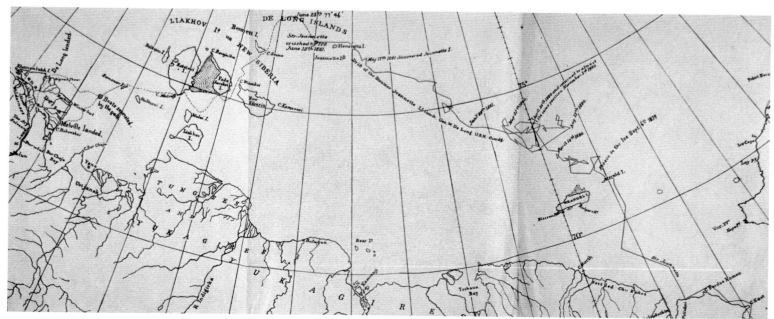

MAP 211.
The map taken from George De Long's journal and published posthumously in his book in 1883. The track and drift of the *Jeannette* is shown by the solid red line, with the point the ship became trapped in the ice marked *Frozen in the Ice Sept. 6th 1879*. The dashed red line is the route of De Long and his men as they made for the Lena Delta, on the mainland.

135

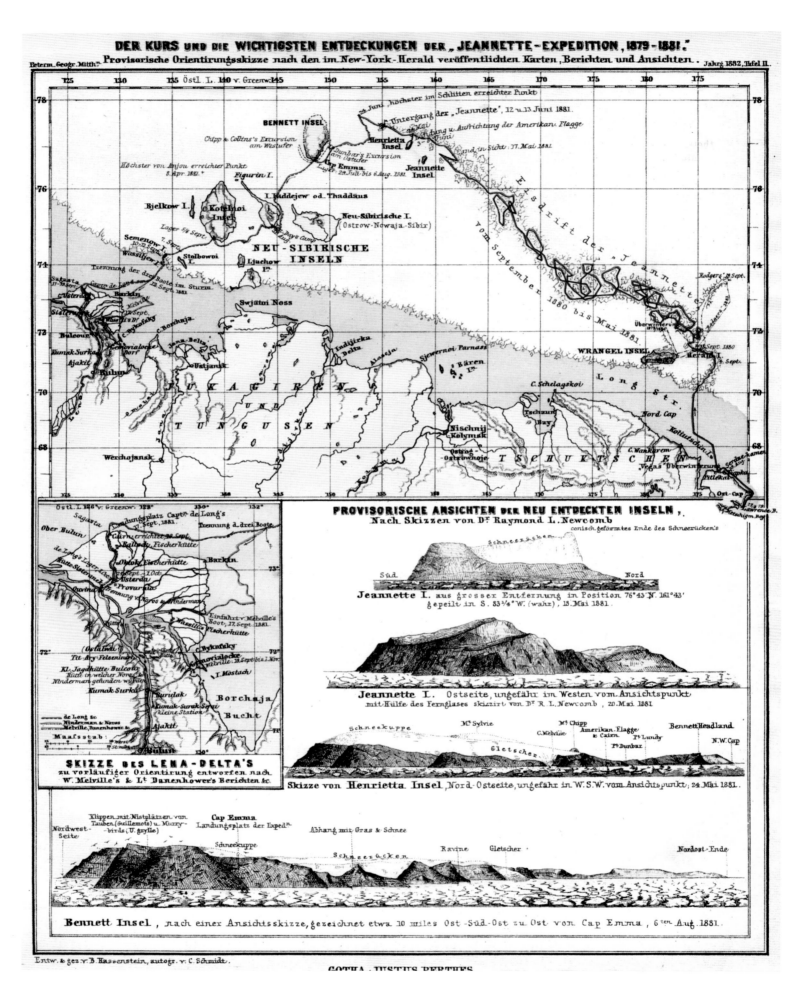

MAP 212.
A map complete with coastal views published in 1882 in *Petermann's Geographische Mittheilungen* showing the track and demise of the *Jeannette* and George De Long in 1879–81.

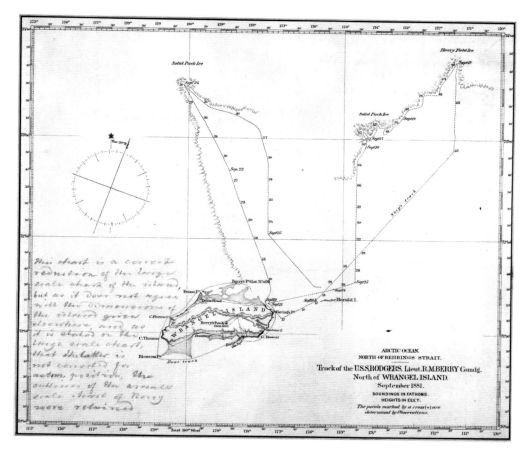

On the face of it, their predicament was not hopeless, for De Long had acted quickly in unloading as much as possible onto the ice, and they had sledges and dogs and boats. As they headed south towards the north coast of Siberia, they encountered another island. This one De Long named Bennett Island (Ostrov Bennetta), after his backer, and they used the newly found island to rest. For De Long discovered to his horror that they were drifting away from the coast. "To work like horses all day for ten or eleven hours, and to make only a mile, is rather discouraging," De Long wrote.

But they made it. On 6 August they reached open water and took to their three boats, only to have a storm separate them. One boat disappeared, one made it to the Lena Delta and was saved. De Long's boat made it to the shore but the men could find no hunters to help and eventually all died from starvation. Two of his party De Long had sent to find help met a language barrier and were unable to communicate with Russian hunters who might have saved the group.

On hearing what had happened, Bennett dispatched reporters to get the story, one of whom dug up bodies to see if they contained any notes. It was an inglorious start to the involvement of newspapers in Arctic exploration.

MAP 213 (above).
A map of the search for the *Jeannette* by the USS *Rodgers*, commanded by Robert Mallory Berry, in 1881. The *Rodgers* was destroyed by fire in December 1881. Berry searched the northern Siberian coast the next year, finding the *Jeannette* survivors.

MAP 214 (below).
A map from Adolphus Greely's book showing the exploration of the northern coast of Greenland. This map shows the section from Fort Conger to Lewis Beaumont's Cape Britannia. MAP 216 (overleaf) continues this map eastwards to show the achievement of a new "farthest north" by Lieutenant James Lockwood and Sergeant David Brainard.

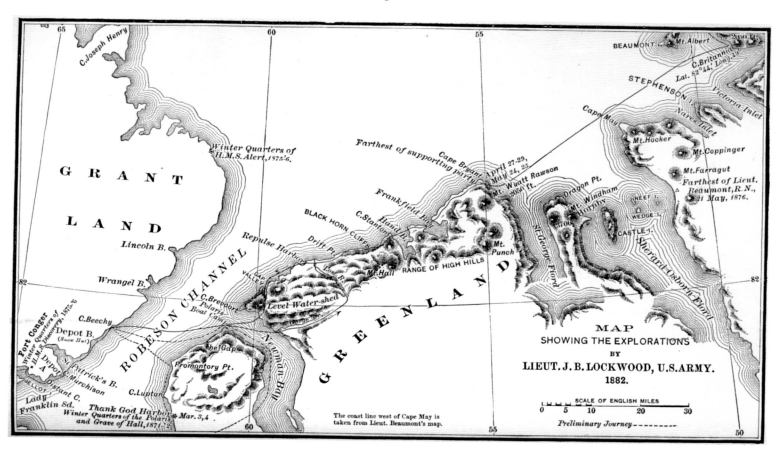

Flotsam from the crushed *Jeannette* drifted clear across the Arctic Basin and was found in 1884 on the shores of southwest Greenland. It was this that gave Fridtjof Nansen the idea for his voyage and drift of the *Fram* in 1893–96 (see page 148).

Another dramatic tale of woe attaches to the next expedition to attempt the Pole, that of Adolphus Washington Greely in 1881. He was sponsored by the U.S. government and was to set up the American contribution to the first International Polar Year in 1882–83. This was a multinational effort to study all things scientific and polar originally proposed by Karl Weyprecht, the discoverer of Franz Josef Land. But Greely also had the Pole on his agenda, or at least the achievement of a higher latitude than the British had in 1876.

Greely and twenty-five men from the Signal Corps of the U.S. army sailed north in 1881 on the steamer *Proteus*, which took them to Discovery Bay, where the British ship *Discovery* had wintered in 1875–76; then it departed. Greely built a base he named Fort Conger, after a U.S. senator, and the various meteorological, magnetic, tidal, and other observations began.

In the spring of 1882 two parties explored north and east. Octave Pavy sledged to Cape Joseph Henry and struck out over the ice to try to determine if any land existed to the north of Ellesmere Island. He was stopped by an open lead and was forced to return to Fort Conger.

James Booth Lockwood and David Legge Brainard sledged to the northwest coast of Greenland. They reached a new "farthest north" at the north end of Lockwood Island, at 83° 23′ 08″ N (Map 216). "We unfurled the glorious Stars and Stripes to the exhilarating northern breezes with an exultation impossible to describe," wrote Brainard in his field notes. They could see that the coast extended yet farther to the northeast, but their food was almost gone, so Lockwood wisely decided to go no farther. They built a cairn and spent some time ensuring that their position was accurately measured, something certain later polar explorers could have learned from.

Greely meanwhile had explored the interior of northern Ellesmere Island, finding and naming the large Lake Hazen. Then, in what would turn out to be an omen of impending disaster, in 1882 their relief ship, sent to resupply them, failed to reach far enough north.

In the spring of the following year Lockwood trekked west across Ellesmere Island to the west coast, where he found a very large inlet he named Greely Fiord. That summer the relief ship again failed to make it to Fort Conger, and so, following his instructions, on 10 August Greely abandoned their base and started southwards by boat and sledge.

After an epic journey of 800 km they reached Cape Sabine, on the northwestern

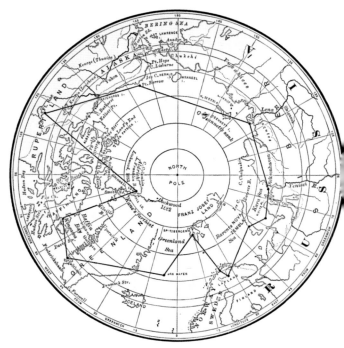

Map 215 (*above*).
Polar map showing the locations of all the stations participating in the first International Polar Year, 1882–83, connected by lines to emphasize how far north they were. Greely's station at Fort Conger was the northernmost station. The intention was to simultaneously record scientific data such as temperatures and magnetic information at all the stations.

Map 216 (*right*).
An eastward extension of Map 214, this map shows the achievement of a new "farthest north" by Lieutenant James Lockwood and Sergeant David Brainard between April and June 1882. Lockwood Island, the farthest point reached, is at top right. The sketch *Next point beyond farthest* shows Cape Washington to the left and Cape Kane in the foreground. The *Sketch of "farthest" from the west* shows Lockwood and Brainard Islands in front of the Cape Kane peninsula.

shore of Smith Sound, where they hoped to winter, using caches of provisions left by the relief ships. They found only one cache, with little in it, for the relief ship, the *Proteus*, had been sunk by ice.

The winter was a terrible ordeal; hunting was a failure and starvation and scurvy claimed the lives of sixteen men during the winter. Another killed himself, and another was executed for repeatedly stealing food.

Seven survivors, one of whom was Greely, were finally rescued on 22 June 1884 by a relief expedition led by Winfield Scott Schley. "In a frenzy of feeling as vehement as our enfeebled condition would permit, we realized that our country had not failed us," wrote Greely later. The survivors were in a very bad condition; one more man died on the way south. Twenty-five went north, and only six returned. Death in the Arctic was becoming all too common.

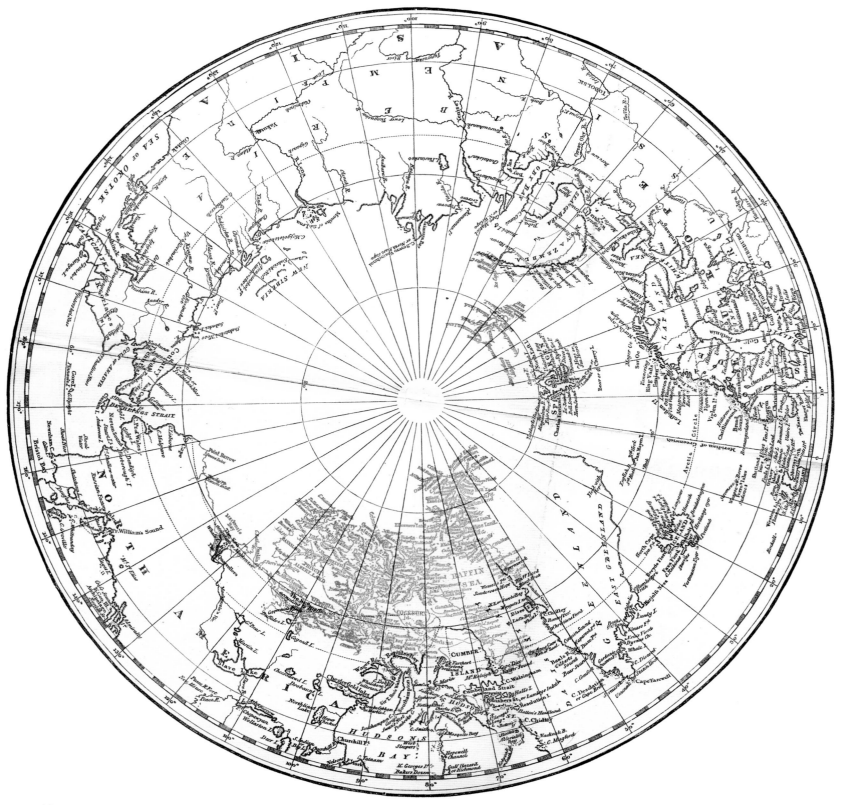

MAP 217.
An updated version of John Barrow's polar map of 1818 (MAP 104, page 60). This map is from Charles Hall's book about his second Arctic expedition, 1864–69 (see page 109), published in 1879, after his death. New information is depicted in red. Despite its relative crudity, the map gives a good impression of the advance of geographical knowledge in the intervening period. The 1818 and 1879 versions of this map can also be compared with that of 1846, shown as MAP 136, page 81.

The Northeast Passage Achieved

The first navigation of the Northeast Passage, that goal of men for centuries, was by a Swedish scientist and explorer, Adolf Erik Nordenskiöld. He had at first been intrigued by the question of the attainment of the Pole, and in fact reached a farthest north at the time, in 1868, of 81° 42´ N, during an attempt by sea from Spitsbergen (MAP 218, *below*). After a failed test run in Greenland in 1870, he tried again in 1872 with reindeer pulling sledges, but the reindeer did not get to his proposed starting point and he had to be content with the first crossing of the ice cap of Nordaustlandet, the easternmost large island of the Svalbard (Spitsbergen) group. He did this with Adolf Arnold Louis Palander, a naval officer.

By 1875 Nordenskiöld, a man of far-ranging interests, had turned his attention to the achievement of the Northeast Passage. In 1874 Joseph Wiggins, a captain of a merchant ship, steamed as far as the mouth of the Ob' River (Obskaya Guba) with the intention of proving the feasibility of trade between Britain and Siberia. Nordenskiöld noticed this and realized that navigation of the Northeast Passage, at least in a good year, should be possible for a small steam-powered vessel.

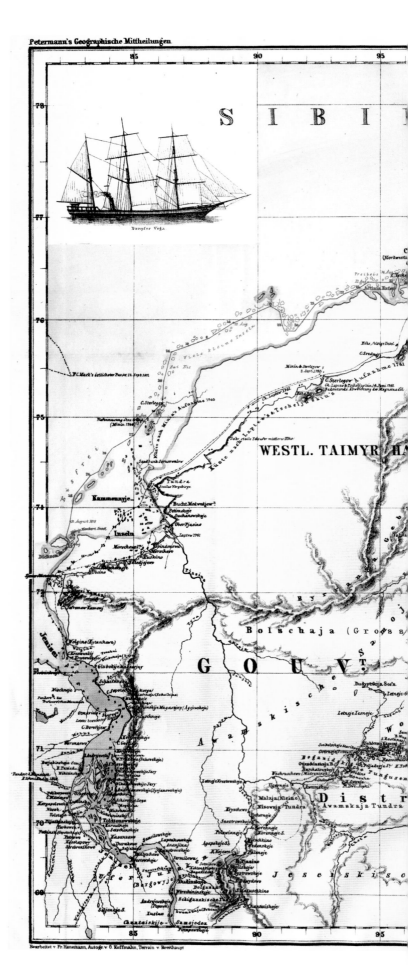

MAP 218 (*above*).
A map from *Petermann's Geographische Mittheilungen* illustrating the voyage of Nordenskiöld to Spitsbergen in 1868. His attempts to push farther north are marked by a maze of tracks at top. His farthest north, near to that of Parry's in 1827 (but by sledge and boat), is also shown.

MAP 219 (*right*).
The western part of a nicely detailed map of Nordenskiöld's voyage in the *Vega* through the Northeast Passage in 1878. This covers the period to 27 August; the map was published in 1879, before Nordenskiöld had completed his transit of the passage.

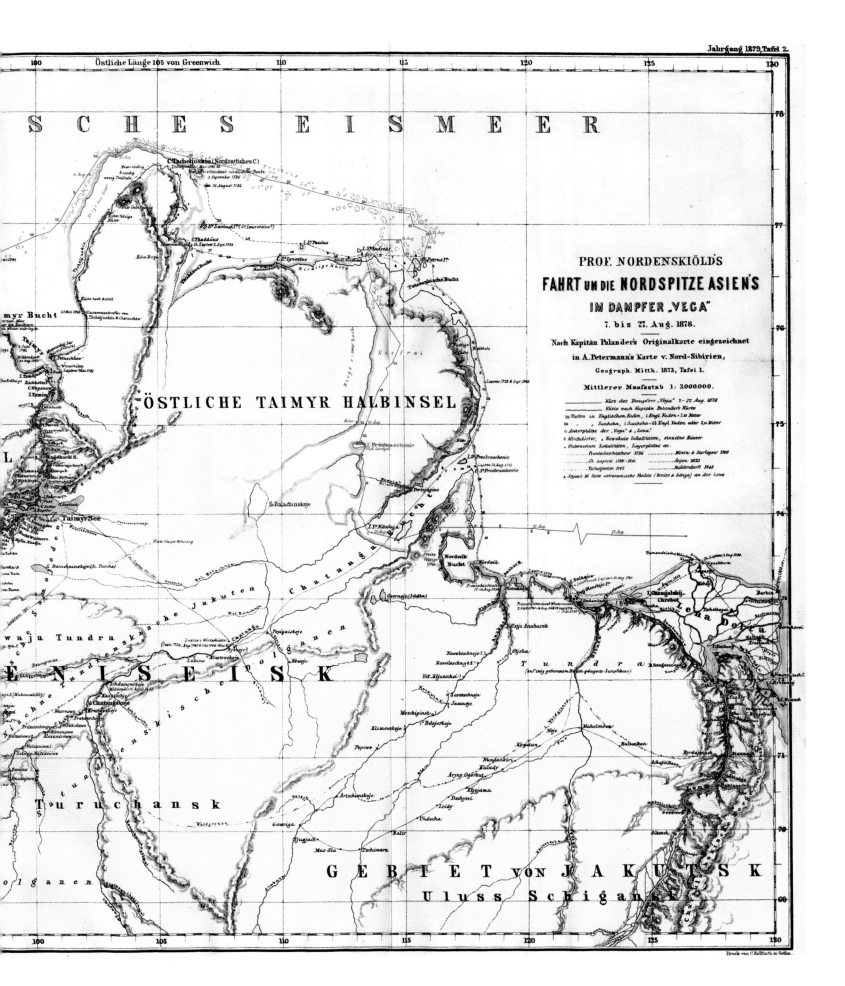

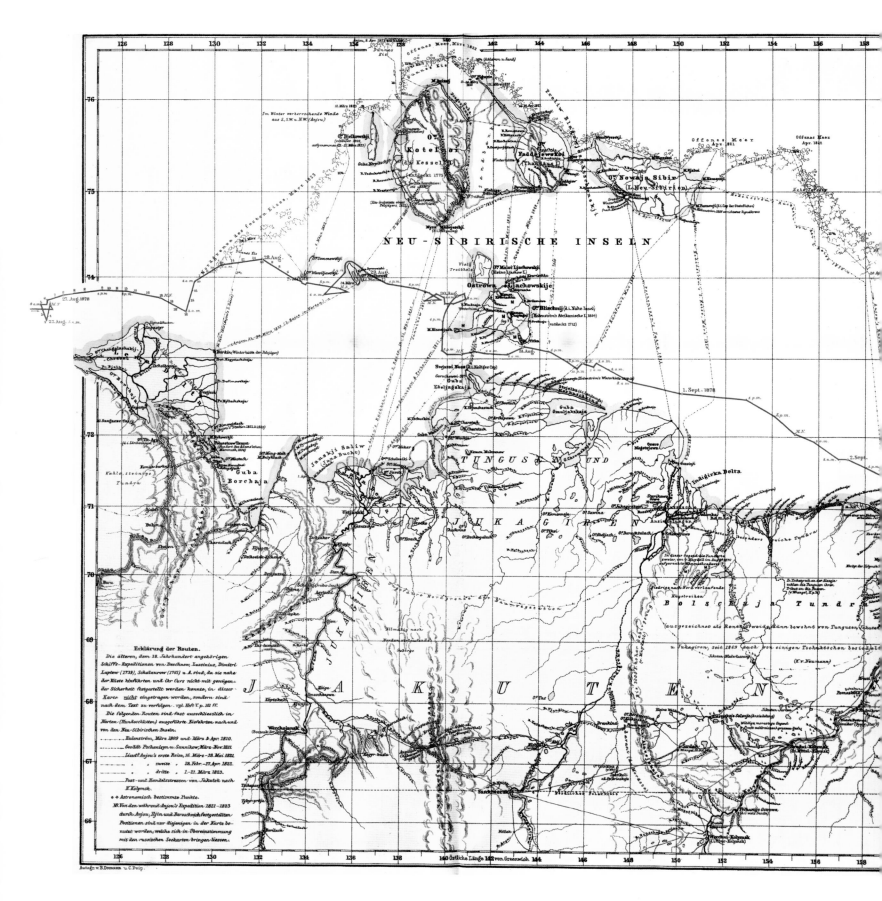

Map 220.
The eastern part of the map of Nordenskiöld and Palander's transit of the Northeast Passage in the *Vega* to the end of 1878. Like Map 219, it was published in *Petermann's Geographische Mittheilungen* in 1879, before Nordenskiöld's completion of the passage. His track here stops at Kolyuchinskaya Guba, the place he wintered in 1878–79. It is marked *Winter-Quartier der Vega* in German on this map. Nordenskiöld was at this point only about 200 km from Bering Strait.

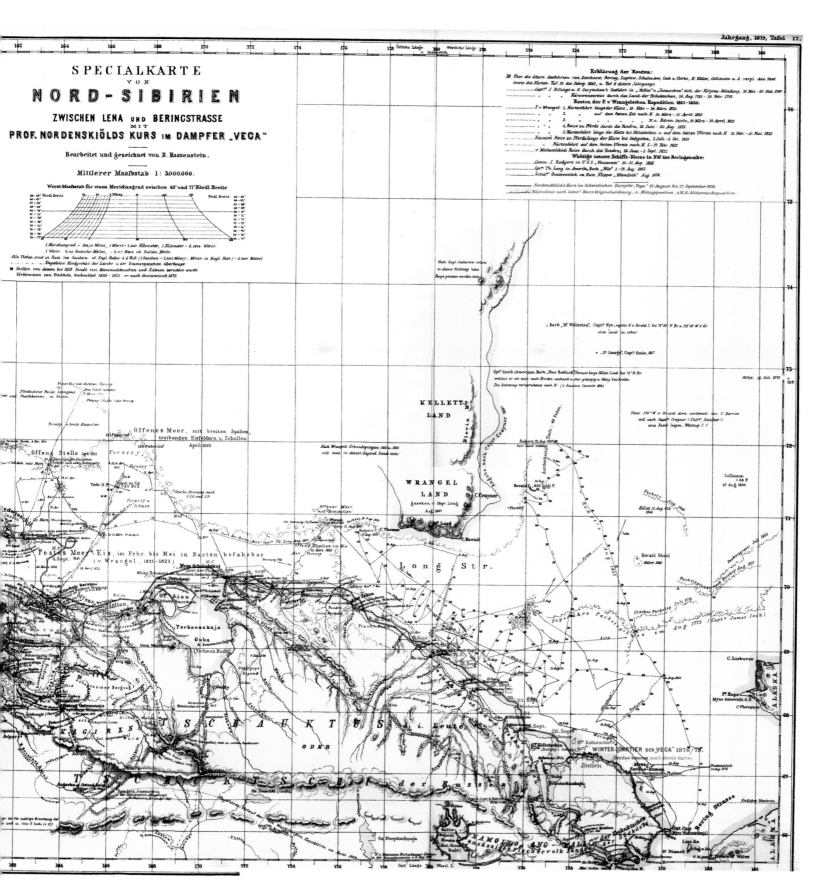

By 1878 Nordenskiöld had organized an expedition. His ship, the *Vega*, was commanded by Louis Palander, and three other ships would play various support roles. Nordenskiöld, the scientist, had worked everything out very precisely. He sailed from Tromsö in the *Vega* with one of the other ships, the *Lena*, meeting the others at Yugorskiy Shar, between Ostrov Vaygach and the mainland. Together they sailed into the Kara Sea.

The *Vega* and the *Lena*, without the other two ships, continued eastwards, rounding the Taymyr Peninsula (Poluostrov Taymyr) at Mys Chelyuskin, the most northerly point of the continent of Asia. They were the first ships ever to do so.

Now late August, they reached the vicinity of the Lena Delta. Here the *Lena* made for Yakutsk, where it was laid up for the winter. The *Vega* continued alone. They were

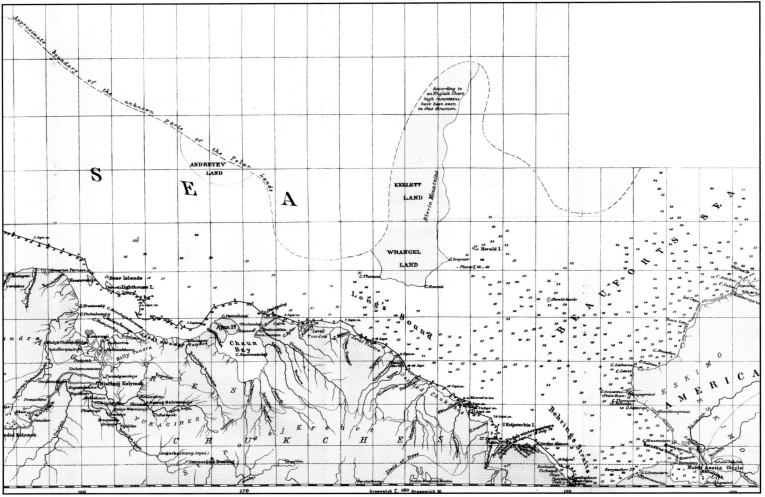

Map 221 (*above*).
This map showing the track of the *Vega* from her winter quarter to Bering Strait also shows the possible land to the north of eastern Siberia.

Map 222 (*right*).
The completion of the *Vega*'s transit through the Northeast Passage in 1879, through Bering Strait and into the Pacific Ocean. It was the dream of centuries finally come true.

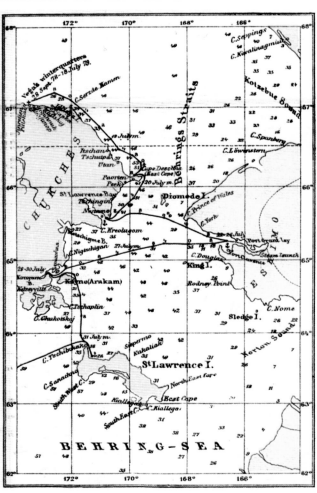

making fast progress east, but after they reached Mys Shelagskiy (*Myss Schelagskoj* in German on Map 220), fog and ice and increasing darkness slowed them down. They made it to Kolyuchinskaya Guba (*Guba Kohutschinskaja* on the map) by 28 September before being stopped by ice.

They were here only about 200 km from Bering Strait. Such proximity must have been frustrating for Nordenskiöld and Palander, but they could rest easy knowing that they were superbly positioned to pass through Bering Strait the following season.

On 18 July 1879 the *Vega* was able to proceed eastwards once again, taking only two days to reach Bering Strait. From there they sailed to Yokohama, Japan, reaching that city on 2 September, and then they completed the circumnavigation of Asia, returning home through the Suez Canal.

Nordenskiöld's Northeast Passage was to become the Soviet, then Russian Northern Sea Route, kept navigable today by massive nuclear icebreakers. It was 1932 before the first ship passed through in a single season, a traverse which has by now become relatively routine. Yet as late as 1938 the icebreaker *Georgii Sedov* became trapped in the ice and was whisked away, *Jeannette*-like, to the Greenland Sea (see page 186).

The Exploration of Greenland

Greenland, correctly now Kalaallit Nunaat, is the world's largest island. It is shaped like a huge bowl, with up to 3 410 m thick ice filling the interior, depressing the ground surface to 365 m below sea level.

Up to the end of the nineteenth century, Greenland was a forbidding and impenetrable land. Native, Norse, and other settlements hugged the coast, the only source of year-round food. Until the beginning of the twentieth century it was not even known for sure that Greenland was an island, for the ice-clogged northeastern coast defied attempts to penetrate it.

In July 1870, Adolf Erik Nordenskiöld led an expedition which pushed 57 km into the interior. He was supposed to be testing dogs for a later attempt on the North Pole, but dogs were unavailable due to disease, so he resorted, unsuccessfully, to man-hauling the sledges.

In 1878 the Danish government sent Jens Jensen to survey the west coast and attempt an extended journey up onto the ice cap. The illustrated map below (MAP 223) shows his track in July 1878, when he took

MAP 223 (below).
A rather nicely illustrated map showing the explorations of Jens Jensen in 1878. The inset map of Greenland also shows the track of Adolf Nordenskiöld's venture onto the ice cap in 1883.

MAP 224 (right).
Map showing the coast of eastern Greenland as discovered by William Scoresby in 1822 and Douglas Clavering in 1823, together with the coast mapped by Karl Koldeway in 1868.

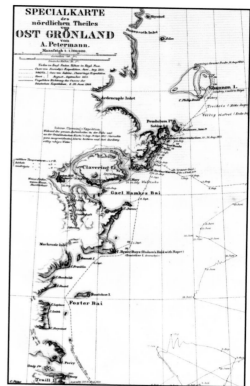

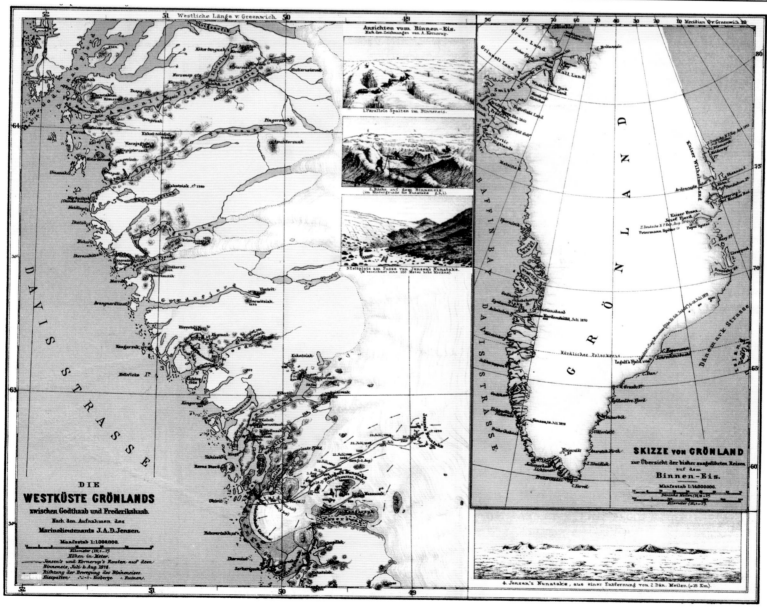

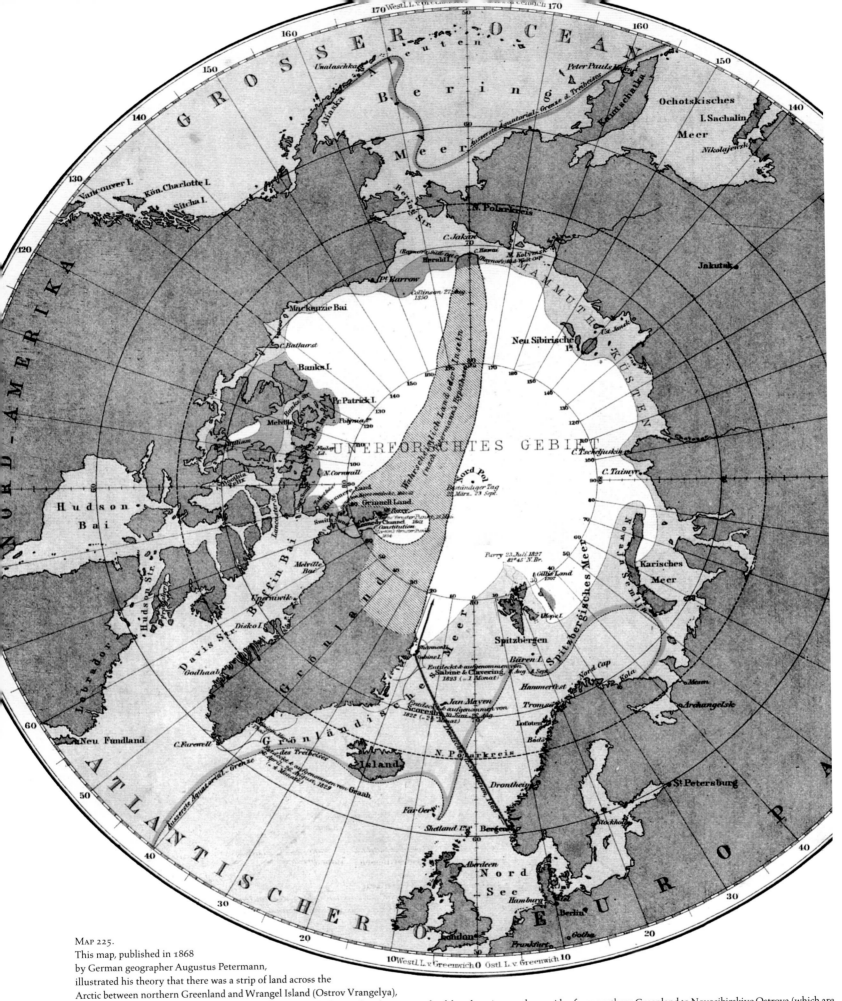

MAP 225.
This map, published in 1868 by German geographer Augustus Petermann, illustrated his theory that there was a strip of land across the Arctic between northern Greenland and Wrangel Island (Ostrov Vrangelya), the latter having been discovered by Thomas Long the year before. In fact there was no land, but there *is* an undersea ridge from northern Greenland to Novosibirskiye Ostrova (which are shown on this German map as *Neu Sibirische In*[Insuln; islands]). This is the Lomonosov Ridge, discovered in 1948 (see page 190); Petermann's idea was not so off base after all. Note that Petermann's land avoids the North Pole. This allowed him to hold the apparently incompatible view of the existence of this vast land at the same time as he believed in an open polar sea, the sea that Karl Koldeway was aiming for, on Petermann's instructions, in 1869 (see pages 130–31). (A proposed route for his attempt the year before is shown in red, marked *Deutsche Expedition, 1868*.) Petermann's ideas were also the impetus for Karl Weyprecht and Julius Payer's expedition that discovered Franz Josef Land in 1873 (see pages 118–21).

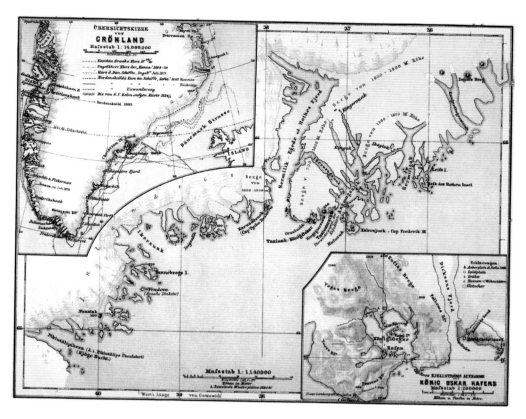

MAP 226 (*above*).
Map of Adolf Erik Nordenskiöld's trek into the interior of Greenland in 1883, shown in the inset. The main map shows part of the east coast of Greenland just below the Arctic Circle, where Nordenskiöld searched for any evidence of Norse settlements, finding none.

squashed two years later by another beginning Arctic explorer, Fridtjof Nansen.

In August to October 1888 Nansen and five others crossed Greenland at about 64° N, from east to west. He was dropped off by his ship on the east coast and deliberately put himself in the position of *having* to cross to the populated west coast. It was a gamble, but it paid off. At the end of the journey the six reached the head of a fjord and had to construct a wooden boat so that two of them, Nansen and Otto Sverdrup, could paddle to Godthåb (now Nuuk) and get a boat to return for the others. Nansen's track in 1888 is shown on MAP 227, *below*.

Nansen had demonstrated that the ice cap was continuous, at least in the southern part of Greenland. He himself admitted later that his crossing of Greenland "was very generally considered to be the scheme of a lunatic." But it established Nansen as an Arctic explorer of note, probably much to the chagrin, one suspects, of the fame-seeking Robert Peary.

MAP 227 (*below*).
Fridtjof Nansen's first crossing of the Greenland ice cap, in 1888.

eleven difficult days to reach some of the mountains, called nunataks, that protrude from the ice sheet. From there Jensen could see an unbroken ice cap extending as far as the eye could see. The length of the journey is put into perspective on the inset map; it was just a short incursion onto the massive ice cap. Nordenskiöld's 1870 trek is also shown on MAP 223.

In 1883 Adolf Erik Nordenskiöld again led an expedition, this time to try to determine if ice covered the whole of the interior of Greenland.

Nordenskiöld set out across the ice cap in July, with nine men hauling sledges. He traveled 117 km, and then sent two Lapp ski experts on farther. They said they covered another 230 km, still seeing no end to the ice. Nordenskiöld concluded that the ice must cover most of Greenland. His track is shown on MAP 226, *above*.

The next explorer of Greenland hoped to become the first to cross it, for he was seeking fame. This was Robert Edwin Peary, whose Arctic adventures loom large twenty years later. Peary, who financed this first of his expeditions himself, managed to penetrate 150 km inland in 1886, not as far as Nordenskiöld's Lapps had claimed, but at a higher elevation, 2 294 m. Any thought of another attempt was

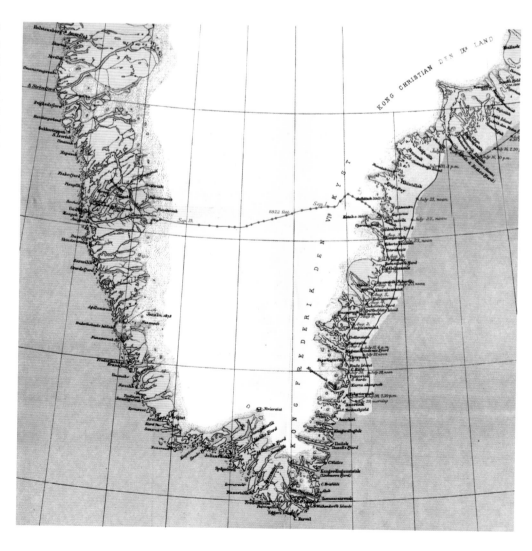

Nansen's Drift

Soon after his epic crossing of Greenland (*see previous page*), the Norwegian explorer Fridtjof Nansen read of flotsam washed up on the southwest coast of Greenland and proven to have come from the ill-fated *Jeannette* expedition of 1879, which had been crushed in the ice in 1881 near Novosibirskiye Ostrova (the New Siberian Islands) far to the east (see page 135).

Nansen's sharp mind immediately thought of the possibility of using this drift to deliberately cross the Arctic Ocean. He proposed to go with the flow, rather than fight against it. Nansen set about organizing an expedition and found sponsors in the Norwegian government, the king of Norway, and other subscribers. Here, Nansen propounded, was an ideal opportunity to explore the unknown with a good chance of success.

A specially constructed ship, the *Fram*, was built with a rounded hull intended to ride up onto the crushing ice rather than be caught in its vice. Its captain was Otto Sverdrup, who had been Nansen's companion for his Greenland crossing.

The *Fram* left Christiania (Oslo) in June 1893, followed the Northeast Passage route recently forged by Adolf Nordenskiöld (see page 140), picked up thirty-four Siberian dogs at Khabarovo, on Yugorskiy Shar, and on 25 September became stuck fast in the ice at about 78° 50´ N to the north of Novosibirskiye Ostrova.

Nansen must have wondered about his theories at first, for the *Fram* hardly drifted at all for the first few months. Then, in January 1894, a northwesterly drift began, slow and irregular at first, but soon unmistakable. By the summer, Nansen was convinced that the ship was indeed drifting towards Greenland, as he had anticipated.

During the ensuing winter, realizing that although the *Fram* would approach the Pole it would not drift over it, Nansen hatched a bold plan to leave the ship and sledge to the Pole. On 14 March 1895, after two previous attempts, Nansen left the ship with one other companion, Frederik Hjalmar Johansen, a stoker on the *Fram*, selected over Sverdrup because the latter was the only other person capable of commanding the ship. The *Fram* was at this time at 84° 04´ N. The pair's dog-drawn sledges proved difficult to handle over the rough, pressure-ridged ice, and it was extremely cold. In addition they found themselves fighting against a southward drift of the ice, a phenomenon which was to plague later attempts on the Pole.

On 8 April 1895 at about 95° E, Nansen and Johansen reached 86° 13´ 06´´ N before deciding that they could go no farther. Nevertheless, this was a record "farthest north."

Returning southwards, they aimed for Zemlya Frantsa-Iosifa (Franz Josef Land). Now, with the advancing season, the problem became leads of open water. But, helped

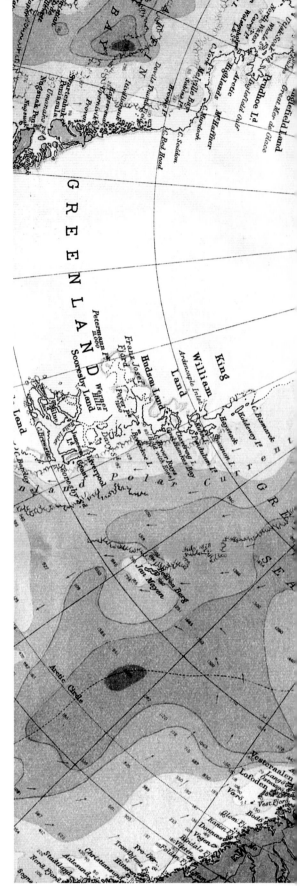

MAP 228 (*left*).
The map from Frederick Jackson's explorations of Zemlya Frantsa-Iosifa (Franz Josef Land) in 1894–97. The northernmost blue cross indicates Nansen's first landfall, at what he named "Hvidetland," now Ostrova Belaya Zemlya, and the other blue cross shows where he met with Jackson's expedition. Nansen's Hut, his wintering place in 1895–96, is shown on the unmarked island (Ostrova Dzheksona) just southeast of the northernmost blue cross. Nansen later named the island "Jackson Island," though at the time he was unsure as to exactly where he was due to malfunctioning chronometers.

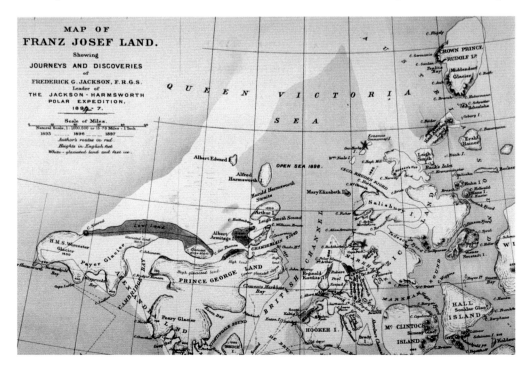

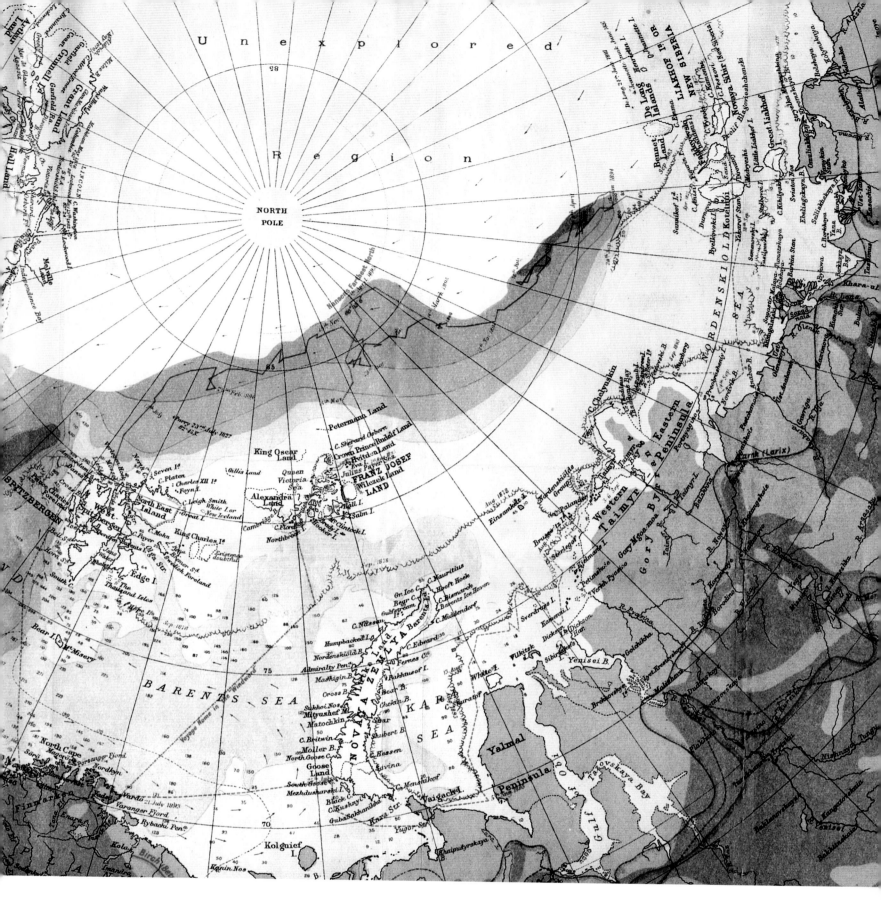

MAP 229 (above).
This map shows the drift of Nansen's ship *Fram* in 1893–96 and his attempt on the North Pole in 1895. The dashed red line is Nansen's route along the northern coast of Asia, and the point at which it becomes a solid line is where the *Fram* was beset in the ice and began its long drift northwards and westwards. The dotted red line is the track of Nansen and Johansen. The island first encountered by them on their trek southwards Nansen named "Hvidetland" (White Land, now Ostrova Belaya Zemlya), shown as *Hvitden Land* on this map, from the English translation of Nansen's book *Farthest North*. The track of Frederick Jackson's supply ship *Windward* as far as Vardo in northern Norway is also shown. The beginnings of a bathymetric map of the Arctic Ocean are shown in the swath either side of the path of the *Fram*, extrapolated from the regular soundings taken from the ship. Nansen collected much scientific data, some of which is still useful today.

by albeit leaky kayaks, they were able to find ways to bridge the waters. Nansen fed the dogs by killing some and feeding the meat to those remaining, a ghastly procedure followed by many polar explorers who used dogs.

At last, they spied land. Nansen's entry for 24 July speaks of his relief: "At last the marvel has come to pass—land, land, and after we had almost given up our belief in it . . . a new life is beginning for us, for the ice is never the same." The land was the northernmost small island of the Franz Josef group, an island that Nansen was the first to

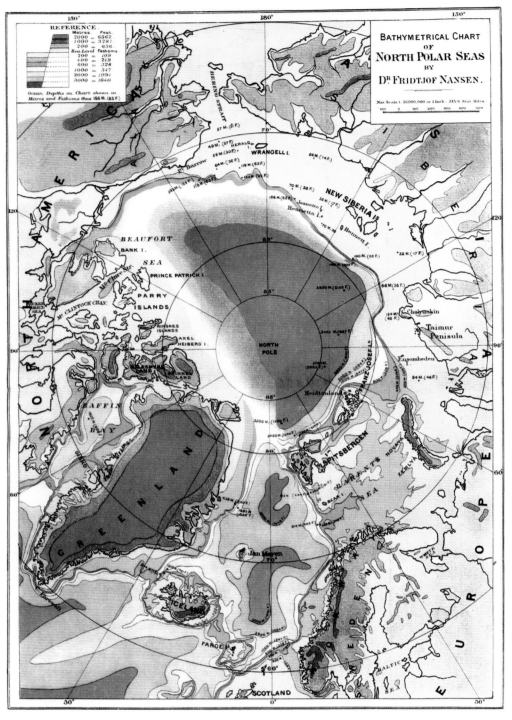

MAP 230 (left).
This first attempt at a comprehensive bathymetric map of the Arctic Basin was published by Nansen in 1907. Data from his 1893–96 drift has been updated with other geographical information such as that from Otto Sverdrup in 1898–1902. Compare this map with the most modern bathymetric map, the *International Bathymetric Chart of the Arctic Ocean,* shown on pages 190–91. The comparison reveals that Nansen mapped the broad shape of the ocean depths quite correctly, although, naturally enough, there were many details left to be filled in.

discover. But still the going was difficult; it took until 7 August for Nansen to be able to write "At last we are under land; at last the drift-ice lies behind us." They were on small islands, but ahead was the edge of the pack ice and open water.

Here they had to kill the last of their dogs and take to their kayaks. They paddled and sailed southwest, but by September were forced to winter on the west coast of Ostrov Dzheksona, later named Jackson Island by Nansen (MAP 228, page 148). Surviving the winter by hunting, the following June Nansen and Johansen fortuitously met the expedition of Frederick Jackson, who was exploring Franz Josef Land, trying to complete the geographical discoveries begun by Karl Weyprecht and Julius Payer in 1872–74 (see page 120). Thus Nansen was able to sail south in August on Jackson's supply ship *Windward*, and he arrived in Tromsø on 21 August.

Since Nansen left the ship in March 1895, Sverdrup and the *Fram* had continued their drift towards the Greenland Sea, reaching 85° 55′ 06″ at 66° 31′ E on 15 November of that year. The *Fram* finally broke free of the ice and reached land at Danskøya (Danes Island), Spitsbergen, on 14 August 1896. Sailing south, *Fram* reached Tromsø on 20 August, one day before Nansen and Johansen. The entire expedition was thus reunited, reaching Christiania on 9 September.

Nansen, ever the scientist, had arranged for many scientific observations to be taken from the *Fram*, which was essentially the first of many drifting stations that would be set up in the Arctic in the next century. In particular, regular soundings were taken, which soon began to demonstrate that the Arctic was a deep basin. Nansen's soundings showed depths of 3 000 to 3 900 m along the line of the drift of the *Fram* and for the first time produced enough data for him to draw the conclusion that it was unlikely that further land would be found towards the Pole, at least on the Russian side of the Arctic Ocean. And he had graphically demonstrated the significant drift of ice in the Arctic Ocean.

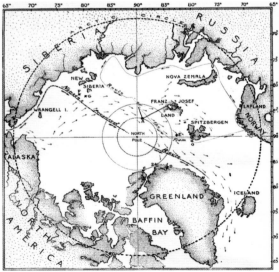

MAP 231 (left).
Map Showing Nansen's Proposed Route to the Pole from a magazine article by Nansen entitled "Future North Pole Exploration," published in February 1898. Nansen was by this time more convinced than ever that the Arctic was largely deep ocean and that by simply choosing the right starting point, he could drift across the Pole as he had intended in 1893. That his hypothesis was correct was proved by the drift of Russian ice stations (see page 186). From May 1954 to April 1957 station SP-4 drifted across the Arctic Ocean, coming within 13 km of the North Pole. The only place that Nansen thought land might exist was north of the North American continent, and indeed, more land was discovered there in 1900 by Nansen's captain, Otto Sverdrup, in his ship, the *Fram* (see page 156).

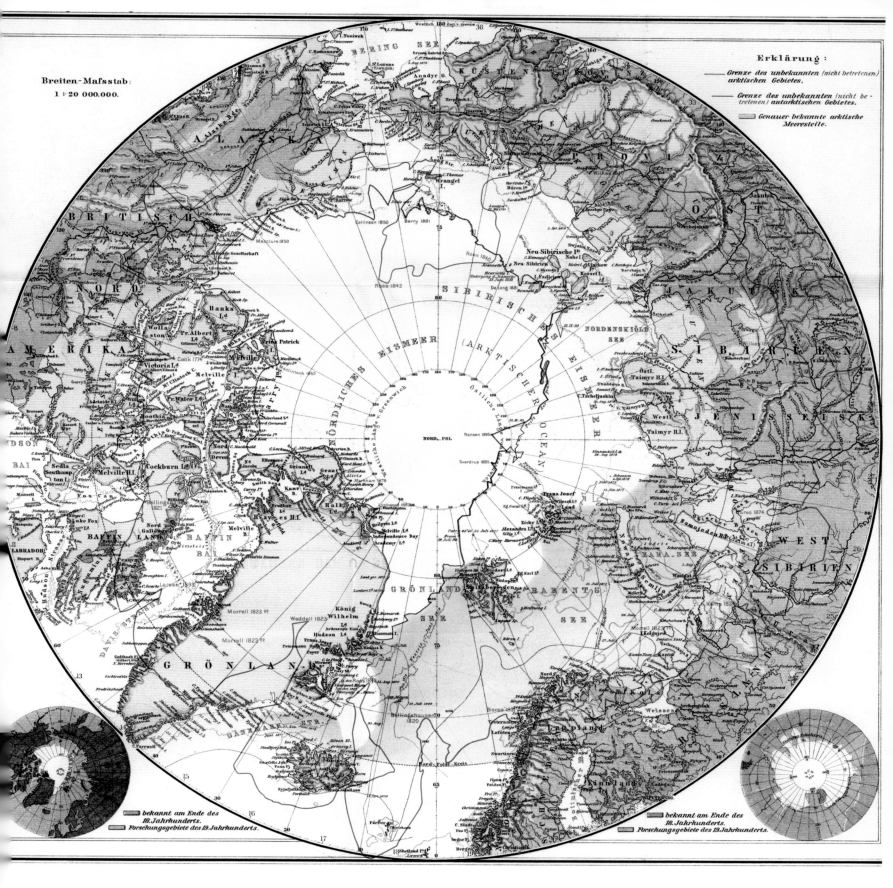

Map 232.
This polar summary map appeared in *Petermann's Geographische Mittheilungen* in 1897. It shows the newly reported drift of Nansen's ship *Fram* together with his "farthest north" on his attempt to reach the North Pole by sledge. The blue line delineates the unexplored area in 1897. The red line, a little strangely, is an attempt to relate the attempts on the North Pole to those on the South Pole; names of "farthest south" explorers are shown on this "farthest north" map. The green area is the well-known part of the Arctic according to Petermann. The smaller polar map at left bottom shows the known area by 1800 in blue and the area discovered from 1800 to 1896 in red, the white area being still unknown. The right-hand smaller map does the same thing for the south polar regions.

In October 1937 a Russian icebreaker, the *Georgii Sedov*, became trapped in ice in the Laptev Sea, just west of the New Siberian Islands, at about 130° E. With a damaged rudder and a crew of fifteen after the first winter, she was left to drift with the ice. It was January 1940 before the ship was freed—in the Greenland Sea, at about the meridian of Greenwich, 0°. The ship drifted first east then west, coming within 340 km of the Pole, following a similar path to that of the *Fram* forty-three years before (see Map 286, page 186). It was Nansen all over again.

Peary's Push for the Pole

American Robert Edwin Peary was to feature large in Arctic exploration, not least of all in his apparent achievement of the North Pole in 1909, still accepted by many as the first (see page 160). But Peary was a dogged and persevering explorer who accomplished a great deal even if his Pole claim is discounted.

Peary was determined to become famous from the beginning. His first exploration was an unsuccessful attempt to cross Greenland in 1886 (see page 147). After this, Peary turned his attention first to northern Greenland, which offered a vast as yet unknown region ripe for exploration—and possible fame—and later he set his sights on the Pole.

In 1891 Peary left New York on an expedition to northern Greenland. He intended to cross the ice cap and locate the undefined northeastern coast of Greenland and perhaps its northern point. Unfortunately for Peary, on the voyage northwards he broke a leg, which limited his activities that year. But the following year, after wintering at MacCormick Fjord, and after some initial forays out onto the ice cap, Peary was off in earnest. Some 240 km northeast, he sent back two of his party (one of whom was Frederick Cook, who would later become Peary's nemesis on his race to the Pole) and continued with only one other person, Eivind Astrup, and thirteen dogs. They arrived at Independence Fjord on the northeast coast at the point at which the Academy Gletscher (Glacier) meets the sea.

Both features were named by Peary, who also thought Independence Fjord was the eastern entrance to a strait which would have made the northeastern part of Greenland another island. At a place he named Navy Cliff, Peary wrote that "as I had for some days suspected, this channel stretched from the Lincoln Sea to the Arctic Ocean on the N. E. coast of Greenland." This he called Peary Channel, and it is shown on his maps. It was found to be non-existent by an ill-fated expedition under Dane Ludvig Mylius-Erichsen in 1907; Mylius-Erichsen perished while trying to map it. The fact that Peary had been wrong was not revealed until the return of a searching expedition under another Dane, Ejnar Mikkelsen, in 1912. He had found Mylius-Erichsen's notes in a cairn in 1910.

Peary returned to his base and sailed for New York in August 1892, but the next year was back to continue his explorations. He

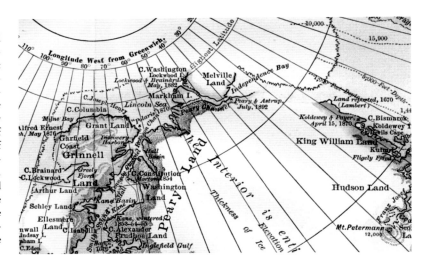

MAP 233.
An atlas map, dated 1897, showing northern Greenland as it was thought to be after Peary's 1892 expedition. *Independence Bay* leads to *Peary Channel*, cutting off the northern part of Greenland, here labeled *Melville Land*, and today renamed Peary Land. (Melville Land is still the name of a strip of land bordering Independence Fjord.)

made his base at Bowdoin Fjord, close to his previous base, and intended, as before, to cross the ice cap to Independence Fjord. But this time he was not as successful. After overwintering he set off in March 1894, but was dogged by bad weather and sickness. After only 237 km, he decided to turn back.

That season, he sledged south in search of the "iron mountain" first reported by John Ross (see page 60) and the Inuit source of iron for centuries before contact with Europeans. He found the iron on an island in Melville Bugt (Bay), at Savigsivik; it was in fact large meteorites. The island is now called Meteorit Ø (Meteorite Island).

In 1895, after wintering again, Peary set out across the ice cap. He again reached Independence Fjord, but due to lack of food, made few additional discoveries. Back on the west coast, he was picked up by a relief ship in August and, after collecting two of the meteorites from Melville Bugt (partly to deflect

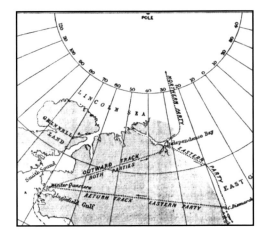

MAP 234.
This map showing Peary's proposed route for 1893 and 1894 reveals that he already had his eye on the North Pole.

criticism that he had not achieved much), he sailed back to New York.

In 1896 and 1897 Peary made two attempts to collect the largest meteorite, weighing perhaps 100 tons. The first attempt failed when ice threatened his ship, but in 1897 it was successfully transported to New York.

Now Peary's attention turned to bigger things yet: the North Pole. Peary was now financed by a group of New York businessmen called the Peary Arctic Club, men whose names Peary would scatter around the capes and bays of northern Greenland and northern Ellesmere Island in the years to come. In July 1898 Peary sailed from New York on his first expedition aimed at the Pole.

His ship, the *Windward*, could go no farther than Allman Bay on Ellesmere Island (shown on MAP 237, *overleaf*, on the west side of Kane Basin). From here Peary made a number of short trips into the interior of Ellesmere Island and also laid depots northwards to Adolphus Greely's old base (see page 137) at Fort Conger (shown on Lady Franklin Bay, at about 81° 40′ N, on MAP 237). This was to be Peary's base for his attempt on the Pole. On this trip, Peary suffered from frostbite and lost all or parts of seven toes. Many lesser men would have given up at this point, but Peary carried on, undeterred. That summer he met a supply ship at Etah (Taseq), and both ships departed, leaving him to winter there.

In March 1900 Peary again went to Fort Conger, and on 11 April he left there on his first attempt on the Pole. The route he took was northeast along the coast of northern Greenland, beyond the farthest point of the Greely expedition (see page 138). At Kap (Cape) Washington, after finding it not to be

MAP 235.
The members of the Peary Arctic Club sent a letter of appreciation to Peary in October 1902 in which they signed themselves on a map of the northern part of Greenland against the features Peary had named for them. U.S. president Theodore Roosevelt's name is prominent amongst them; Peary named a range of mountains after him.

the northernmost point of Greenland, Peary made an interesting comment, in light of his future claims. "It would have been a great disappointment to me, after coming so far," he wrote, "to find that another's eyes had forestalled mine in looking first upon the coveted northern point." Peary was clearly very concerned about being first; he wanted fame. The "coveted northern point" of mainland Greenland was reached on 14 May; this was Kap (Cape) Morris Jessup, named by Peary after the president of the Peary Arctic Club. From here he launched northwards over the ice.

After two days of "frightful going" Peary was defeated in his northward progress by water, at the edge of what he called "the disintegrated pack." He had reached 83° 50′ N, well south of Nansen's 1895 mark of 86° 13′ N and, unknown to Peary, Umberto Cagni's 86° 34′ N, reached twenty-two days before (see page 154). He returned to land and instead decided to explore farther east, reaching Clarence Wyckoff Island (now known to be part of the mainland) near Kap Wyckoff on 20 May (see MAP 237, *overleaf*). Here he left a record of his achievement in a cairn, as he had at Kap Morris Jesup: "With me are my man Henson; Ahngmalokto, an Eskimo; sixteen dogs and three sledges," and a list of the members of his Peary Arctic Club.

Peary arrived back at Fort Conger on 10 June. He prepared to winter here, making a trip to Lake Hazen to find musk oxen for a winter food supply, as he realized that his ship was not going to reach him. *Windward* had again sailed north, now with Peary's wife and daughter on board, but it was trapped in the ice at Payer Harbour, on the west side of Smith Sound, and had to winter there. In the spring of 1901 Peary trekked south to meet the ship, which sailed home in August, leaving Peary to winter once more.

In March 1902 Peary set out to make another attempt on the Pole. He stopped briefly at Fort Conger, then traveled round the coast to Cape Hecla, not the northernmost point of Ellesmere Island, but close to it. From here he began his trek northwards over the ice. On 21 April Peary reached 84° 17′ 27″ N before being stopped by difficult conditions, "old rubble and deep snow." In his journal for that day he wrote, "The game is off. My dream of sixteen years is ended . . . it is impracticable . . . I have made the best fight, I knew. I believe it has been a good one. But I cannot accomplish the impossible."

Returning to Fort Conger and then to Payer Harbour, Peary met the *Windward* and sailed home in August. It would be three years before he would try again.

Money was raised by the Peary Arctic Club to build the *Roosevelt*, specially strengthened to withstand the ice and, like Nansen's *Fram*, designed to ride up over the ice instead of being crushed. Peary left New York in July 1905, and the *Roosevelt* forced a way through the pack to Cape Sheridan, at the northeast tip of Ellesmere Island. There the expedition wintered.

In February 1906 Peary set out from Cape Sheridan to Point Moss, just west of Clements Markham Inlet on the north coast of Ellesmere Island (shown on MAP 237). From here, on 6 March, he struck out northwards over the ice once more. It was difficult travel; leads and pressure ridges caused numerous detours and delays, graphically represented on Peary's map by the zigzag track. He lost seven days trying to cross the "big lead"—Peary termed it his "Hudson River"—where the ice attached to land meets the true sea ice. He sent back his support parties to Point Moss for supplies, but before they returned, the water froze sufficiently to allow him to cross. So he continued north with only Henson and six Inuit men. It was only while waiting to cross the "big lead" that Peary had first taken a longitude reading, and so only at this point did he have any real idea where he was.

Three days after crossing the "big lead" Peary was stopped again—this time by the weather. A fierce storm lasted seven days, breaking the ice floe on which they were camped right under Matthew Henson's iglu. Peary's journal entries show he was now coming to the realization that, once again, he would not make it to the Pole. So instead of the North Pole he would bring back a new "farthest north" for his Peary Arctic Club supporters, who, like most businessmen, demanded results for their money.

MAP 236.
Peary's map of his (here unnamed) Peary Channel, cutting off the northern tip of Greenland. The dashed line is Peary's route in 1892.

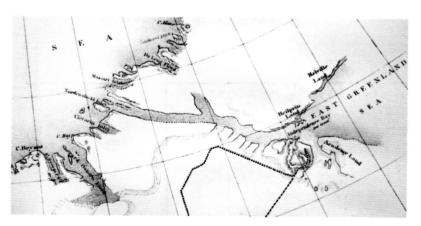

But in a pattern that was becoming all too familiar, Peary's continued journey north involved claims of dubiously high speeds over the ice, coupled with astonishingly poor documentation. Peary's map shows he thought he traveled east before turning north again, although even then more north-northeast than north, yet neither Peary's journal nor his book mention this apparent anomaly.

Peary claimed to have reached 87° 06´ N, a new "farthest north," on 21 April 1906, having traveled at three times his normal speed after leaving the camp where he was pinned down by the storm. His return journey, shown on his map as mainly due south along the 50° W meridian, apart from a short jog to the east to find a place to cross the "big lead," is curiously devoid of dates. Peary reached Kap Neumayer on the northern Greenland coast on 9 May. Here he met a lost Charles Clark, in charge of one of his support parties. Clark had been unable to cross the "big lead" and had returned to land at the same time as Peary set off northwards. Yet here he was, in the same place at the same time having traveled less than half the distance Peary claimed. The fact that Clark made a landfall here demonstrates the eastward movement of the ice, yet Peary's route is shown as due south.

Peary and Clark returned to the *Roosevelt*, but Peary, by now desperate to bring home some new discovery, set off again only a week later to survey the remaining unmapped coast between the farthest west point of Pelham Aldrich in 1876 at Alert Point (see page 134) and the farthest north and east of Otto Sverdrup in 1900 of Lands Lokk, now the Kleybolte Peninsula, at the northwestern tip of Ellesmere Island. On 16 June he passed Alert Point, and in his book he makes a statement that is very revealing about the nature of the man. "What I saw before me in all its splendid, sunlit savageness," he wrote, "was *mine*, mine by right of discovery, to be credited to me, and associated with my name, generations after I have ceased to be." That Peary would write this in a book shows the importance he attached to being *first*, not to mention a rather overwrought vanity.

Peary mapped the coast westwards, reaching Cape Stallworthy, the northern tip of Axel Heiberg Island, on 28 June. He reflected that this could be the end of his "Arctic work." "Twenty years last month since I begun," he wrote, "and yet I have missed the prize . . . It seemed as though I deserved to win this time."

It was from here that Peary thought he saw a land to the northwest, "the snow-clad summits of the distant land in the northwest, above the ice horizon," to which he later gave the name Crocker Land. Not until 1914 would the land be shown to be non-existent (see page 163). It was likely an Arctic mirage like the one that had fooled John Ross in 1818 (see page 60).

Peary returned to the *Roosevelt*, which, although damaged, was sailed by its captain, Robert Bartlett, in a brilliant feat of seamanship, back to New York. But this was not the end; Peary would, thanks to the deep pockets of the members of the Peary Arctic Club, get yet another chance at the Pole (see page 161).

While Peary was in the North in 1900, a rival attempt on the North Pole was made by Italian Umberto Cagni. He was second-in-command of an expedition led by Luigi Amedeo di Savoia, the duke of Abruzzi. He wanted to reach the North Pole by dogsled from Franz Josef Land. The expedition reached Franz Josef Land in August 1899, but by September their ship was nipped by ice, and the expedition had to spend the first part of the winter on shore, during which period the duke's fingers were frostbitten. Hence the leadership of the attempt on the Pole passed to Cagni.

After they set off north in February 1900, an accident forced a return, and it was not until 11 March that the sledge expedition got going. There were 10 men, 102 dogs, and 13 sledges. Three men acting as a support party left Cagni to return to the ship on 23 March; they were never seen again. The second support party of three turned back on 31 March, returning safely. Cagni and three others continued north. They got as far as 86° 34´ N on 24 April before being forced to return, reaching their ship on 23 June. But it was 16 August before the ship could finally be extricated from the ice.

Cagni achieved a new "farthest north" for the time, one which would stand until Peary's 1906 claim. Cagni's achievement of his farthest north has, like Peary's, been seriously questioned, for his final dash north was remarkably different from the rest of the expedition, encountering, according to Cagni, unexpectedly smooth ice conditions allowing a much faster speed. No sextant observations were shared by Cagni, and his compass variation figures, difficult to fake, are off by over 20° *only* on the final-dash portion of the expedition. Thus, it has been suggested that Cagni faked his farthest north in order to save face for the otherwise lacklustre performance of the Abruzzi expedition. Much later, it was Cagni who in 1928 headed the Fascist court

that disgraced Italian airship aviator Umberto Nobile, who didn't fake anything and was likely one of those first to really reach the Pole, on the airship *Norge*, in 1926 (see page 176).

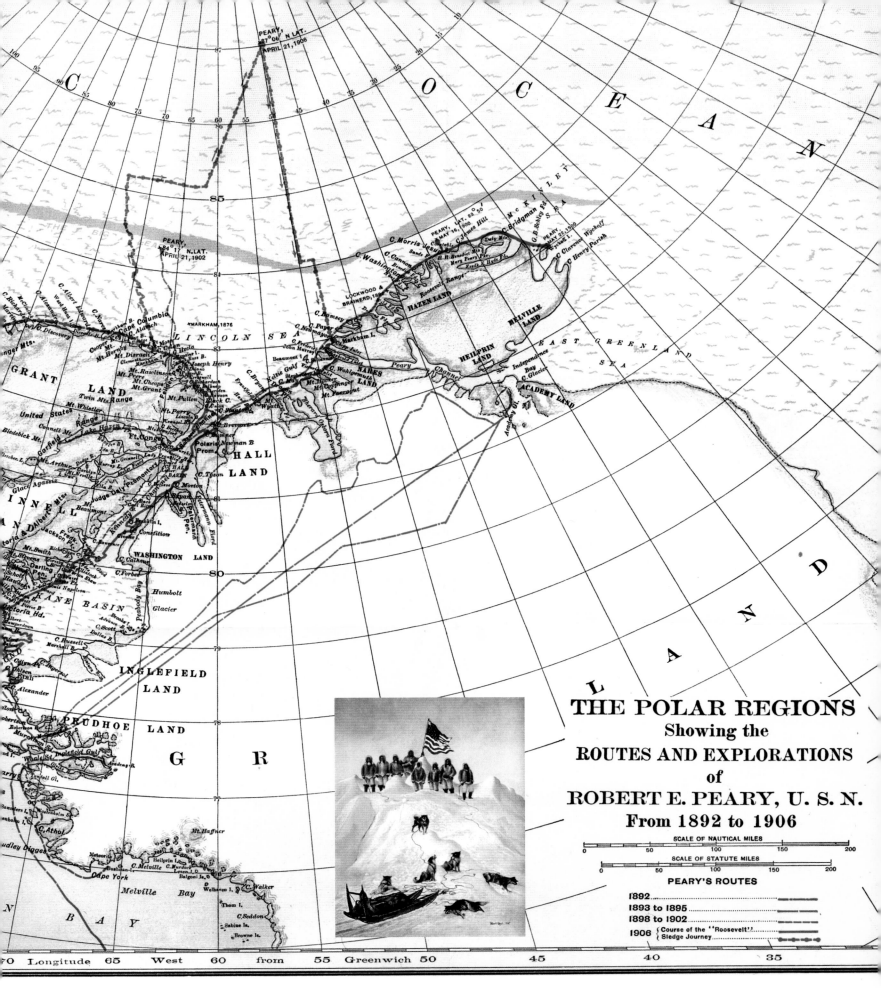

MAP 237.
The map published by Peary in his 1907 book *Nearest the Pole*. The northernmost and southernmost of the red lines across northern Greenland are Peary's outbound and return journeys in 1892. The middle line, when, one suspects, Peary was not so sure of his route, is that of 1895. Peary's three forays northwards over the ice are shown: the shortest in 1900, the bolder 1902 attempt, and finally the 1906 attempt for which a new "farthest north" was claimed. Peary's zigzag initial route over the ice in 1906 contrasts with the straight one shown in 1909 (MAP 247, page 162). Note Peary's mapping of the "big lead," a "river" apparently separating the land-attached continental ice from the sea ice pack, and his imagined Crocker Land, shown to be non-existent by the MacMillan expedition of 1914. Inset is a *painting* from Peary's book—notably not a photo—of the 1906 "farthest north."

New Land

Soon after Otto Sverdrup brought Nansen's ship *Fram* back to Oslo in 1896 he was asked to lead another expedition, this time to discover the land that was widely thought to exist in the Arctic Ocean north and west of Ellesmere Island. As there were "still many white spaces on the map," wrote Sverdrup, "I was glad of an opportunity to fill some of them in with Norwegian colors."

The expedition was financed by the Norwegian consul Axel Heiberg and Ellef and Amund Ringnes, brothers who ran a beer company. Sverdrup sailed in his old ship *Fram* in June 1898, proceeding northwards through Smith Sound. Here he was caught in the ice and spent the winter at the Bache Peninsula, on the east coast of Ellesmere Island. The following spring Sverdrup set off overland, west from the head of Flagler Bay (the southern side of Bache Peninsula), reaching the head of Bay Fiord, on the west coast. From here they could see land that Sverdrup named Axel Heiberg Island; they were the first Europeans to see it.

Intending to take the ship round the southern end of Ellesmere Island, they wintered in 1899–1900 on the north side of Jones Sound at Harbour Fiord (MAP 238). Late in the winter the south coast of Ellesmere Island was mapped. From here in 1900, and from Goose Fiord, farther west, in 1901 and 1902 (the *Fram* could not be moved in the summer of 1901 due to ice), Sverdrup and his crew explored the vast region to the north and west by sledge.

In April 1900 Amund Ringnes Island was seen from the west coast of Axel Heiberg Island and duly explored; the following year Ellef Ringnes Island was found and mapped (although it had been seen by Edward Belcher in 1853—see MAP 158, page 93). In 1902 Sverdrup reached the northwest tip of Ellesmere Island, at a point he named Lands Lokk—Land's End.

Sverdrup, with his cartographer Gunnerius Isachsen and the rest of his small crew of fifteen, methodically surveyed and mapped for the first time some 250 000 km^2, and 2 800 km of new coastline, using their ship *Fram* as a base. It was a superb achievement.

But Sverdrup was Norwegian (Norway becoming independent of Sweden in 1905). Following the normal rules of exploration and discovery, the new land should have belonged to Sweden or Norway. Neither government seemed very interested, but Sverdrup kept putting his case before the Norwegian government, which made the Canadian government nervous. In 1903 none other than Leopold M'Clintock noted to the Royal Geographical Society that the Norwegians had "stopped the advance of the British Empire," remarks that alarmed Ottawa so much that an expedition was hurriedly commissioned, and DGS *Neptune*, with geologist Albert Peter Low, left in August 1903 on an Arctic patrol to establish Canadian sovereignty. These sovereignty patrols were continued by Joseph-Elzéar Bernier a few years later (see page 165).

Sverdrup died in 1930, fifteen days after seeing his years of pressuring of officialdom finally yield some results. With the agreement of the Norwegian government, the Canadian government purchased Sverdrup's maps and diaries for what was then the huge sum of $67,000, and with them tacit sovereignty over the region Sverdrup discovered, now named the Sverdrup Islands. Despite their importance and cost, Sverdrup's maps (though not his diaries) now appear to be lost.

The names bestowed by Sverdrup show up on all the coasts and islands—Colin Archer Peninsula at the eastern end of Jones Sound, named after the builder of the *Fram*; Cape Maundy Thursday and Good Friday Bay on Axel Heiberg Island, named after their day of discovery; Isachsen Peninsula, on Ellef Ringnes Island, after his crew member and cartographer; and dozens more. And the Canadian Arctic added Ringnes's beer to its Booth's gin.

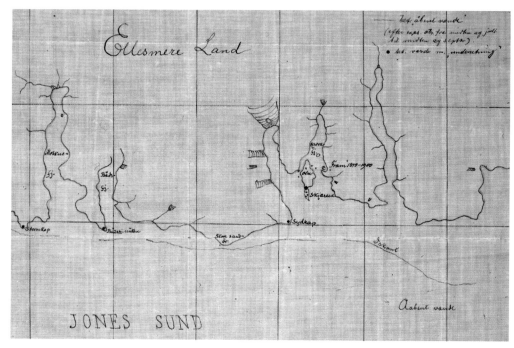

MAP 238.
This original map from the Sverdrup expedition shows the wintering place in the second year, 1899–1900, in Harbour Fiord, on the south shore of Ellesmere Island. This map was found in a cairn by Joseph-Elzéar Bernier on 12 August 1907. Bernier was attempting to restore—or confirm—Canadian sovereignty in the Arctic Archipelago (see page 165).

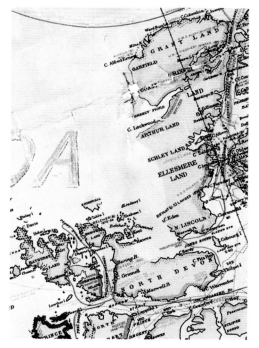

MAP 239.
This is how this part of the Arctic appeared on maps just before Sverdrup left in 1898. It shows the "white spaces" which he was to fill with "Norwegian colors."

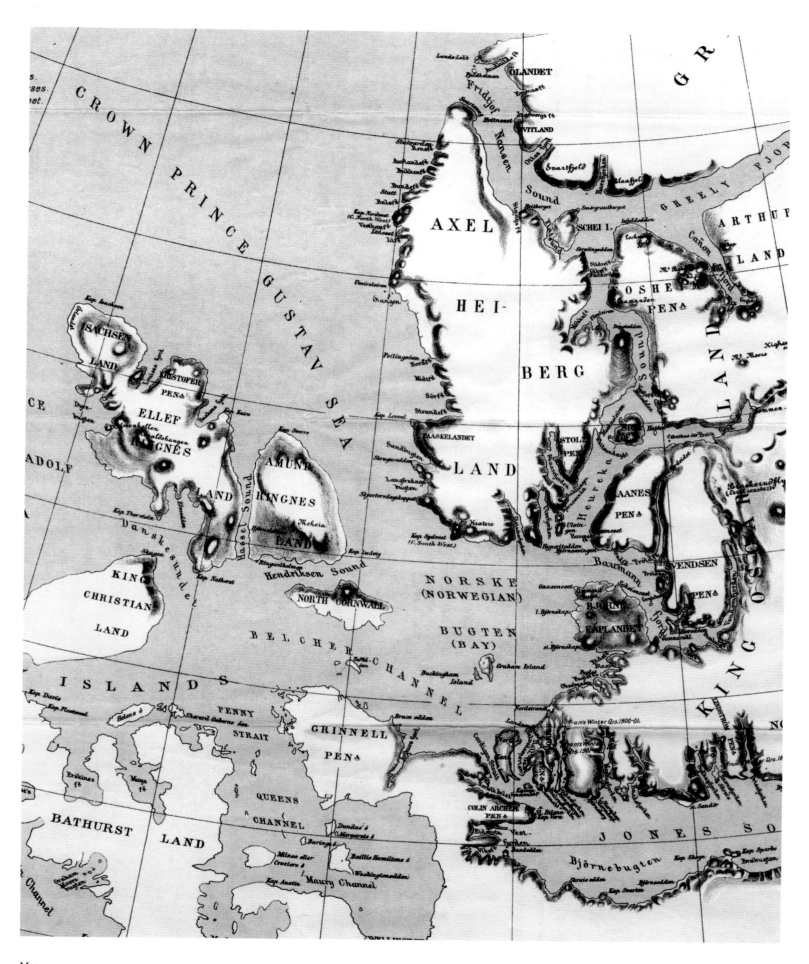

MAP 240.
Gunnerius Isachsen's summary map of the discoveries of the Otto Sverdrup expedition in 1898–1902. The map was published in his book *New Land* in 1904, an English translation of the Norwegian edition *Nyt Land,* published the year before.

The Northwest Passage at Last

After Robert M'Clure's transit of the Northwest Passage minus his ship (see pages 97–99), it remained for someone to do it *with* their ship. Allen Young, a well-heeled British adventurer, tried it in 1875, being stopped by ice in Peel Sound, and again in 1876; the Admiralty requested him to divert to take communications to the Nares expedition. He did not try again, but might well have succeeded if he had, for Young had intended to hug the coast of mainland North America just as Roald Amundsen was to do. Young's ship, the *Pandora*, was that bought by George Washington De Long and renamed the *Jeannette* (see page 135).

Roald Amundsen, like Nansen and Sverdrup also a Norwegian, was arguably the most successful and certainly the most versatile polar explorer of all time. As well as sailing through the Northwest Passage, Amundsen would become the first to reach the South Pole in 1911, and in 1926, by air, one of the first to attain the North Pole—if not the first, since Robert Peary's and Frederick Cook's achievements are by no means certain, and the same applies to Richard Byrd in 1925 (see pages 160 and 175).

But Amundsen was not well heeled and was to lurch from one expedition to the next, always seemingly underfunded. In 1903 Amundsen received support from the king of Sweden, Norwegian shipowners, and others for an attempt on the Northwest Passage. Still without enough funds, he slipped away from Christiania (Oslo) in June 1903 in a specially strengthened little herring boat with sails and a small engine named *Gjøa*. Amundsen intended to keep close to the coast and realized that a small boat was best. He had only six crew members and six dogs, although he picked up more dogs in Greenland.

Amundsen visited Beechey Island, then headed south into Peel Strait. After the crew dealt successfully with a fire in the engine room and got the boat off some shoals where she grounded, a wintering place was found at the southern end of King William Island, at a snug little harbor they called GjøaHavn.

Here two years were spent, carrying out magnetic and other scientific observations, and trying (unsuccessfully) to locate the North Magnetic Pole. In 1905 a sledging party surveyed what was by then the only major uncharted coast in the Arctic Archipelago, the eastern coast of Victoria Island, on M'Clintock Channel. The islands at the southern end of Victoria Strait were mapped and named after the Royal Geographical Society, which had contributed to the expedition. And the sea to the south was named Queen Maud Sea, now Queen Maud Gulf. Amundsen also found skulls and bones apparently from the Franklin debacle; these were carefully buried in a cairn.

Finally, on 12 August 1905, *Gjøa* left King William Island and continued westwards. By 17 August they passed Cambridge

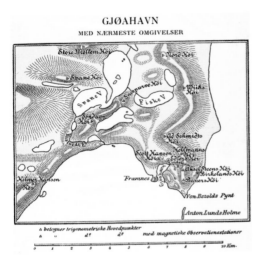

MAP 242 (*above*).
Amundsen's map of his GjøaHavn—now Gjoa Haven—on the south coast of King William Island, where the winters of 1903–04 and 1904–05 were spent.

Bay, where Richard Collinson had wintered—with a far larger ship—in 1852–53. Amundsen considered the day significant, for they had now sailed through what he called the "hitherto unsolved link in the North West Passage." Surely, he must have thought, if Collinson's *Enterprise* could escape west from here, then the little *Gjøa* should be able to also.

And so it could. By 26 August they were near Nelson Head, the southernmost tip of Banks Island—in the water body later named Amundsen Gulf—when they saw another ship. It was the *Charles Hanson*, a whaler from San Francisco, one of many whaling ships in the region.

But they were not destined to escape that easily, for the season was closing down. The *Gjøa* was stopped by ice near the Mackenzie Delta at King Point, and here the crew spent the winter of 1905–06, their third.

Amundsen took the opportunity to sledge to Eagle City, Alaska, where there was a telegraph, to send out news of his triumph to the *Times* newspaper, with whom he had contracted an exclusive story. Unfortunately for him, the telegraph was not private, and others printed the story; the *Times* refused to pay.

The ship was released on 10 July 1906 and passed through Bering Strait on 30 August to a great welcome at the bustling gold city of Nome. Then Amundsen sailed to San Francisco, where *Gjøa* was seized by creditors. She was to remain in San Francisco

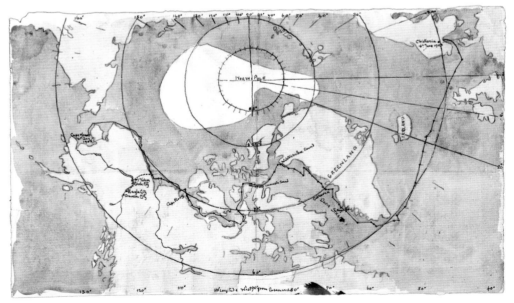

MAP 241.
A contemporary hand-drawn map of the voyage of Roald Amundsen and the *Gjøa* from Christiania (Oslo) to Nome, Alaska. Also shown is Amundsen's sledge journey in the winter of 1905–06 to Eagle City, Alaska, to reach the telegraph.

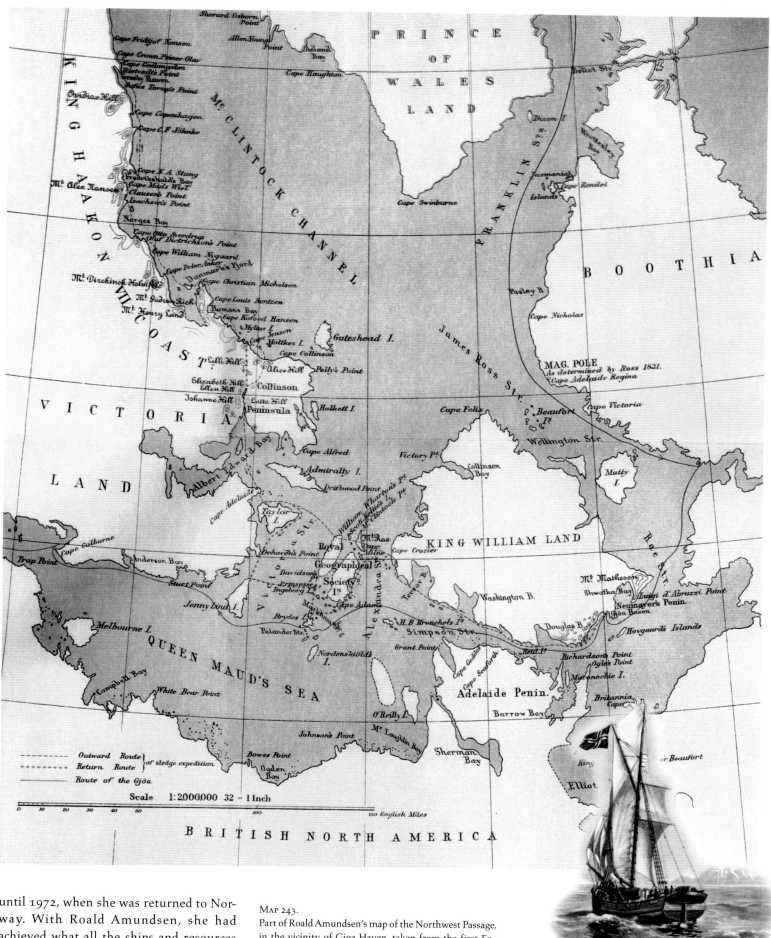

until 1972, when she was returned to Norway. With Roald Amundsen, she had achieved what all the ships and resources of the British Royal Navy could not, and the Northwest Passage had finally been conquered.

MAP 243.
Part of Roald Amundsen's map of the Northwest Passage, in the vicinity of Gjoa Haven, taken from the first English edition of his book. The western shore of M'Clintock Channel (the east coast of Victoria Island), which was explored and surveyed by a sledge party from *Gjøa*, is covered with Norwegian names.

Amundsen's *Gjøa* in the Northwest Passage, from his book.

The Pole Achieved?

No geographical controversy has raged longer than the question of who reached the North Pole first—Frederick Cook in 1908 or Robert E. Peary in 1909. But if one accepts that the burden of proof lies with the explorer then the likely conclusion is that neither reached the Pole. Nevertheless there are staunch supporters of each side who still do battle to this day. Certainly both explorers ventured north onto the ice towards the Pole; the only question that remains is how far they got.

Both Cook and Peary claimed they had reached the Pole, with their claims being made public within four days of each other in September 1909, and both wrote books with accounts of their treks, including maps shown here.

Never in the history of exploration has there been written such a voluminous amount of apparently authoritative material produced for two opposing cases. Reading one account, one case, is convincing—until the contradicting account is read. Books with titles like *The Noose of Laurels,* or the over 1,100-page *Cook and Peary: The Polar Controversy, Resolved,* seem to be the end of the matter until the next volley is fired.

Frederick Cook's trek, according to his own account, began with the transfer of supplies to Etah (now Taseq), on the Greenland coast of Smith Sound, followed by a sledge journey to Anoritoq (he called it Annoatok, or Anortok on his map), just north of Etah. The effort was financed by John R. Bradley, a wealthy American big game hunter. He was to be rewarded for his support of an apparently illusionary expedition with his name on a chimerical land north of Peary's equally delusionary Crocker Land.

On 19 February 1908 Cook began his trek, accompanied by his assistant, Rudolph Francke, and a number of Inuit. He crossed Smith Sound to Flagler Bay, crossed Ellesmere Island to Bay Fiord and traveled north through Eureka Sound to Nansen Sound, between Ellesmere and Axel Heiberg Islands, reaching Cape Stallworthy (at the time called Cape Svartevoeg), at the northern tip of Axel Heiberg Island.

Here he established a base where he left Francke and most of the support party. On 18 March, he set out northwards over the ice with only two Inuit companions (E-tuk-i-shuk and Ah-wehlah), two sledges, and twenty-six dogs. On his way to the Pole he claimed to

MAP 244.
Map showing the farthest north positions of previous polar explorers, from Cook's book *My Attainment of the Pole.*

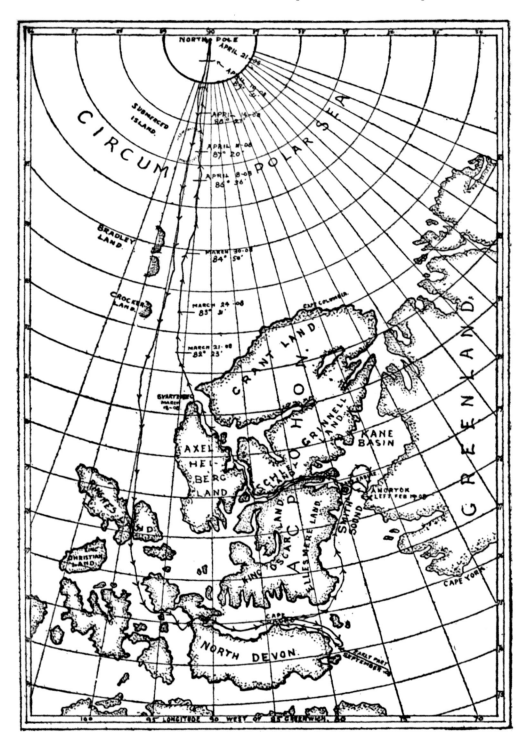

MAP 245.
Frederick Cook's own map, published in his book *My Attainment of the Pole* in 1911. Note his *Bradley Land;* this could have been placed on the map because Cook believed the theories of Rollin Harris of land in the Arctic Ocean (see MAP 250, page 164). Farther north, the "submerged island" had more credibility as a possible ice island (see page 186), until Wally Herbert found a photograph that showed the ice with *land* attached (see page 164).

have discovered new land. "As well as I could see," he wrote, "the land seemed an interrupted coast extending parallel to the line of march for about fifty miles, far to the west." This he named Bradley Land, after his sponsor.

Farther north, he encountered hard ice which he thought was land ice above a "submerged island," and this is also shown on his map (MAP 245). Cook claimed to have attained the North Pole on 21 April 1908, having taken thirty-five days on the ice to reach it.

On his return, they missed the depot on Axel Heiberg Island, with the result that they trekked farther south, but were unable to turn eastwards to land because of "fifty miles of small crushed ice and impassable lines of open water." They somehow missed Meighen Island and reached land only at Amund Ringnes Island, discovered only a few years before by Otto Sverdrup (see page 156). Several years later Vilhjalmur Stefansson, who first reported the existence of Meighen Island, was to consider the fact that Cook had not reported land there as "incontrovertible proof" that he had not passed this way. Cook's path is shown on his *own map* (MAP 245) as passing through the position of Meighen Island (about 80° N, 100° W), but his path is *plotted* east of his *stated* position. On MAP 246, said to be a copy of the route traced by Cook's two Inuit companions, the island *is* shown.

Continuing south, Cook entered Penny Strait, crossed the northwestern part of Devon Island, and by sledge and then folding canvas boat got as far east as Cape Sparbo (Cape Hardy), on the north coast of Devon Island. Here, barely surviving, he wintered.

In February, Cook set out again, reaching Anoritoq at the beginning of April after another journey of near starvation. From there he traveled to Upernavik, and thence by ship to Copenhagen, where he presented his discoveries to the Royal Danish Geographical Society. He was back in New York by 21 September 1909, hailed as conqueror of the Pole.

Robert Peary, meanwhile, had set off from New York in July 1908 with two ships, one of which, the *Roosevelt*, wintered in 1908–09 at Cape Sheridan, at the northeast tip of Ellesmere Island. This was Peary's base for his attempt on the Pole. In February 1909, using the so-called Peary system, support parties set out from the ship to carry provisions and make a trail for Peary and his assistant Matthew Henson, who were the last to leave the ship, on 1 March. The route was along the coast to Cape Columbia, at 70° W, then north on the 70° line of longitude, what Peary called the Columbia Meridian. They were delayed by the "big lead," a river of open water supposedly marking the boundary of the land-attached ice and the true sea ice, until 11 March.

The last support party, that led by Robert Bartlett, captain of the *Roosevelt*, turned back at 87° 47´ N on 1 April, leaving Peary and Henson, together with four Inuit men, Ootah, Egingwah, Seegloo, and Ooqueah, to continue to the Pole.

Peary claimed to have reached the North Pole on 6 April and stayed there thirty hours to verify his position. Unfortunately for him, if indeed he was at the Pole, he did not do this very well. He then took only sixteen days to reach Cape Columbia, a *straight-line* distance

MAP 246.
A copy of Otto Sverdrup's 1904 map on which Cook's two Inuit companions, E-tuk-i-shuk and Ah-wehlah, were persuaded by Peary's men to draw the routes they took with Cook. This appeared in the *Chicago Daily Tribune* on 13 October 1909. It shows no northward trek by Cook over the ice. It does show Meighen Island, however, marked here as *small low island*. Cook made no mention of the island in his account, and it is not shown on his map (MAP 245, *opposite*), despite a route that would have taken him close to it (101° 22´ W, about 20 km east of the island; the route on his map is not plotted accurately). Meighen Island was formally discovered by Vilhjalmur Stefansson in 1916 (see page 166). One of the most damaging arguments against Cook has always been the so-called Eskimo Testimony, the supposed admission by the two Inuit that they had never been out of sight of land. But there is ample evidence that the interview of the two men was biased, so that they agreed to what was suggested to them. And those present seem not to have had much understanding of their language in any case. Cook admitted that he encouraged the Inuit to believe that they were near land as they apparently ventured far out onto the ice only with trepidation. But how to account for the appearance of what is now Meighen Island on this map? A mystery indeed.

161

of 765 km (475 miles), and a further two days to reach the *Roosevelt*.

Many analyses of Peary's speed have been written, for the assumed impossibility of maintaining the speeds required over the often broken polar ice is a major argument against him. Conservatively, and based on much data from polar explorers before and since, one must add about 25 percent to the total distance to allow for necessary detours due to rough ice and leads. From the point where he left Bartlett to the Pole and back again, Peary claimed to have covered 363 km (225 miles) in four days, which, allowing for the 25 percent added for detours, is 113 km (70.2 miles) per day. This is a speed never approached by any expedition, even modern-day ones much better supported than Peary's. The typical *maximum* for a dog and sledge is about 60 km (35 miles) in a day (the speed of Naomi Uemura in 1978, over the same route), about *half* of the claimed speed of Peary. For the same reason Peary's 1906 Pole attempt has now come under scrutiny.

The claim of Peary to have gone straight to the Pole along the Columbia Meridian and to have returned along it, hitting the iglus built on the outward journey, also defies belief, for the ice, it has been demonstrated many times, is constantly moving, so that the position of the iglus, even if built on the meridian, would have moved by the time of the return journey. Wally Herbert, who himself made a trek to the Pole with dogs and sledge in 1968–69 and has traveled extensively over both Peary's and Cook's routes, meticulously calculated that Peary would have ended up at least 80 km from the Pole. He did this by considering prevailing winds and currents and their effect on the ice. The point, however, is that whatever the movements of the ice, it is highly unlikely it would have remained in the same position, with everything—including Peary—right on the 70° meridian.

There is no doubt that Peary was alarmingly slack in his navigation, particularly considering that he knew full well that he might have to prove his claim over Cook's. He made his final trek without witnesses able to corroborate his position—Matthew Henson was not able to use a sextant—even leaving Robert Bartlett at the last depot. He may well have done this so as not to have to share "his" Pole with anyone else. (In those days, Henson, a black person, and the Inuit would not have counted in Peary's mind.)

Shadows in some of Peary's photographs appear to be in the wrong places for the claimed place and time, and he did not take enough soundings to have enabled his route to be verified once the topography of the ocean floor was known in more detail. All this, yet some still claim his attainment of the Pole as genuine. In 1990, the Foundation for the Promotion of the Art of Navigation, at the

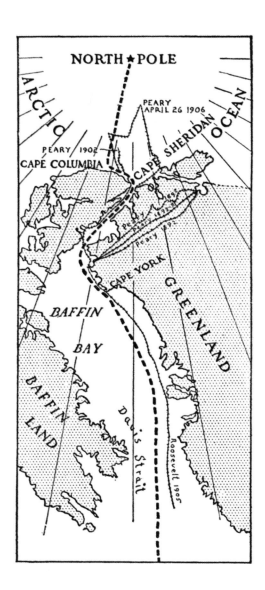

MAP 247 (*left*).
The only map Peary himself drew of his claim to the North Pole in 1909 was this very generalized map, published in his book.

MAP 248 (*right*).
Who would have thought the Arctic could be so full of information? This superb map summarizes attempts on the North Pole during the nineteenth century. It was originally published in 1907, then hastily overprinted in red to show Cook's and Peary's claimed achievements of the Pole and reissued to an eager public in 1909. This map is a "specimen map," presumably a proof. The identical map, under the byline of Gilbert Grovenor of the National Geographic Society, accompanied Peary's book *The North Pole*, published in 1910.

Blatantly nationalistic, the "coasts discovered by" and the explorers' routes (key, bottom right) are in the color of their country according to the key at top right.

SPECIMEN MAP

COMPILED, ENGRAVED AND PRINTED BY
THE MATTHEWS-NORTHRUP WORKS, "THE COMPLETE PRESS"
NEW YORK OFFICE   BUFFALO, N.Y.   CLEVELAND OFFICE
MADISON SQUARE GARDEN TOWER   517 CITIZENS' BUILDING

FRANZ JOSEPH LAND

Explorers of Franz Josef Land:
Payer and Weyprecht (Austrian) 1872-4
Leigh Smith (British) 1880-2
Frederic Jackson (British) 1894-7
Fridtjof Nansen (Norwegian) 1896
Walter Wellman & Evelyn B. Baldwin (American) 1898-9
Duke of Abruzzi (Italian) 1899-1900
Anthony Fiala (American) 1903-1905

Coast Explored by
United States..........
British..........
Scandinavian (Norwegian, Swedish and Danish)..........
German and Austrian..........
Dutch..........
Italian..........
Russian..........

The Routes of the Explorers have the color of their native country.

The Circumpolar Stations of 1881-1883 are indicated thus:
✦ C. P. Sta. with name of nationality following.

THE ARCTIC REGIONS
Showing Explorations towards the
NORTH POLE

COPYRIGHT, 1907, BY THE J. N. MATTHEWS CO., BUFFALO, N.Y.

Explorers' Routes:
Parry (British) 1819-..........
Franklin and Richardson (British) 1821 & 1826..........
Franklin (British) 1845-'(6)..........
McClure's (Br.) Northwest Passage, 1850-'53..........
Second German North Pole Expedition, 1869-'70..........
Austro-Hungarian Expedition 1872-'73..........
Nordenskjöld Northeast Passage in "Vega" (Swedish) 1878-'79..........
DeLong in "Jeannette" (United States) 1880-1881..........
A. W. Greely (United States) 1881-84 (Cape Sabine to Cape Washington)..........
Nansen in "Fram" (Norwegian) 1893-'96..........
Duke of Abruzzi in "Stella Polare" (Italian) 1900..........
Sverdrup in "Fram" (Norwegian) 1898-1902..........
Peary (United States)..........
Ziegler Polar Expedition of 1903-'05 (A. Fiala)..........
Amundsen's Northwest Passage in "Gjöa" (Norwegian) 1903-1906..........
Duke of Orleans in "Belgica" 1905..........
Journeys by foot, sledges or boats .... with color of expl......

Telegraph Lines:..........
Railroads..........
Steamship Routes..........
Winter Harbors +
Average Limit of Icepack and solid Drift Ice ———
Tundra

Cook's Route thus: ------

instigation of the National Geographic Society, produced a seemingly conclusive report analyzing Peary's photographs and soundings in which it was concluded he had reached the Pole, a report which was then roundly attacked by a number of navigation-savvy people who concluded that the photographs showed that he did not reach the Pole and the soundings were inconclusive. And so the argument continues.

While Peary's claim was from the beginning widely accepted, that of Cook was not. Supporters would say that Peary won the public relations battle, but not the argument.

Cook's other exploits have been used to show he was of doubtful character. He was later convicted of mail fraud in connection with an oil stock promotion, and a previously claimed ascent of Mount McKinley, the highest point in North America, was also attacked as fraudulent, but these are circumstantial and do not in themselves mean anything. Clearly Cook's "discovery" of Bradley Land was false, although one could attribute this to an Arctic mirage, as, presumably, one must Peary's Crocker Land. But Cook went a step further. In his book there is a photograph of Bradley Land; Herbert says this is a photograph of the west coast of Axel Heiberg Island. Herbert found the original plate of Cook's photograph of a "submerged island of Polar Sea," which is shown in his book and said to be close to the North Pole (and is shown on MAP 245). The whole plate showed that it was in fact a glacier *with land clearly visible to one side*. However, even this is not by itself proof positive of Cook's deception, for it is possible that his publisher—they do this sort of thing, you know—substituted a photograph of something else when none was available of the submerged island. However, does it also seem likely that Cook would have let that go when he knew people reading his book would be looking for evidence he did not reach the Pole?

Ultimately, there probably will never be any clear resolution of the question of whether Cook or Peary reached the Pole. The onus of proof was with them, and neither provided it. But the supporters of each will battle on; of that there is no doubt at all.

In 1909, Roald Amundsen had been planning an expedition to the North Pole, and its departure had been set for January 1910. Then on 1 September, the news broke of Cook's apparent achievement, followed a few days later by that of Peary. Naturally, this threw Amundsen's plans into disarray and his sponsors started pulling out. Amundsen, never one to give up easily, changed his goal and made for the South Pole instead, finally reaching it—just ahead of Robert Scott—in December 1911. But given Amundsen's record before and after, it is quite within the realms of possibility that if Cook and Peary had not made their bogus announcements, then Amundsen would have become the first to achieve the North Pole in 1910 rather than in 1926, as he did, by air (see page 176). And Scott, presumably, would have been first to the South Pole.

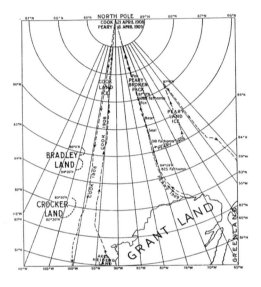

MAP 249 (*above*).
Map by Edwin Swift Balch, published in 1913, in a book in which he discussed the "new land" discovered by Peary and Cook.

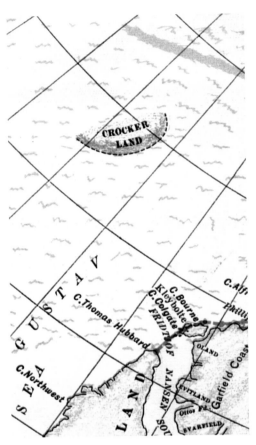

MAP 251 (*above*).
Crocker Land, from Peary's 1906 map. In 1914 one of Peary's assistants, Donald Baxter MacMillan, with one companion, set out northwards across the ice from Cape Thomas Hubbard, at the northern tip of Axel Heiberg Island, in an attempt to find Peary's Crocker Land. Of course, he did not find it, and the belief in land in this area largely died once MacMillan reported his search. MacMillan also led a later expedition to explore the islands west of Ellesmere Island (see page 174).

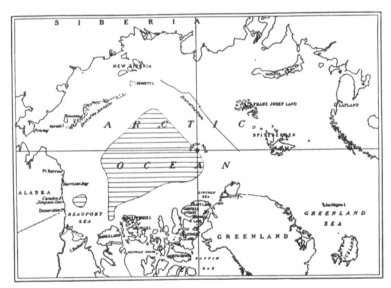

MAP 250.
Rollin Harris's 1904 map showing his hypothesized land in the Arctic Ocean. Harris was a scientist with the U.S. Coast and Geodetic Survey. The idea of land was based on a study of tides, as well as the drift of the *Jeannette* and the *Fram*, and presumed sightings. Harris wrote that "the tides clearly prove that there can be no large and deep polar basin." He was right. In the 1950s, not land, but an undersea ridge crossing the Arctic Basin was detected by Russian scientists making soundings from drifting stations. Today it is the Lomonosov Ridge (see page 190). At the time of Cook and Peary, the reports of the existence of Crocker Land and Bradley Land seemed to corroborate the scientific predictions of Rollins, and well into the 1920s many believed land must exist in this area of the Arctic. Hubert Wilkins (see page 178) spent a lot of time searching for it.

Claiming the Arctic

In 1880, Britain, which had explored a great deal of the Canadian Arctic, transferred its territorial claims to Canada, and in 1895 (with an amendment in 1897) provisional districts were established which included that of the District of Franklin, covering the Arctic Archipelago. When Otto Sverdrup reported new land in the region the Canadian government suddenly considered it important to establish or confirm its sovereignty more directly. The DGS *Neptune* was hurriedly commissioned to patrol the North (see page 156), and this was followed by three voyages of the DGS *Arctic*, commanded by Joseph-Elzéar Bernier, specifically to establish sovereignty by erecting cairns and taking formal possession of most of the larger islands of the archipelago.

While Bernier was in the Arctic in 1907, a Canadian senator, Pascal Poirier, proposed the "sector principle" of territorial ownership right up to the North Pole, in pie-like slices (MAP 252). Although the proposal was not adopted then, it has since been adopted by Canada and other polar countries. It was not until 1925 that Canada officially claimed the territory, land or water, "right up to the Pole," but this is still a claim rather than an internationally recognized boundary. (The boundary between Greenland and Canada was agreed to in 1974, however.)

Bernier was a staunch proponent of Canadian sovereignty in the Arctic. In 1906 he landed on Somerset, Cornwallis, and Griffith

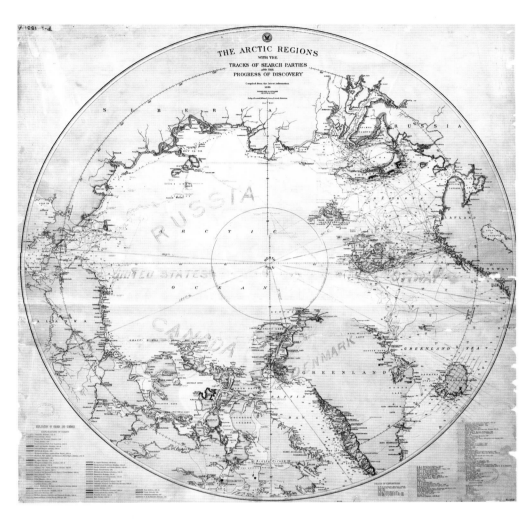

MAP 252 (*above*).
The sectors proposed for international control of the north polar region have been drawn on this 1896 printed American map. Bernier is probably the person who drew in the sectors, for his name is written on the map.

In the nineteenth century gentlemen had calling cards, and Bernier was no exception. This is his calling card, dating from the early years of the twentieth century and already espousing the sector principle.

MAP 253 (*right*).
A pencil sketch-map made by Bernier in 1911, with the aid of an Inuit person he called "Cornow." It shows Bylot Island and Pond Inlet. Bernier later set up trading posts and took up mining claims in the region, and these were added later to this map.

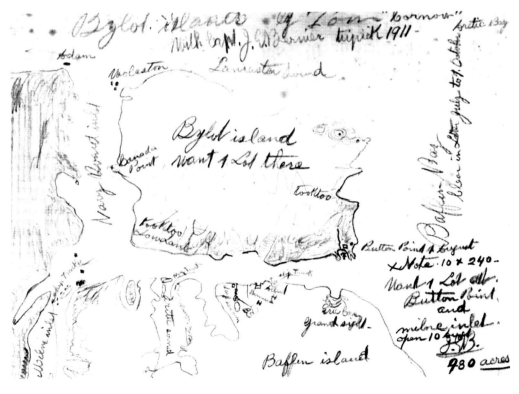

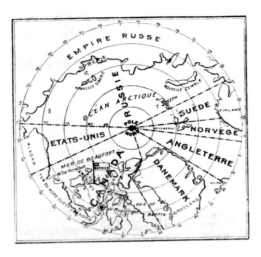

MAP 254 (*above*).
The sector principle as put forward by Canadian senator Pascal Poirier in 1907. This map was published in an Ottawa newspaper, *Le Temps*, on 18 October 1909, illustrating a report on a speech Bernier had given to the Canadian Club two days earlier. The speech had outlined the results of his sovereignty assertions during his voyage of 1908–09, from which he had just returned.

Islands, the latter being a small island just south of Cornwallis; and, sailing west, Bathurst, Byam Martin, and Melville Islands. In 1908 Bernier wintered at Winter Harbour on Melville Island, Parry's original wintering place. In the spring of 1909 sledging parties set out for Banks Island and Victoria Island, with possession ceremonies taking place on each. In 1910 Bernier was given permission to attempt a transit of the Northwest Passage. In 1908 he had reported ice conditions so good that he felt he could have done it then. But ice conditions change from year to year in the Arctic, and this year he met an impenetrable barrier of ice; instead, northern Baffin Island was explored and surveyed (MAP 255). The inlets of northern Baffin Island had tended to be ignored up to this point for they clearly did not lead to a Northwest Passage.

In 1913 the Canadian government sponsored an expedition to the Arctic to carry out

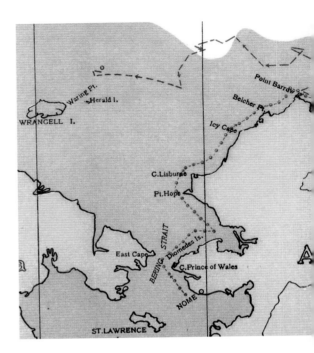

MAP 256 (*above*).
The voyage (red dashes with circles) and drift (red dashes) of the *Karluk* in 1913, to the point north of Herald Island (Ostrov Geral'd) where the ship was crushed in the ice on 11 January 1914. Stefansson had left the ship early, ostensibly to hunt, but this is disputed. The survivors of the wreck walked to Wrangel Island (Ostrov Vrangelya). As their situation deteriorated, the ship's captain, Robert Bartlett, walked with an Inuit guide 1 120 km to the coast of Siberia to get help. Meanwhile those on Wrangel Island fought over provisions and survived on eggs and roots. One man died from a gunshot; whether it was murder or an accident has never been determined. The wreck of the *Karluk* is one of the many epics of Arctic survival and hardship. It led to an ongoing dispute between Bartlett and Stefansson lasting almost fifty years.

further exploration, mapping, and scientific work. Its leader was Vilhjalmur Stefansson. The expedition was in two sections, and three ships were initially involved, one of which, the *Karluk*, left Victoria, British Columbia, in June. In mid-August the ship became trapped in the ice off the north coast of Alaska and began an inexorable drift north and then west. Stefansson left the ship, which drifted east until it was crushed (MAP 256, *above*).

Stefansson joined another part of the expedition which, over a period of four years, ranged by sledge over a vast area of the western Canadian Arctic. He was out of touch with Ottawa, which was preoccupied with the war in Europe, and was thought likely to have perished. But Stefansson was able to live off the land in a way few Europeans have. He was very skilled at both ice travel and survival, skills he taught to one of his companions, George Hubert Wilkins, who would go on to become a pioneer Arctic aviator (see page 178).

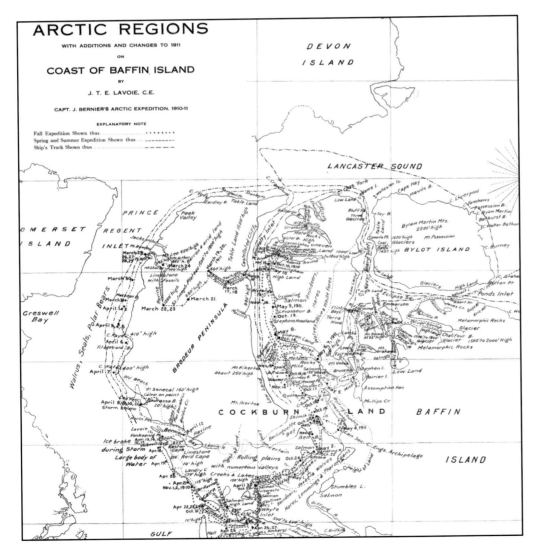

MAP 255.
The survey of northern Baffin Island made in 1910 and 1911 by J. T. E. Lavoie, a member of Bernier's scientific staff. The Brodeur Peninsula was named after L. P. Brodeur, the Canadian minister of fisheries.

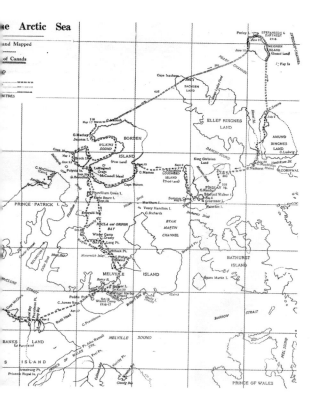

MAP 257 (*above*).
A detailed map of Stefansson's discoveries in 1916, published in his book *The Friendly Arctic* in 1921. It shows his Meighen, Brock, Borden (in reality two islands), and Lougheed Islands.

Stefansson discovered more Arctic islands for Canada. North of the Ringnes Islands in 1916 he found an island he named Meighen Island, after a minister in the Canadian government, later briefly prime minister. This was the island marked on a map by two Inuit describing the route Frederick Cook took in 1908 (see MAP 246, page 161). He also found Brock Island and what he called Borden Island, after the Canadian prime minister. In 1947 aerial survey determined that this is in fact two islands. Borden's name was retained for the northernmost island, and the name Mackenzie King Island was given to the one to the south, after the prime minister of Canada in 1947. Between them is a strait named for Hubert Wilkins. Stefansson also discovered Lougheed Island, although this had been shown as part of a larger King Christian's Island by Otto Sverdrup.

Stefansson was a controversial figure, and the *Karluk* episode together with his claim of Wrangel Island for Canada in 1921, which embarrassed the government, and the fact that he moved to the United States, meant he was never employed by the Canadian government again. In 1921 Stefansson published a book he called *The Friendly Arctic*, in which he maintained that the Arctic was easy to survive in—as long as you knew how.

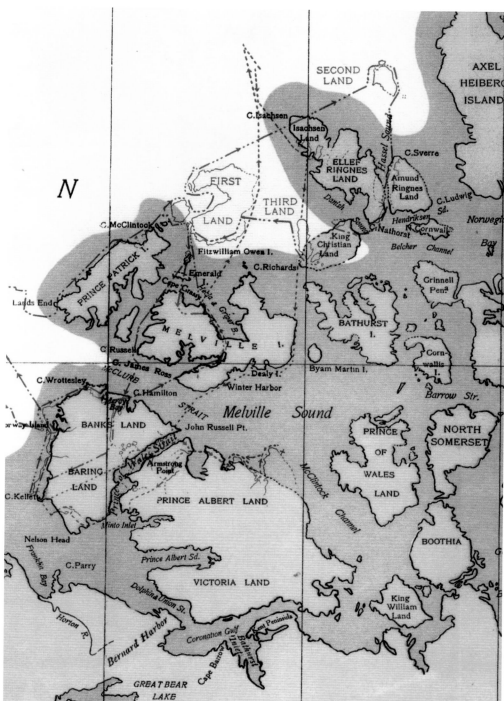

MAP 258.
Part of the summary map of Stefansson's expeditions from 1914 to 1917. The red lines show his tracks. The long dashed lines with one dot represent the 1914 expedition, those with two dots, 1915, and those with three, 1916. The 1917 track is shown with short dashes only. *First Land* is Borden Island and Mackenzie King Island, with an unnamed Brock Island just to its west; *Second Land* is Meighen Island, probably the land seen by Frederick Cook in 1908 (see page 160); and *Third Land* is Lougheed Island, in reality detached from a much smaller *King Christian Land* (Island), shown in red outline to the east of Lougheed Island.

In 1940 police officer Henry Larson began a long command of the Royal Canadian Mounted Police ship *St. Roch*, which had been the government's Arctic supply ship and police presence—and protector of Canadian Arctic sovereignty—since 1928. That year the *St. Roch* began what was to become the first west-to-east transit of the Northwest Passage, emerging into an enemy submarine–infested Atlantic in 1942 and sailing to Halifax. Later the ship traveled in the reverse direction, becoming in 1944 the first ship to sail through the Northwest Passage in both directions. Still later, in 1950, it became the first ship to circumnavigate North America.

New Land in the Russian Arctic

Russia had long had its eyes on the idea of using its northern coastal waters for military or even commercial purposes. The world's first seagoing icebreaker was the Russian *Yermak*, built for the Imperial Russian Navy in Newcastle, England, in 1898, which reached 81° 28´ N near Spitsbergen on its maiden voyage in 1899, under the command of its conceptor, vice admiral Stepan Makarov.

Largely as a result of the disastrous Russo-Japanese War of 1904–05, it was decided that efforts should be made to develop the northern coastal route to Bering Strait—the Northeast Passage. Accurate maps would first be required, and two icebreaking survey ships were specially built to carry out this work. They were the *Taymyr* and the *Vaygach*. During the period 1910 to 1915 the ships undertook detailed surveys and mapping of the entire route, culminating in 1915 with a complete transit from Vladivostok to Arkhangel'sk.

In 1913, as part of this Arctic Ocean Hydrographic Expedition, as it was known, the *Taymyr*, commanded by Boris Andreyevich Vil'kitskiy, and the *Vaygach*, under Per Novopashennyy, found hitherto unknown new land, only about 60 km from the mainland coast at Mys Chelyuskin (Cape Chelyuskin), the northernmost tip of mainland Asia.

The discovery happened almost by accident. They were trying to survey the coast of the Taymyr Peninsula (Poluostrov Taymyr) eastwards when heavy ice along the shore forced the ships northwards. On 20 August a new island was found, which they called Ostrov Tsesarevicha Aleksaya (now Ostrov Malyy Taymyr). Continuing northwest, they found the next day what was clearly a major new landmass. Vil'kitskiy mapped the coast from his ship to its northern point, Mys Arkticheskiy, at 81° 07´ N. A landing was made on 22 August, possession declared for Russia, and the archipelago was named Zemlya Imperatora Nikolaya II, after the tsar.

In 1928 Umberto Nobile in his airship *Italia* tried to reach the western coast of this "Northern Land," as its name, given to it in 1926, means in English, but didn't quite get that far due to the weather (see page 179).

Severnaya Zemlya was not fully delineated until 1930–32, when an expedition planned by Otto Shmidt and led by Georgiy Alekseyevich Ushakov explored it. The expedition spent three years exploring and mapping Severnaya Zemlya, finding that it was comprised of three large islands and several smaller ones. New islands including Ostrov Shmidta (shown on MAP 263, *far right*) were found. A new strait was discovered between the revolutionary-named southernmost large island, Ostrov Bol'shevik, and the largest, central island, Ostrov Oktabr'skoy Revolyutsii, significant as an alternative route from the Kara Sea to the Laptev Sea.

MAP 260 (*right*).
This American map dated 1929 shows Severnaya Zemlya just as it was discovered and mapped by Vil'kitskiy in 1913. The following year a new Russian expedition would complete the discovery and mapping of the entire island group.

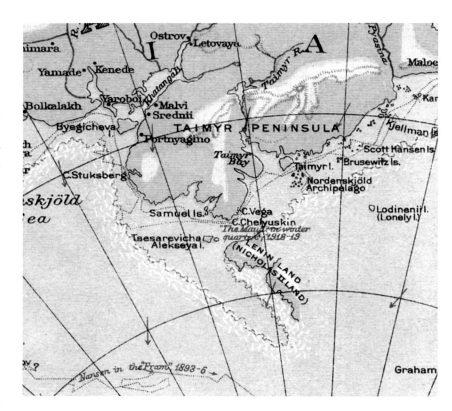

MAP 259 (*above*).
This 1925 American atlas map shows the east coast of Severnaya Zemlya as discovered by Vil'kitskiy in 1913, but *Nicholas II Land*, named after the tsar of Russia, has been shown on this map also as *Lenin Land*. It was officially renamed Severnaya Zemlya—Northern Land—in January 1926. The notation "The Maud" in winter quarters 1918–19 refers to Roald Amundsen's ship *Maud*. From 1918 to 1921 Amundsen tried repeatedly to get his ship beset in the ice pack north of Russia, with the intention of drifting to the Pole, as Nansen had intended with *Fram* in 1893. Despite overwintering three times, he was unable to persuade the fickle pack to envelop his ship, a strange thing, really, when you consider how many ships had unwillingly become trapped in the ice. He did end up making the third transit of the Northeast Passage, however. In 1921 Amundsen returned to San Francisco, where he became enthralled by the idea of Arctic aviation (see page 171). The *Maud*, seized by creditors just as his *Gjøa* had been fourteen year earlier, was sold to the Hudson's Bay Company and renamed *Baymaud*. It proved unsuitable for company purposes and was abandoned in Cambridge Bay in 1930, where its rotted hull is still visible.

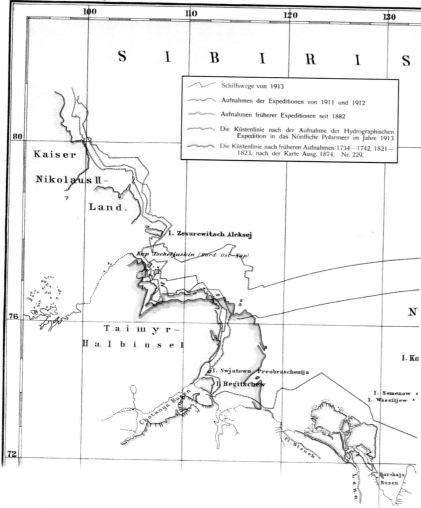

MAP 261 (left).
A German map from *Petermann's Geographische Mittheilungen* showing the voyage of Boris Andreyevich Vil'kitskiy in the *Taymyr* in 1913 and the discovery of Severnaya Zemlya, here shown as *Kaiser Nikolaus II–Land*. The coastlines represented on the maps of some earlier explorers are also shown. The thin red lines are the tracks of Vil'kitskiy's *Taymyr* and Novopashennyy's *Vaygach*. The red coastline is the one Vil'kitskiy mapped in 1913, the blue that mapped by similar Northern Sea Route hydrographic expeditions in 1911–12. The green coastline is that shown on maps in 1874, and was itself the result of earlier expeditions from the Great Northern Expedition (see page 38) forward. Vil'kitskiy was in the process of surveying the northern coast of Asia for a Northern Sea Route when he discovered hitherto unknown land not far off Mys (Cape) Chelyuskin (on this map, in German, *Kap Tscheljuskin*), the northernmost point of mainland Asia. *I. Zesarewitsch Aleksej* is Ostrov Malyy Taymyr, the first small island found by Vil'kitskiy. The German key has been moved from its original position on this map.

Other discoveries in 1930 derived from the voyage of the *Georgii Sedov*, the ship that had taken Ushakov to Severnaya Zemlya. In examining the drift of a ship from an earlier Northern Sea Route expedition that had been caught in the ice in 1913, Russian explorer Vladimir Yul'yevich Vize had predicted the existence of land in the northern part of the Kara Sea, and even plotted it on a map published in 1924. In 1930, this island was found exactly where it had been predicted to be; it was, not surprisingly, named Ostrov Vize. Further exploration found shoals and an island that was named Ostrov Ushakova. All are shown on MAP 263.

Between July and September 1932 an ex-Newfoundland sealer, the *Aleksandr Sibiryakov*, with Otto Shmidt on board, sailed from Arkhangel'sk to the Pacific around the northern tip of Severnaya Zemlya. Although towed towards the end of the voyage because ice had ripped off the ship's propellers, its voyage marked the first transit of the Northeast Passage—transformed into the Northern Sea Route—by a regular, non-icebreaking ship. But the success was short-lived; the following year ice trapped and crushed the *Chelyuskin*, attempting the same passage. Sea route or not, the ice would always be a danger.

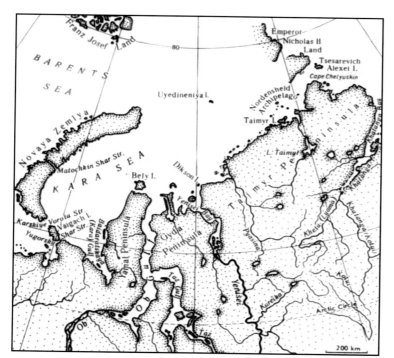
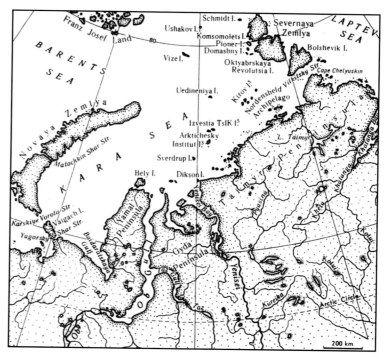

MAP 262 (below) and MAP 263 (below right).
A copy of a 1916—pre-revolutionary—map of the western Russian Arctic (MAP 262) and the same region from a 1974 map (MAP 263), showing the islands found in the intervening period and the complete delineation of Severnaya Zemlya. The one island shown in the middle of the Kara Sea in 1916 was Ostrov Uyedinaniya—Lonely Island—discovered by Norwegian walrus hunter Edvard Johannesen in 1878. Following the Russian surveying efforts since the Revolution, Johannesen's island was not so lonely after all. These two maps were published in Soviet Russia in 1974 to demonstrate the prowess of Soviet, as opposed to just Russian, explorers.

The Arctic by Air

The coming of the airplane was to radically change the way exploration was conducted in the Arctic, as elsewhere. Suddenly journeys that would have taken years could be carried out in a matter of hours or days. But the Arctic is an unforgiving place to fly if your craft is unreliable; failure often meant death.

The undoubted pioneer of Arctic aviation was the Swedish engineer Salomon August Andrée. Pursuing his interest in aerial navigation, he had taken up ballooning as a hobby in 1893, and experimented with steering a balloon by means of guide ropes which dragged along the surface of the sea. Andrée considered that he could steer a balloon up to thirty degrees either side of the wind using this method, plus the use of sails. And if it worked on the surface of the sea, why not on the surface of ice?

Andrée thought he would either drift to the Pole and back or go right over the Pole. However absurd the scheme appears to us today, in 1895 it appeared sensible enough that he was able to secure the support of the Swedish Academy of Sciences, the king of Sweden, and Alfred Nobel, of dynamite and prize fame. So the project, now well financed, went ahead.

A silk balloon capable of holding 4 530 m³ of hydrogen was made and named

This remarkable photograph of Andrée's balloon after its final crash onto the ice north of Spitsbergen on 14 July 1897 was taken by one of the three doomed men aboard. The unexposed film lay on the shore on Kvitøya for thirty-two years before being found and developed. Out of a total of 192 exposures found, some 30 were able to be developed successfully.

Ornen (*Eagle*). The balloon was equipped with many of Andrée's ideas. A sleeping bag was to double as a darkroom. Photos were to be taken en route, developed in the sleeping bag, and dispatched back to civilization with carrier pigeons. Cooking was to be achieved with a spirit stove hung 8 m below the basket, lit with a string and extinguished by blowing through a tube, so as to keep the flame away from the hydrogen in the balloon.

In June 1896 Andrée sailed for Spitsbergen, where he found a suitable site for a balloon shed on Danskøya, Danes Island, at the northwest tip of Spitsbergen. It took a month to get the hydrogen generator assembled and the shed built. Then the wind blew persistently in the wrong direction, and on 20 August the ship's insurance against ice expired, so all had to leave.

The next year, 1897, Andrée tried again, a little earlier. But this time although he was successful at getting away, it cost him his life. As he was leaving, the balloon lost all three lower sections of the dragline ropes, and with them went most of Andrée's chances of a safe return, for without them he had little control of the balloon's direction. Instead of abandoning the balloon right away, Andrée for some inexplicable reason kept going; he was never seen again.

We only know what happened to Andrée and his two companions because in 1930 a Norwegian scientific expedition going to Franz Josef Land called at tiny Kvitøya (White Island), probably the "Gilles Land" seen first by Dutchman Cornelius Giles in 1707 (see Map 193, page 121). There on the shore were the remains of a small encampment and a boat

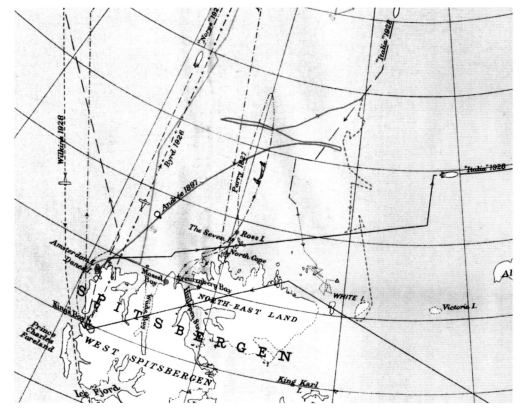

Map 264.
The drift of Andrée's balloon, shown in red. This map is from the printed version of Andrée's diaries, published in 1930. The balloon's track has been reconstructed from positions in the diary. The dashed red line is Andrée's trek to White Island (Kvitøya), where the remains of his encampment were found in 1930. The red dots are depots laid down before Andrée's flight. By the time this map was published, many later expeditions had traversed the area north of Spitsbergen. The light-blue line is the drift of Nansen's ship *Fram* in 1895–96; others marked on this portion of the map are later aerial expeditions described in this section. The dotted line towards White Island is the drift of the wreck of the airship *Italia* in 1928 (see page 180).

bearing the words "Andrée Polar Expedition 1897." Diaries were found; the last entry was dated 17 October 1897. Andrée's balloon had crashed onto the ice 600 km from Danes Island and 775 km short of his goal, the North Pole. The three men had managed to trek and boat back to land, but nobody found them, and they were ill-equipped for the coming winter.

It seems that Andrée continually failed to appreciate the dire straits he was in, for a photo was even taken of the crashed balloon. The exposed film had lain on the shore at Kvitøya for thirty-two years, yet was able to be developed.

After Andrée's disappearance, it was understandably some time before anyone made another attempt to fly in the Arctic. But in 1907 an American journalist, Walter Wellman, having made several unsuccessful attempts to penetrate the ice in ships, decided that air was the way to go. Realizing that Andrée's problem was his lack of ability to control his balloon, Wellman thought that this could be solved by using one of the new airships being developed at this time by Ferdinand Graf von Zeppelin and others.

Wellman, now financed by the *Chicago Record Herald*, purchased a Godard airship in 1906, had a hangar built on Danskøya, and sailed there to begin a flight to the North Pole. But it became evident that the airship had too many defects and the attempt was not even begun that year. Two further attempts were made by Wellman in his airship, named *America*, in 1907 and 1909, but both were dismal failures; the first time he was forced to land over a glacier, and the second the airship plunged into the sea not far from the hangar. This time it sank, and with it went Wellman's fortunes for, as they were sailing back to Norway, news was received that Frederick Cook had reached the Pole (see page 160). "There is no more honor to be had in reaching the Pole and I shall not try again," said Wellman.

The first use of an airplane in the Arctic came in 1914, when Russian navy lieutenant Yan Iosifovich Nagurskiy flew a seaplane some 1 060 km on five separate flights. He was searching for Georgii Sedov, who had set off on a private expedition to walk to the Pole in 1912–14. Nagurskiy flew over the ice off the northwest coast of Novaya Zemlya, about 100 km from land.

The First World War intervened at this point, and exploratory aviation was put on hold. But the war did improve the technology of the airplane and made it evident to many that aviation had a future.

Walter Wellman's airship *America* leaving its hangar on Danskøya (Danes Island) in 1907 for the first of his two abortive attempts to fly to the North Pole.

In 1920, at the instigation of the Canadian and American governments, an air expedition from New York to Nome, Alaska, was planned, and the return trip was carried out by four biplane bombers between July and October of that year, covering the astounding distance of 14 500 km over often difficult country with hardly any prearranged landing facilities.

On his return from the South Pole in 1912, Roald Amundsen had set about planning a drifting scientific expedition into the pack ice through Bering Strait. To this end he designed another ship, the *Maud*, similar in concept to Nansen's *Fram*, with a rounded bottom which would lift up in the ice rather than be crushed by it.

Delayed by World War I, the expedition finally sailed in June 1918. It was at the time the largest and best equipped geophysical expedition ever. But the ship was trapped in the ice for two years (see MAP 259, page 168), taken back to Seattle for repairs, and in 1922 again froze fast in the ice, this time near Wrangel Island. For three years it moved with the ice off northwestern Siberia. The time was not wasted, however, for considerable amounts of valuable oceanographic, meteorological, and magnetic data were collected. Meanwhile, Amundsen turned his attention to exploring by air.

It was Amundsen who first proposed to use the airplane for exploration in the high Arctic. He intended to fly a plane fitted with skis from Point Barrow in Alaska towards Spitsbergen. Towards, because he realized that he would not have the fuel capacity to make the entire trip. He therefore planned to have a fuel depot set up on the ice at which he could refuel. This not very practical idea was undoubtedly the worst that Amundsen ever had.

To set up the fuel depot, an expedition to Spitsbergen was undertaken by Amundsen's friend, H. H. Hammer. With him was Walter Mittelholzer, a well-known Swiss photographer. Amundsen did not get very far, canceling his flight due to mechanical problems, but Hammer decided to turn his expedition into a photographic survey of northern Spitsbergen. Walter Mittelholzer took numerous aerial photographs on several flights, producing the first aerial photographs of the Arctic.

MAP 265.
The method used by Walter Mittelholzer to convert aerial photographs into maps is shown here. A tracing paper with a perspective grid of two kilometer squares was laid over the photograph. Then the information shown in each square could, with a little judgment, be transposed to an undistorted square on the map. This example is on an air photograph of Hinlopen Strait, Spitsbergen.

The following year, 1924, Briton George Binney took a small seaplane to Spitsbergen as part of an Oxford University surveying and mapping expedition, taking more aerial photos to aid with mapmaking.

Amundsen, meanwhile, feeling the financial pinch of his failed expedition from Alaska in 1923, met Lincoln Ellsworth, son of a multimillionaire mine owner, who persuaded his father to underwrite another major effort to conquer the North Pole by air. Two new Dornier Wal flying boats were purchased in Europe, and to save shipping them to Alaska they decided to attempt the Pole from Spitsbergen. Amundsen, as well as reaching the Pole, wanted to prove that there was no land north of Spitsbergen.

The plan was to fly the two airplanes with only three men in each, so that if one failed there was a possibility of them all making it back in the one remaining. And, as it turned out, this was a sensible plan (unlike Amundsen's previous unworkable plan to refuel on the ice), for this was precisely what happened. But neither plane had the fuel capacity to fly to the Pole and return, and so this plan *had* to be carried out. The glitch in the plan was that it demanded a suitable landing place be found.

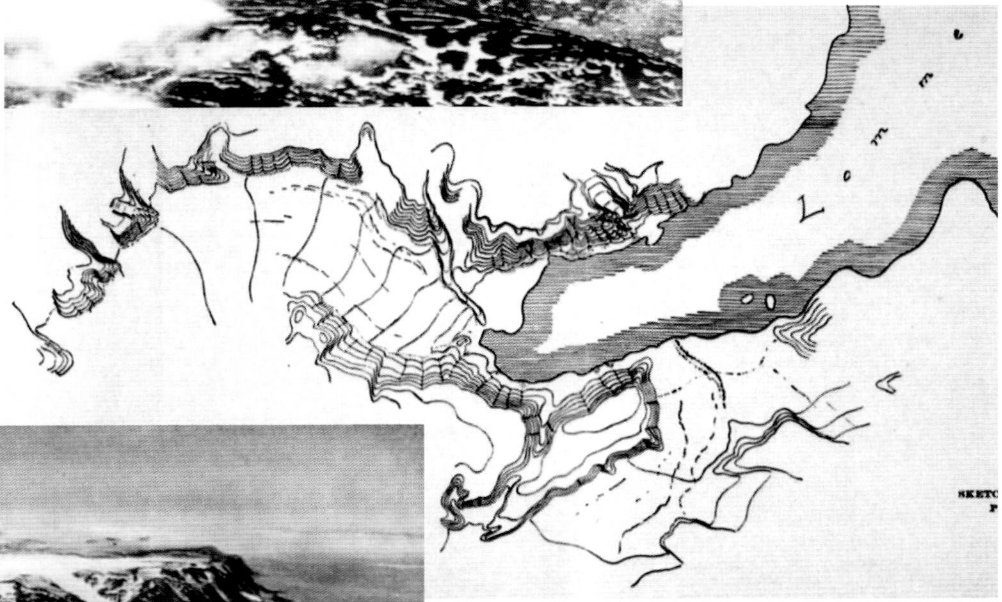

The two planes took off from King's Bay (now Ny Ålesund), Spitsbergen, on 21 May 1925. Amundsen as navigator, Hjalmar Riiser-Larsen, pilot, and Ludwig Feucht, mechanic, were in one plane; Lincoln Ellsworth, navigator, Lief Dietrichson, the pilot, and Oskar Omdhal, mechanic, in the other. After a relatively uneventful flight north they were nearing the Pole when a polynya—open water—was seen, and a landing attempted. This was because half the fuel had been used up, and Amundsen had lost track of his longitude. The first plane, flown by Riiser-Larsen, missed the water but was not badly damaged on landing, and Dietrichson landed safely some distance away, although one of his engines was burnt out. It took four days, with repeated attempts over the difficult ice interlaced with open leads, for Ellsworth, Dietrichson, and Omdhal to reach Amundsen's plane. A few days later, Omdhal and Dietrichson returned to their plane to retrieve a gas tank.

Now remained the task of getting the plane into the air once more. Amundsen calculated that they would have to take off before 15 June or the ice would have broken up so much as to make it impossible, and the men set to work to clear a runway. But the moving ice kept foiling their best efforts time and time again. The temperature was rising, meaning that wet snow or slushy ice had to be removed or trampled down. And the men were becoming weaker. Attempts to take off bogged down. Finally, on the morning of 15 June, the very last day Amundsen had calculated they could leave, the

Map 266.
Lommes Bay, Spitsbergen, mapped by Mittelholzer from several air photos, including the two shown here.

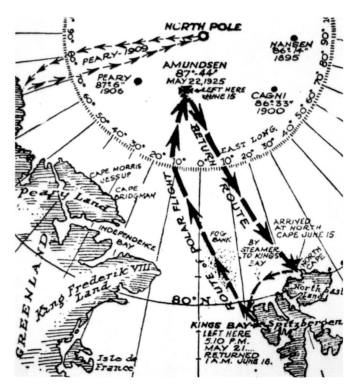

Map 267 (below).
Amundsen and Ellsworth's route is shown in this map from the *New York World* newspaper. Also shown are the "farthest north" points of Nansen in 1895 and Cagni in 1900, together with Peary's route to the Pole in 1909.

THE NEW YORK TIMES, FRIDAY, JUNE 19, 1925

AMUNDSEN BACK SAFE IN SPITZBERGEN; PLANES LANDED 150 MILES FROM POLE; SHORTAGE OF FUEL FORCED RETURN

FLEW BACK OVER THE ICE

Explorers Dropped Into the Sea and Were Rescued by a Fishing Boat.

PLANES ICELOCKED IN NORTH

Water Lane Froze After Landing, Balking Efforts to Reach Pole and Endangering Return.

ONE MACHINE LEFT THERE

Expedition Sighted No Land in the Far North—New Attempt May Be Made.

By EDWIN L. JAMES.

Copyright, 1925, by The New York Times Company.
Special Cable to THE NEW YORK TIMES

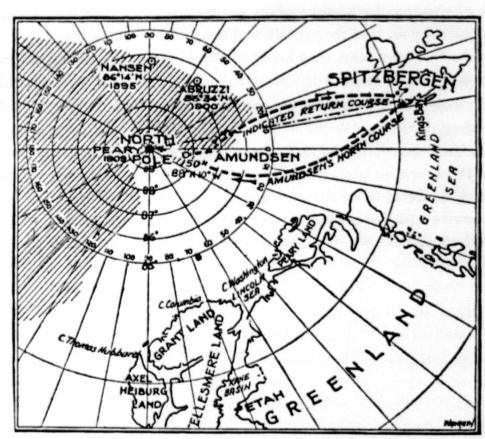

AMUNDSEN'S ROUTE NORTH AND BACK.
The Course Taken by the Norwegian Explorer's Two Airplanes on the Trip North From Spitzbergen to Within About 130 Miles of the North Pole The Shaded Portion Marks the Area MacMillan Plans to Study.

MAP 268.
Map and headlines from the *New York Times* newspaper on 19 June 1925. The unexplored area of the polar region is shown also. This was the area that the MacMillan expedition (*see overleaf*) planned to explore, although they never got any farther than the west coast of Ellesmere Island.

ground was frozen hard. Now was their chance. Riiser-Larsen opened up the engines and the six men held their breath as the Dornier finally lumbered into the sky. It had to be a moment of great satisfaction for Amundsen and his crew, one that had taken them twenty-four days of desperate labor to achieve.

Their trials were not quite over; approaching the coast the elevators on the plane seized up, forcing an immediate landing in the rough sea. Luckily they were able to taxi to a small bay and were soon relieved by the sight of a ship, whose captain agreed to tow them back to King's Bay. Hence they arrived, seemingly back from the dead, to find rescue seaplanes which had been brought in to start a search for them.

On this flight, Amundsen achieved a latitude of 87° 43´ N, a mere 260 km from the North Pole.

In 1924, a U.S. government–sponsored plan to fly an airship from Point Barrow to Spitsbergen had been canceled because of technical difficulties. Naval officer Richard Evelyn Byrd, the navigator assigned to this flight, was disappointed. The following year, determined to fly in the Arctic, he got himself appointed as one of the naval volunteers on a private expedition to Ellesmere Island organized by Donald MacMillan and the National Geographic Society. The intention was to carry out initial depot-laying activities for an attack on the area to the north and west that was as yet unexplored (Map 268, *previous page*) and where, it was thought, land might be found. The expedition had three navy-loaned amphibious biplanes.

The ship carrying the expedition arrived at Etah, in Smith Sound, on 1 August 1925, and from this base Byrd and the other fliers searched for open water or smooth ground on which they could land supplies, for the expedition was trying to establish a forward base on Axel Heiberg Island for continued exploration the next year. Despite numerous flights, shown on Map 269, few landing places were found. One that was found, on Flagler Bay, disappeared the next day.

The expedition itself did not achieve a great deal, although it began the aerial exploration of northern Greenland and Ellesmere Island and made the Canadian government concerned about its sovereignty over the region. But it did ingrain a love of Arctic flying and navigating into Richard Byrd. The next year he and one of the expedition's mechanics, Floyd Bennett, were to achieve much greater things. For Byrd saw no reason why he should not fly to the North Pole by air, and such were the times that he quickly found himself a sponsor willing to finance the attempt, Edsel Ford, son of automaker Henry Ford.

On 9 May 1926, Richard Byrd and Floyd Bennett took off from King's Bay, Spitsbergen, bound for the Pole, a sixteen-hour flight of over 2 500 km for the return trip. Their plane was a Fokker three-engine monoplane named *Josephine Ford* after Edsel Ford's daughter. They carried a sled and a rubber boat and food for ten weeks—not enough if they were forced down onto the ice. If this happened, they would have to walk to Greenland, for the drift of the ice would not let them return to Spitsbergen.

Map 269 (*above*).
The various flights of Richard Byrd and the two other planes in the MacMillan expedition of 1925 are shown in this National Geographic Society map. The expedition *de facto* challenged Canadian claims to sovereignty over the area.

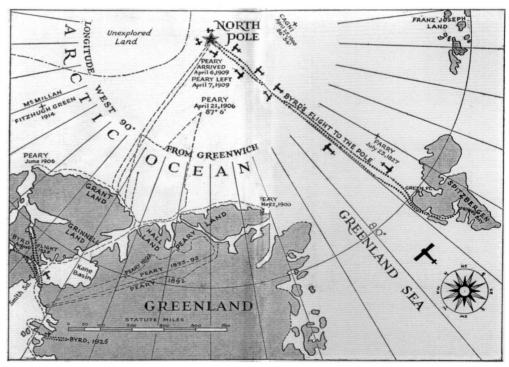

Map 270.
Map of Byrd's flight to the North Pole in May 1926, from his book, *Skyward*. Note that Peary Channel is shown cutting off the northern tip of Greenland. This was despite this channel having been shown not to exist by the ill-fated Mylius-Erichsen expedition of 1906–08 (see page 152).

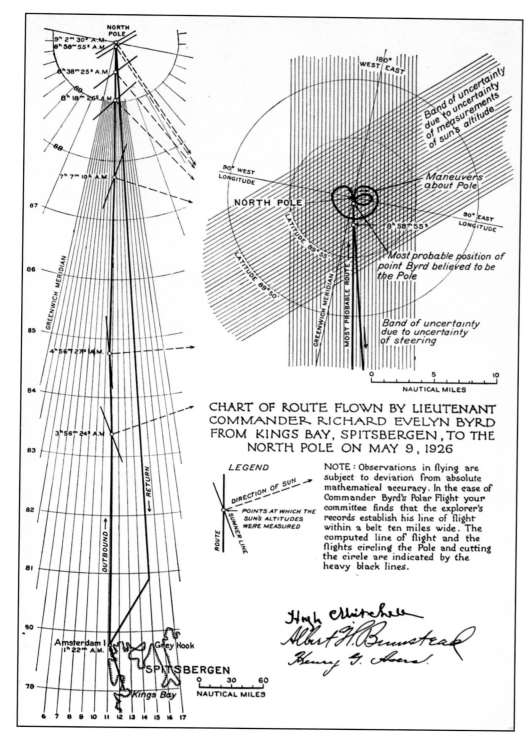

MAP 271.
The chart drawn by the committee of the National Geographic Society which examined Byrd's navigational record. The committee concluded that he had indeed made it to the Pole, *or very close to it*. Today there is renewed speculation that Byrd did not actually reach the Pole, and the subject draws heated arguments and supposedly learned papers.

Byrd and Bennett's plane, the *Josephine Ford,* named after the daughter of Edsel Ford, Byrd's financial backer.

Byrd's navigation was superb. He wrote: "We were now getting into areas never before viewed by mortal eye. The feelings of an explorer superceded the aviators. I became conscious of that extraordinary exhilaration which comes from looking into virgin territory. At that moment I felt repaid for all our toil." But then, near disaster. "When our calculations showed us to be about an hour from the Pole, I noticed through the cabin window a bad leak in the oil tank of the starboard motor." Bennett suggested a landing to try to fix it, but Byrd decided to keep going. "There was no doubt in my mind that the oil pressure would drop any moment. But the prize was actually in sight. We could not turn back," he recorded.

Next, the Pole. Byrd wrote: "At 9.02 a.m., May 9, 1926, Greenwich civil time, our calculations showed us to be at the Pole! The dream of a lifetime had at last been realized." After making wide circles to ensure that they had made it to the Pole even if there had been some errors in navigation, they headed back to Spitsbergen, having abandoned on account of the oil leak plans to return via Cape Morris Jesup. Soon after they left the Pole, Byrd's sextant slid off the chart table and its horizon glass broke, making it necessary for him to navigate by dead reckoning.

Byrd aimed for Grey Point, Spitsbergen. "Finally when we saw it dead ahead, we knew we had been able to keep on our course," he wrote, in a classic understatement, and "it was a wonderful relief not to have to navigate any more." Luck was with them. Their leaky engine was still running. The leak, they found out later, was caused by a rivet in the oil reservoir jarring out of its hole; when the oil reached the level of the hole left by the rivet it stopped leaking.

On his return to King's Bay Byrd sent a cable to the United States in which he reflected: "The elements were surely smiling that day on us, two insignificant specks of mortality flying there over that great, vast, white area in a small plane with only one companion, speechless and deaf from the motors, just a dot in the center of 10,000 miles of visible desolation."

A later check by a committee of the National Geographic Society confirmed Byrd's navigation, and that he had made it to the Pole. "The feat," they wrote in their report, "of flying a plane 600 miles from land and returning directly to the point aimed for is a remarkable exhibition of skillful navigation and shows beyond a reasonable doubt that he knew where he was at all times during the flight."

Nevertheless, as with Frederick Cook and Robert Peary, there have been a number of

The airship *Norge*.

attempts to prove that Byrd did not in fact reach the North Pole. The evidence remains inconclusive and the controversy continues. If Cook and Peary did not reach the Pole and Byrd did, he would have been the first to reach that coveted point.

At King's Bay at the same time as Byrd was a large airship designed and piloted by an Italian aviator, Umberto Nobile. He was accompanied by Roald Amundsen and Lincoln Ellsworth. Although Ellsworth was financing the whole thing, he was certainly becoming an aviator-explorer in his own right. After his disappointments with planes, Amundsen had conceived of an airship flight right across the Pole. Again he wished to disprove the existence of any unmapped land as well as achieve the Pole. Now, with Ellsworth and Nobile, he was waiting with the airship *Norge* for the right conditions to begin the attempt.

Early on 11 May, just two days after Byrd's return had meant that they would not be first to the Pole by air, conditions were good and so off they went. After fifteen hours of relatively uneventful flying—by Arctic standards, at any rate—they reached the Pole, dropping the flags of Norway, the United States, and Italy onto the ice below. A message was transmitted on a new Marconi radio. Then they kept going, heading for Nome, on the Bering Strait coast of Alaska.

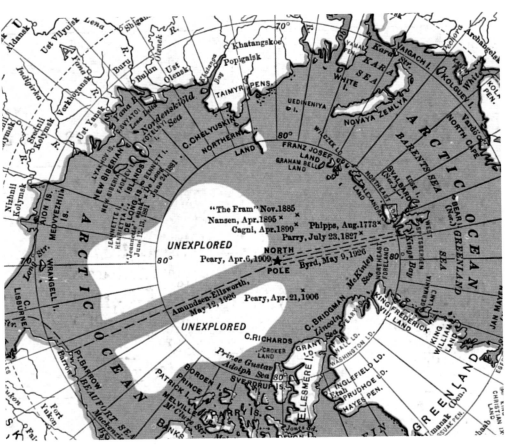

MAP 273.
This atlas map from 1930 shows very clearly the swath of "known" territory cut through the previously unexplored region. The height at which an airship flew enabled the "discovery" of a considerable area either side of the flight path.

The attainment of the North Pole by Amundsen, Ellsworth, and Nobile is the first *uncontested* achievement of the vaunted goal. All those previous—by Cook, Peary, and Byrd—have been seriously questioned.

Amundsen considered that he had the easiest task of any of the sixteen men aboard, acting as the explorer-observer, watching particularly to see if there appeared any signs of a possible Arctic continent, for now they were in completely virgin territory, never before traversed by man. As can be seen from MAP 273, the bulk of the unknown area at the time was north of Alaska and eastern Siberia. The airship cut a swath of "known" area through it; "theoretically seen" would be a better term, for the frequent presence of fog is ignored here.

On this stage of their journey, ice and fog became a major problem. Ice is particularly dangerous for an airship because ice buildup increases the weight that has to be supported by the hydrogen gas in the bag. It

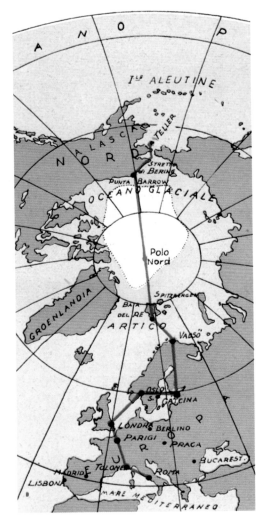

MAP 272.
An Italian map showing the entire route of the *Norge* from Rome to Teller, Alaska, from Umberto Nobile's account of the flight, in Italian, published in 1926. Nobile and Amundsen had a falling-out over who should be credited with the success of the flight. Amundsen had organized it and led it with Ellsworth, but Nobile had designed and piloted the airship. Hjalmar Riiser-Larsen had navigated; and in the Arctic navigation is most certainly as important as piloting. In truth, it was a joint achievement of all the participants.

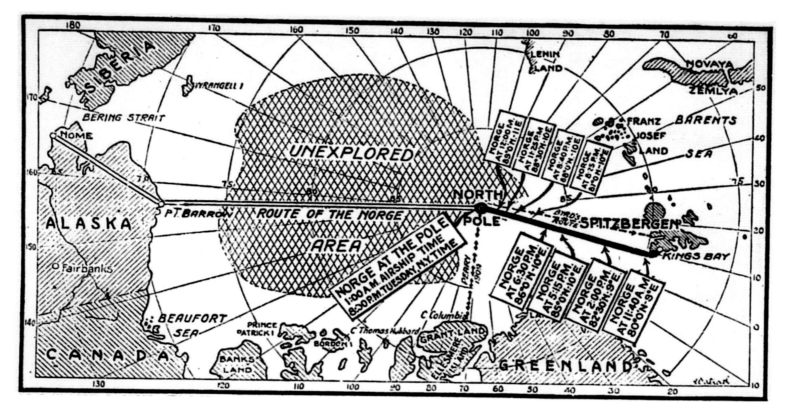

MAP 274 (*above*).
The progress of the *Norge*, reported by radio, was followed by newspapers all over the world; it was a major media event, as we would say today. This map is from the *New York Times* on 12 May 1928. The airship has reached the Pole and it is about to continue through the still considerable region of unexplored terrritory between the Pole and the northern coast of North America. It was the combination of flying and radio that made flying in the Arctic safer, whether by airship or airplane.

MAP 275 (*below*).
Amundsen's chart of the flight of the *Norge* in May 1926 from *Kingsbay* (Ny Ålesund), Spitsbergen, to Nome, Alaska, via the North Pole. This is the map from Amundsen's 1928 book about his flight.

also increases drag. In addition, ice can be flung off the propellors and pierce the envelope, with disastrous results. One of the advantages of an airship, however, was that the engines could be stopped in midflight for checks and repairs, and this was done several times.

Despite nerve-wracking bangs of splintering ice, the airship stayed aloft, and land was spotted after forty-five hours in the air. Soon they were able to recognize that they were near Wainwright, Alaska, about 120 km west of Point Barrow. They were considerably elated by their success, but the weather deteriorated, and strong winds sprang up. They eventually landed at Teller, a small village some 145 km from Nome. But no matter. They had achieved the first crossing of the Arctic from one side to the other. Inuit from the village took them to the edge of the pack ice, where they were picked up by a coast guard cutter sent from Nome.

Navigation had been aided on this flight by a Marconi radio direction-finder, carried for the first time. The flight had taken 70 hours 40 minutes, and they had traveled 5 118 km, at an average speed of 72 km per hour.

Flying in the Arctic must have been popular in 1926, for in addition to the Byrd and Amundsen-Nobile flights, the first sorties

were made by George Hubert Wilkins, who would go on to make many pioneer flights both in the Arctic and Antarctic. Although Wilkins was then a newcomer to flying, he was an Arctic veteran; he had been with Vilhjalmur Stefansson on his explorations of the Canadian Arctic in 1913–16 and had been one of the four who had left the doomed *Karluk* in 1913 (see page 166). In 1926 Wilkins flew more than 200 km out across the pack ice north of Barrow, searching to see if there was any land in this region. Soon after, both planes he had with his expedition crashed, curtailing the season.

The next year he was back. With pilot Carl Ben Eielson he flew northwest out over the Chukchi Sea. When they were 800 km away from land the engine failed and they had to land on the ice. Despite the immediate danger, Wilkins took the opportunity, much to Eielson's horrified bemusement, to bore a hole through the ice and take a sounding. He was looking for evidence of land and was not going to be sidetracked.

The two worked in appalling conditions—Eielson's fingers were frostbitten—but the engine was repaired and they managed to take off again. However, that was not the end of the story, for fuel was being used up faster than calculated, and 100 km from Barrow, they ran out. Crash-landing on the ice, in the dark, both men miraculously survived. Now Wilkins' Arctic survival skills, well learned with Vilhjalmur Stefansson, came into play, for he led Eielson over the ice to Beechey Point, just east of Barrow, where there was a radio station. At one point the ice cracked under Wilkins and he fell into the icy water beneath. Managing somehow to climb out, he astonished Eielson once again by stripping naked, rubbing the inside of his fur suit in dry snow, and at the same time prancing around wildly—and this at –30° C. Eielson only found out later that this was a determined and correct attempt by Wilkins to prevent himself freezing to death, as he would have surely done if he had not somewhat dried his suit in the snow and restored his circulation with his wild dance.

Wilkins' greatest Arctic achievement came the following year, 1928, when he flew from Alaska to Spitsbergen, again with Eielson. Wilkins had managed to scrape together enough money by selling his two existing planes to buy a new Lockheed Vega, the second one ever made.

With Eielson piloting, Wilkins took off from Point Barrow on 16 April 1928, after a

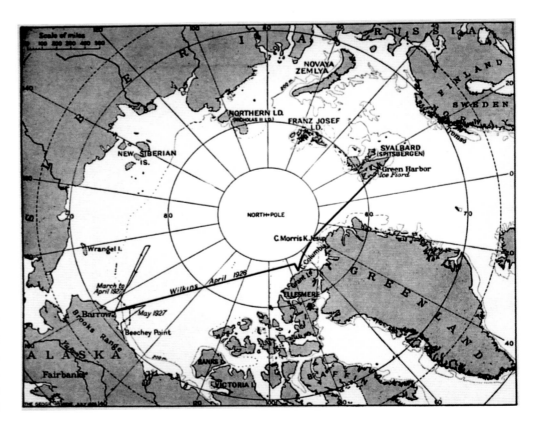

MAP 276.
The map from Wilkins' book *Flying the Arctic*, published in 1928. In addition to the track of the Alaska-Spitsbergen flight of that year, the tracks of his shorter flights the previous year are also shown.

number of failed takeoff attempts, and headed for Cape Columbia, at the northern tip of Ellesmere Island. Wilkins was a very skilled navigator who was also able to read information from the condition of the ice to tell him how far from land he was, and he knew from obterving the clouds where to fly to pick up tailwinds. These skills proved invaluable, for after thirteen hours of flying—with not the slightest glimpse of land—they perceived the glaciers of northern Ellesmere Island, showing that Wilkins' navigation was correct after flying about 2 250 km (1,400 miles).

Then Wilkins set a course eastwards to Spitsbergen, which they reached after 20 hours 20 minutes. But landing was not easy. Despite running into a storm, they found a landing spot on a little island, but Eielson's windshield had become covered with frozen oil, with snow sticking to it, reducing his visibility to essentially zero. They managed to land with Wilkins looking out of the side windows, giving written directions to Eielson, because the noise inside the plane was too loud for them to hear each other.

But where were they? Consulting his charts, Wilkins decided they were on an uncharted island in King's Bay. After they slept, the storm abated temporarily and Wilkins was able to take a reading with his sextant. He found he was indeed near King's Bay. But the stormy weather prevented a takeoff attempt for another four days.

The plane had settled into the snow and would require a push to begin a takeoff. But it took three attempts to get airborne, at least with both men in the plane, for Wilkins found he could not get into the hatch fast enough after pushing. At the first attempt the plane took off leaving Wilkins behind. At the second attempt Wilkins was knocked down by the tailwing, and it was only after a last desperate attempt that both men were able to take off together. But soon after takeoff they spotted Green Harbor (Gronfjorden), where they were able to land safely.

The transpolar flight of Wilkins and Eielson was the longest nonstop flight ever attempted until the advent of jets. Roald Amundsen paid tribute by writing, "No flight has been made anywhere, at any time, which could compare to it."

Wilkins and Eielson went on, later in 1928 (the seasons, of course, being reversed, allowed this), to become the first to fly in the Antarctic, where Wilkins achieved his aim of discovering land.

In 1926, Roald Amundsen and Umberto Nobile had quarreled over who should re-

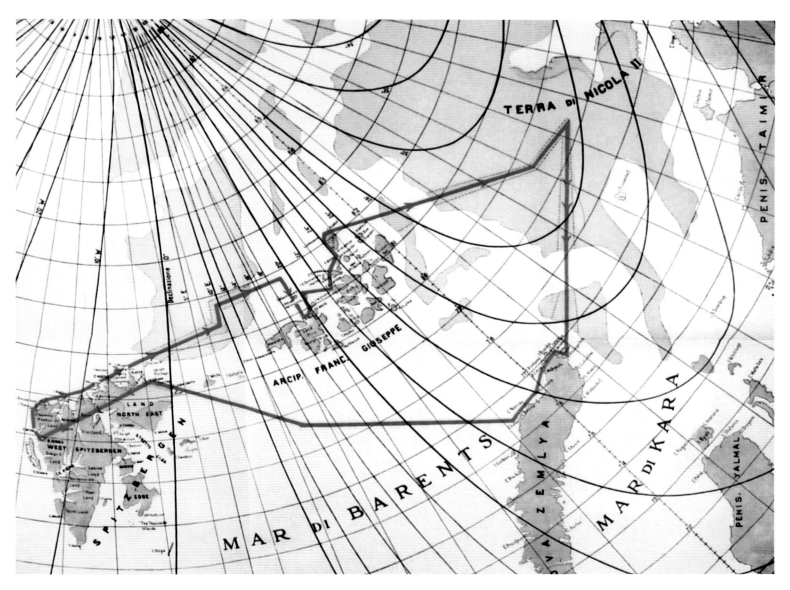

MAP 277.
The flight of the *Italia* from Spitsbergen nearly to the west coast of Severnaya Zemlya, here shown as *Terra di Nicola II* (Nicholas II Land), and the northwestern coast of Novaya Zemlya. The path of the airship was marked by Nobile; the map was presented to the American Geographical Society.

ceive the credit as leader of the *Norge* expedition. Thus in 1928 Nobile, the airship designer and pilot, organized another airship expedition to the Arctic. The purpose was to try to find land. To this end flights were proposed along the Siberian and North American coasts where the water was shallowest and the probability of land being found was higher. Studies of magnetism and other scientific subjects were to be made. In addition to these scientific excursions, a flight was to be made to the North Pole, this time to set down. This latter trip made the gathering of support for the expedition easier, for it was to be an *Italian* polar expedition.

A new airship was built and named—predictably—*Italia*. All of the crew were Italian save two scientists. The base ship would be supplied by the Italian navy. Nobile was received by the king of Italy and Mussolini. Flags, and a cross from the Pope, were taken to be dropped on the Pole.

Nobile flew the *Italia* from Milan to King's Bay, where the mooring mast and hangar built for *Norge* were still standing (as the mast does to this day).

On 10 May 1928 *Italia* set off on the first of the planned flights, eastwards towards Severnaya Zemlya. But the weather deteriorated and they had to return after a few hours.

By 15 May the weather had improved and they set out again. After thirteen hours they reached Franz Josef Land, and after another twenty-one they were approaching the then unknown western coast of Severnaya Zemlya. But severe winds made Nobile decide to turn back. They headed for Novaya Zemlya, exploring its western coast, then headed back to Spitsbergen, crossing the island in the process. North East Land (Nordaustlandet), the easternmost island of the Svalbard group, which had up to this time been shown as an area of continental ice, was seen to be rock covered with only thin ice in places.

Italia had airlifted sixteen people over some 3 850 km of ocean and land, most of which was still unknown at the time, in sixty-nine hours of flying. The stamina of the airship compared to a plane was evident, and the prospects for the flight to the Pole seemed very good.

Five days later, again with sixteen people on board, *Italia* took off for the Pole. A northwesterly course was taken initially, turning north as they approached the northern tip of Greenland. Thus the route was about halfway between that taken on the ground by Peary in 1909 and that of the *Norge* in 1926, converting a vast swath of unknown territory into known. Conditions were good, with a tailwind, and they arrived over the Pole, where with appropriate ceremony the Italian

MAP 278 (*above*).
The manuscript version of a chart by Nobile showing the path of the *Italia* to the North Pole in 1928. Nobile has marked the track of his airship on a printed base map, *Carta delle Regioni Artiche*. This map was presented to the American Geographical Society in January 1929.

flag and the papal cross were dropped from a height of 140 m.

Then the return journey began. The wind that was a tailwind now became a headwind, and although the onboard meteorologist, Finn Malmgren, had predicted it would veer, it did not. The weather deteriorated, and it was difficult to know if their heading was correct; only a radio signal from the base ship was available to them, and thus no means of a cross-fix, though later they were able to get a fix using the sun by climbing higher. This showed their position to be about 290 km northeast of King's Bay. Then,

disaster struck. Ice started to build up on the airship, making it too heavy, and it plunged towards the ice. Although Nobile stopped the engines to reduce the speed, *Italia* hit the ice with considerable impact, breaking the gondola away from the main body of the airship, which, now lightened, immediately shot into the sky, carrying six crew members in an uncontrolled flight to their doom.

The nine men on the ice assessed their situation. They found an emergency radio and began transmitting to the base ship on the 55th minute of each hour, as had been previously arranged. These messages went strangely unanswered; the ship's captain had decided that the radio operator must have been killed in any crash, so inexcusably did not bother to listen in. A small four-man red tent was in the emergency supplies. In it the nine men huddled. A sextant in the supplies was used to establish that their position was 81° 14´ N, 28° 14´ E, just 80 km north of North East Land. But the ice was drifting east and came within view of a small island, touching off a debate as to whether they should attempt to march there. This was decided against, because Nobile and another of his men had broken legs and would have been difficult to move, but when they drifted to within 11 km of Foyn Island (Foynøya) Malmgren and two naval officers decided to

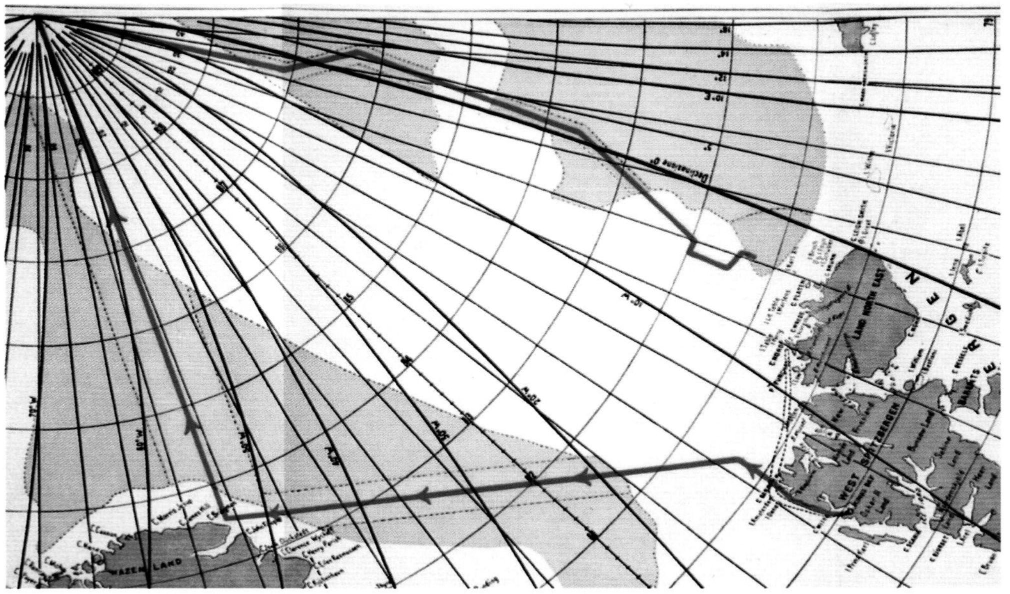

MAP 279.
The version of the map printed in Nobile's book on the flight of the *Italia*, published in 1930. The yellow area is the zone "explored" —or rather, seen—by the crew of the *Italia*; the green area is unexplored.

try it. In the attempt, Malmgren was later abandoned by the two officers because of his inability to keep up due to his injuries.

After nearly two weeks on the ice, on 3 June Nobile's radio signals were picked up by a radio operator in Arkhangel'sk, which considerably revived everyone's spirits, and on 8 June contact with the ship was also made.

At this time Roald Amundsen was in Oslo and, on hearing of Nobile's plight, and despite his previous differences with Nobile, borrowed a plane from the French government and headed north to join the search. With him were Lief Dietrichson, his mechanic on his attempt to reach the Pole in 1925, a French pilot, and three others. His plane was never heard from again. Thus the senior polar veteran explorer of his time met his end. He had been the first to sail through the Northwest Passage, the first to reach the South Pole, and the first—with Nobile and Ellsworth—to cross the Arctic by air.

Several nations rushed aircraft to the Arctic to assist in Nobile's rescue. Nevertheless, it took many more days for the airship crew to be located, as it was difficult to spot the small red tent and a few men in a wilderness of ice. Once they were found, the main problem was that the ice was starting to break up, and landing was difficult. Finally a small ski plane manned by a Swede, Einar Lundborg, landed with orders to take Nobile himself to the ship to act as rescue co-ordinator. Protesting, Nobile agreed to go, but when Lundborg returned for others, his plane overturned on landing, and the would-be rescuer joined the ranks of those to be rescued.

Lundborg was later brought out by another Swedish plane, and the rest of the group were finally rescued on 12 July by the Russian icebreaker *Krassin*. So bad was the visibility that the ship steamed right past them before being brought back by a bonfire lit on the ice.

Nobile was denounced by the fascist Italian government of the day, which held him responsible for the crash of the *Italia* initially, and for leaving the marooned party first, despite having been ordered to go by the Swedes so that he could act as rescue co-ordinator. Nobile was stripped of the honors he had received after the 1926 flight of the *Norge* and compelled to resign. He took up an offer to design airships in Russia. It was not until 1945, and the end of fascism in Italy, that Nobile was exonerated.

In 1931 the by now renowned Arctic (and Antarctic) aviator Hubert Wilkins came up with the idea of a submarine expedition to the North Pole, which would also gather scientific information. He leased a submarine from the U.S. navy. It had been built in 1916 and decommissioned in 1924. He persuaded Lincoln Ellsworth to lend his name and influence to the venture. He also persuaded Hugo Eckener, now in charge of the airship *Graf Zeppelin*, back from a round-the-world flight in 1929, to co-operate in a spectacle which was to have his submarine meet the airship at the North Pole and exchange mail. With this agreed, William Randolph Hearst, owner of a chain of newspapers, was prevailed upon by Eckener to offer $150,000 if the two met at the Pole and exchanged mail and passengers, and $30,000, still enough to finance the expeditions, if they merely met somewhere in the Arctic.

Stripped of armaments and filled with scientific equipment, the submarine was christened *Nautilus*. But the vessel was to be dogged with mechanical failures. Heading for Norway, its engines failed in the Atlantic, and it had to be towed to England for repairs. When the submarine finally reached the edge of the ice north of Norway in August, it was found that the submarine lacked its diving rudders, making diving virtually impossible. Wilkins soon decided it was no longer safe to remain at sea, and he returned to Bergen. The submarine was returned to the U.S. navy, which towed it out to sea and sank it, thus ending a very early attempt to explore the Arctic under the ice. Interestingly, it was a submarine of the same name that finally made it to the North Pole; the nuclear-powered USS *Nautilus* arrived at the Pole on 3 August 1958 (see page 185).

Eckener, meanwhile, had determined that the Russian icebreaker *Malygin* could be substituted for the *Nautilus*, and mail could be exchanged, although no monies would be forthcoming from Hearst. But it is a measure of changing times that Eckener thought his largely scientific venture could now be financed by philatelists.

The *Graf Zeppelin*, with a crew of thirty-one and a scientific team of fifteen, and joined at the last moment by Lincoln Ellsworth, took off from its home base at Friedrichsafen, on Lake Konstanz in Bavaria, on 24 July 1931 and reached Hooker Island (Ostrov Hooker) in Franz Josef Land three days later.

Here the exchange of mail took place. Lincoln Ellsworth met here, unexpectedly, with Umberto Nobile, who was on the *Malygin*. "He had aged visibly," recorded Ellsworth. "The *Italia* disaster had made a different man of him."

The *Graf Zeppelin* continued eastwards and was able to confirm that Severnaya Zemlya was two large islands, not one, as had been thought likely until then. In fact it is

MAP 280.
Hubert Wilkins wrote a book about the voyage of the submarine *Nautilus*. This is the map from that book, showing the proposed route of the submarine both on the surface and under the ice. The zigzag course was to search for land. In fact Wilkins' submarine could not dive safely, and the North Pole was not reached under the ice until 1958, when another *Nautilus*, this time nuclear-powered, traveled there in a relatively uneventful voyage.

not two but four. The airship then returned to Franz Josef Land and conducted an aerial survey, mapping the islands of the archipelago. They reached Ostrov Rudolf, the northernmost island, 800 km from the Pole. *Graf Zeppelin*'s tour of the Arctic was a whistle stop by previous standards, arriving back at Friedrichschafen on 31 July after completing a 13 000 km flight.

During and after the 1930s, almost all Arctic expeditions were supported in some way by air, and as aircraft became more and more mechnically reliable, long flights, including those to the Pole, became almost routine. In particular, Russia came to realize that aviation would significantly improve communication in the icy North.

In 1937, the Russians mounted a large scientific expedition to the Pole, with five aircraft, four four-engine planes and one two-engine scout plane. From an advance base on Ostrov Rudolf, the latter flew to the Pole on 5 May 1937 to determine if landing places were possible. Then on 21 May expedition leader Otto Shmidt and pilot Mikhail Vodopyanov landed at the Pole in one of the four-engine planes. This was the first landing on the ice at or near the Pole in an airplane. The three other planes followed, landing five days later.

The landing is significant for another reason, for if we accept that the claims of Frederick Cook and Robert Peary to have reached the North Pole in 1908 and 1909 were false, then it is Otto Shmidt and Mikhail Vodopyanov who were the first to actually stand at the Pole. Even this has been rejected by purists, for it is now thought that the Russians actually landed not at the Pole but at 89° 43´ N, about 30 km short of that elusive point.

The incoming aircraft carried teams of scientists, wireless operators, mechanics, and support personnel, thirty-five of whom stayed at the Pole for up to sixteen days. When the planes left, they had set up a scientific station at the Pole on which three scientists and a radio operator would remain. This became SP-1, a drifting scientific research station (see page 186).

Ever since Nansen's epic 1893–96 drift in the *Fram* (see page 148), it had been known that the ice moved long distances. The four men, led by Ivan Papanin, were left to man the drifting station on 6 June 1937. They busied themselves with all manner of scientific experiments and reported their changing position by radio. The drift

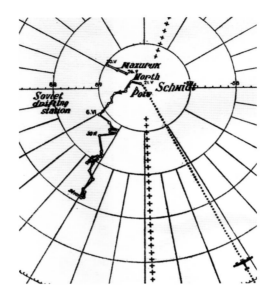

MAP 281.
Map of the drift of Russian scientific station SP-1 from 21 June to 31 July 1937. The complete drift is shown on MAP 287, page 186.

lasted until 19 February 1938, by which time the station was in the Greenland Sea at 70° 54´ N, and ice pressure was threatening to break up the floe. The four men were rescued by the icebreaker *Taimyr*. The drift of the station until the end of July 1937 is shown on MAP 281, and the complete drift on MAP 287, page 186.

Concurrently with the Russian effort at the Pole, several transpolar flights were made. On 18 June 1937 Valery Chkalov took off from Moscow in a single-engine plane specially designed for long-distance flying. It was filled to the brim with fuel and taken to the top of a small hill to assist the takeoff. Chkalov and two others flew from Moscow to Vancouver, Washington, a distance of 10 000 km in 62½ hours.

This success was followed up less than a month later when M. M. Gromov took off from Moscow on 12 July in the same type of single-engine plane. Again with two others, he had intended to fly until his fuel ran out, and he made it to San Jacinto, near Los Angeles. This flight of 11 400 km in 62 hours 20 minutes established a new nonstop long-distance flying record. There were some that doubted whether such a long flight was possible, and thought that Gromov must have landed at the Russian drifting station near the Pole to refuel. But there is no reason to think this was the case.

In 1937, the Russians thought they would try a transpolar flight using a four-engine plane, the same type as the four that had flown to the Pole to establish the drifting station earlier that year. Sigismund Levanevsky took off from Moscow bound for Fairbanks, Alaska. He passed the Pole, then disappeared.

When Hubert Wilkins heard of Levanevsky's disappearance, he immediately volunteered to search for him. He purchased a flying boat and by 21 August was at Coppermine, on Coronation Gulf. Over the next six months Wilkins scoured the Arctic north of the American continent, to no avail. Levanevsky was never found. The four-engine plane in theory was safer than the single-engine, for it could fly on two or three engines. We shall likely never know what happened.

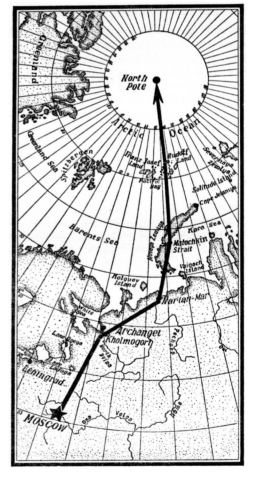

MAP 282.
Route taken by the Russian planes flying to set up the drifting station at the North Pole in 1937.

The Last New Land

As the Arctic continued to be explored in the twentieth century, the last islands that had remained hidden were gradually discovered, and, just as importantly, many phantom lands were relegated to a watery grave.

After World War II, few islands of any size were discovered. A notable exception was the discovery in 1948, by air, of Prince Charles Island, Foley Island, and Air Force Island, in Canada's Foxe Basin, off the west coast of Baffin Island, quite large islands relatively far south. Prince Charles Island, the largest of the three, is 120 km from north to south and 95 km wide. The islands had probably remained undiscovered because they are very low; the highest point, on Prince Charles Island, is only 75 m high. As far as is documented, the islands were unknown at the time even to the Inuit, though this is perhaps doubtful given their southern location. Land in the position of Prince Charles Island had been reported in 1932 by a tug captain, W. A. Poole, but his information had never made it onto any map.

The islands were finally discovered and mapped by an RCAF Lancaster of 408 (Photo) Squadron, flying out of Frobisher Bay. Britain's Prince Charles became the last person to have a major newly discovered part of North America named after him; 1948 was the year of his birth. Foley Island was named after the RCAF navigator on the discovery flight, who was killed in a flying accident early in 1949. Air Force Island, of course, was named after its discoverer, following the time-honored tradition.

The advent of aerial survey eased the task of geographical exploration in the Arctic, but the survey was still by no means comprehensive until the coming of the satellite. Many apparent islands were found to be nothing more than islands of ice, raised upwards by underlying pressures, thus going a long way to explain the myriad sightings of "land" that had taken place over the centuries.

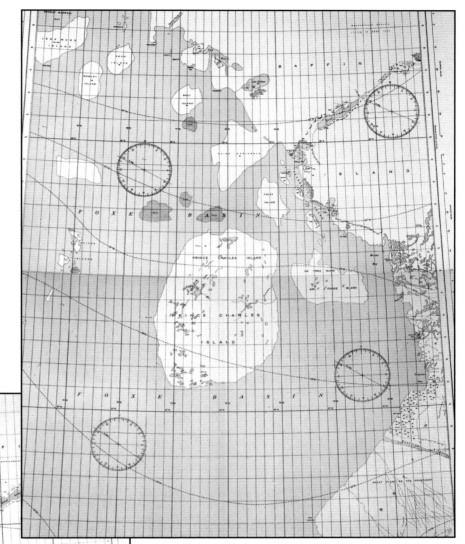

MAP 283 (*left*) and MAP 284 (*above*).
Canadian aeronautical topographic maps of the western part of Foxe Basin dated 1946 (MAP 283) and 1949 (MAP 284). Between the two dates, three islands have appeared, including Prince Charles Island, 9 500 km² in area, an island almost twice the size of the province of Prince Edward Island. Found in 1948 by aerial survey, the latter island was the last-found major land in both North America and the Arctic, a perhaps surprisingly recent conclusion to the centuries-long search for new lands. Each of these maps is composed of two sheets, causing the slight discontinuity halfway down each map. The land to the right is the southeastern coast of Baffin Island.

The Pole by Many Means

USS *Nautilus*, the world's first nuclear submarine and the first vessel to reach the North Pole under the ice, in 1958.

We have already seen that the claims of Frederick Cook in 1908 and Robert Peary in 1909 to have reached the North Pole over the surface of the ice are disputed. Likely the first to actually stand at the Pole were Otto Shmidt and Mikhail Vodopyanov, who flew there in 1937; three other Russian planes landed a few days later (see page 182). Beginning with another Russian, Pavel Gordiyenko, in 1948, many planes have landed at the Pole. But this still left open the achievement of the Pole over the ice.

Realizing this, an insurance agent from Duluth, Ralph Plaisted, gathered sixteen people together in 1967 for the first surface attempt since Peary—but by snowmobile. They reached 84° N before giving up and being airlifted out. However, Plaisted did not give up easily. The following year, having learned a great deal from his experience the season before, he tried again, this time with fewer people and more

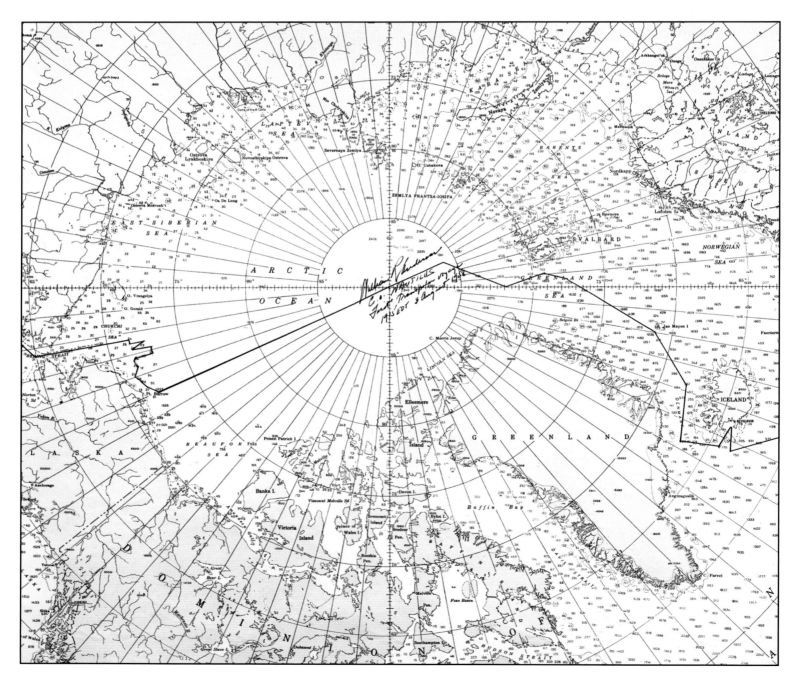

powerful, Bombardier snowmobiles. And he succeeded. On 20 April 1968, Plaisted stood at the North Pole—and nobody contested his achievement. He was airlifted out but received surprisingly little attention from the world's press, who regarded his attainment of the Pole as little more than a stunt, so ingrained was the notion that Peary had reached the Pole in 1909. Nevertheless, it is very likely that Ralph Plaisted, insurance saleman from Duluth, was the first to reach the Pole over the ice.

The next year, the over-ice trek was done "properly." Wally Herbert, a British explorer with considerable experience in the Arctic, sought to reach the Pole using dogsleds. With three others, he left Point Barrow, Alaska, on 21 February 1968 and reached 85° N before being stopped by the end of the season. After wintering on the ice from 6 October to 24 February—and being supplied by air—he continued northwards and reached the North Pole on 6 April 1969. Herbert kept going, now southwards, and reached Spitsbergen on 29 May. His effort was the first crossing of the Arctic without mechanical help *on the ground*.

In April 1978 the first solo attainment of the Pole was made by Japanese adventurer Naomi Uemura, using a dogsled.

USS *Skate* in 1959, the first submarine to surface at the North Pole.

Eight years later, an American explorer, Will Steger, achieved the Pole under the same conditions as the attempts by Cook and Peary, that is, without being supplied by air. It was also the first time that a woman, Ann Bancroft, reached the Pole over the ice, as one of Steger's eight-person team. They reached the Pole on 2 May 1986, but then they were airlifted out. Only nine days after Steger, Jean-Louis Etienne arrived at the Pole, the first to make the trip on skis.

One could certainly argue that going only one way, knowing that you do not have to trek back again, would make a world of a difference. But these are impressive achievements nonetheless, for the Arctic can be a relentless, unforgiving place, as attested to by the number of failed expeditions and deaths.

Finally we should mention the expedition of Richard Weber and Mikhail Malakhov; in 1995 they trekked to the Pole from Ward Hunt Island, near Cape Columbia, man-hauling their own sledges. This achievement, which took them 123 days, from 14 February to 16 June, made them the first to get to the Pole *and back* unsupplied and *without mechanical assistance*.

MAP 285 (*left*).
This is a "souvenir" map of the achievement of the North Pole by the *Nautilus*. On a printed map of the Arctic Ocean William Anderson, the submarine's captain, has drawn the track of the vessel and signed the map.

Two other methods of achieving the Pole merit attention: by submarine and by ship. We have seen how Hubert Wilkins' pioneering attempt in 1931 to use a submarine to travel under the ice failed before it really got started (see page 181). He was simply ahead of his time.

The American nuclear submarine USS *Nautilus*, under the command of William Anderson, reached the Pole *under* the ice on 3 August 1958. Several attempts in 1957 had been aborted due to failure of gyroscopic compasses. In 1958, equipped with a new inertial navigation system, *Nautilus* traversed the Arctic from Alaska to the Greenland Sea under the ice, via the Pole. It was found that some of the pressure ridges extended perhaps 30 m down, presenting an obstacle, particularly in shallow waters, similar to that presented to the surface traveler.

Later the same year, James Calvert, captain of the USS *Skate*, also nuclear-powered, was assigned the task of determining safe ways to surface through the ice. *Skate* surfaced nine times in various locations in 1958. Calvert visited Hubert Wilkins and Vilhjalmur Stefansson later that year, and Wilkins advised him to try the same thing during the winter, through recently frozen leads where the ice would not be too thick. On 17 March 1959 *Skate* surfaced at the Pole. Calvert scattered the ashes of Wilkins at the Pole; the Arctic pioneer had died in December 1958 and Calvert was carrying out his wishes.

Such was the development of powerful propulsion systems that in 1977 the first surface ship reached the North Pole. This was the Russian nuclear-powered icebreaker the *Arktika*. In 1991, another Russian nuclear-powered icebreaker, the *Sovetskiy Soyuz*, crossed the Arctic Ocean, from Murmansk to Provideniya, in July and August. It is now possible to take a trip as a tourist aboard a Russian icebreaker to the North Pole.

One such tourist trip, that of the *Yamal* in August 2000, was surprised to encounter water at the North Pole. This spawned news reports of global warming melting the Pole but the water was simply a polynya—an area of open water—similar to those found all over the Arctic during the summer. As much as 10 percent of the Arctic Ocean may be water at any given time, though the specific areas of water are shifting and changing all the time with the movement of the ice.

USS *Skate* at the North Pole in 1962.

Science, Sonar, and Satellites

After 1945, the combination of radio and more-reliable long-range aircraft allowed the beginning of a more comprehensive gathering of scientific data in the Arctic.

The drift of the ice station SP-1 away from the North Pole in 1937–38 (see page 182) and a similar, but involuntary drift by an icebreaker, the *Georgii Sedov*, from 1937 to 1940, had shown the feasibility, first demonstrated by Nansen, of manned drifting research stations.

Beginning in 1948, the Russians began a program of scientific reconnaisance by air, landing on the ice where possible for a few hours, where a series of scientific observations were made. MAP 287 shows the distribution of these data collection points, and demonstrates the beginnings of comprehensive coverage of the Arctic Ocean. Also shown on this map are the paths of several drifting stations set up on the ice. The first (after the pioneer SP-1) was SP-2, set up in 1950. (The stations were given the SP designation in 1954. The term refers to *Severnaya Polyus*, North Pole.) Not only was the drift tracked, but properties of the ice, temperatures, geomagnetism, and gravity were measured.

Despite some hair-raising incidents where the ice floe split right under buildings, SP-2 was so successful that others followed. One, SP-4, which had been set up north of Ostrov Vrangelya in April 1954, vindicated Nansen's idea of reaching the North Pole by drifting; it reached a point 13 km from the

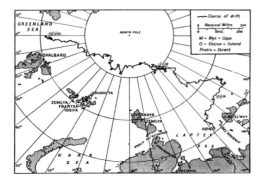

MAP 286 (*left*).
The drift of the trapped Russian icebreaker *Georgii Sedov* over 812 days between 1937 and 1940, covering a total distance of 6 115 km. Soundings were taken from the ship. A number of these were over 4 500 m and the deepest failed to find a bottom at 5 180 m. This compares with the maximum of 3 850 m measured by Nansen, and 4 395 m by drifting station SP-1.

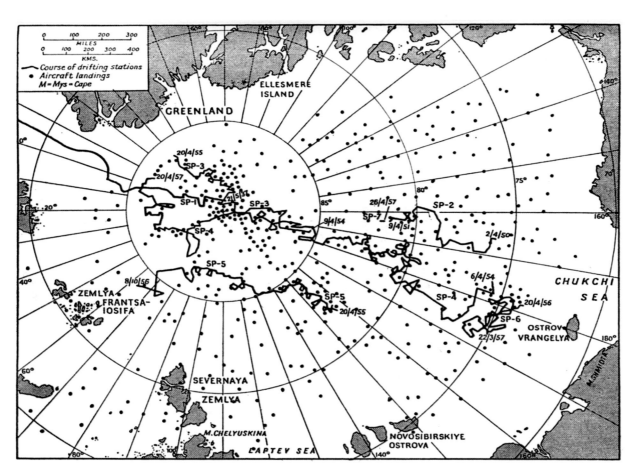

MAP 287.
The courses of Russian drifting research stations and aircraft landings for research purposes, 1948–57. SP-1, the first drifting station, set up at the North Pole in 1937, is also shown. Comprehensive coverage of the whole Arctic Ocean is beginning, but still the landings are but one point in space and time. The drifting station SP-4 demonstrated the truth of Nansen's hypothesis of ice drift (MAP 231, page 150).

MAP 288.
The 1979 edition of the *General Bathymetric Chart of the Oceans* (GEBCO) for the Arctic Ocean, published by the Canadian Hydrographic Service. In 1903 Prince Albert of Monaco, one of the early benefactors of the science of oceanography, initiated the task of gathering data from the world's oceans to produce a series of bathymetric maps. Although these maps looked impressive, the reality was that in most locations, generalizations hid any details of undersea topography. It was not until the development of multibeam or sidescan sonar that more accurate maps could be produced, using more continuous data covering a wider area at one time.

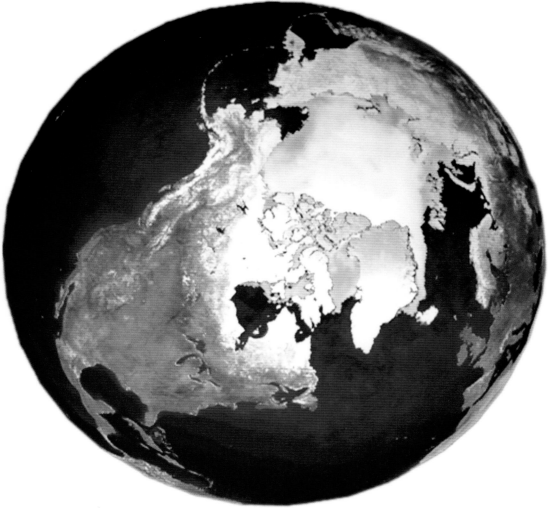

MAP 289 (above).
A top-of-the-world view from a RADARSAT satellite image, showing sea ice in the Arctic Basin. Although this looks like a single image from a high-orbit satellite, in reality it is a composite of many images. Using radar, the satellite can obtain Earth images day and night, without regard to cloud cover. Radar can measure the height of an ice surface, and continuous monitoring will reveal if the height is becoming lower due to melting, information of considerable value in studies of global warming.

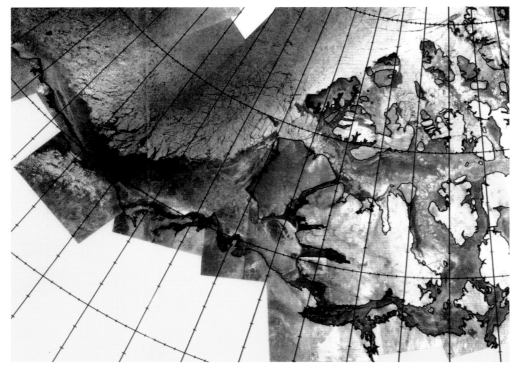

MAP 290.
This map was made by superimposition on a RADARSAT satellite image of the western Canadian Arctic. It shows the state of the sea ice between 31 January and 4 February 2001. The difference in coloration between the multi-year ice pack (lighter gray) and the single-season ice (darker gray) is immediately evident. The multi-year ice was farther offshore this year than it is normally, but the surge of ice into M'Clure Strait and as far east as M'Clintock Channel is still visible. RADARSAT is a sophisticated Earth observation satellite developed by Canada and launched in 1995.

Pole on its way clear across the Arctic Ocean on a three-year drift.

The first American drifting station was set up in March 1952 on an ice island, ice platforms that have broken off of the ice shelves surrounding coasts—in effect a huge super-iceberg. This was the floating research station known as Fletcher's Ice Island T-3, after Joseph Fletcher, the commanding officer of the U.S. air force plane that found it. The island, with a research station on it that was occupied for some time, drifted in the Arctic Ocean for twenty-five years. Other American research facilities have been placed on other ice islands from time to time.

With the end of the Soviet era, there has been much more sharing of information than was the case in the past. In 2000, a new *Arctic Meteorological and Climate Atlas* was released, based on data from the United States, Russia, and other countries. It included much historical data that had previously been unavailable during the Soviet era.

One very practical use of this data is for the forecasting of ice conditions. The Arctic, it must not be forgotten, is, after all, an ocean, and as such it is a living entity, with its ice in flux, constantly on the move. MAP 297, page 193, shows how a lead can form from movement of the ice; it can just as readily close up again, as many Arctic explorers in the past found out to their cost.

Ice forecasting began in 1939, using charts of current ice conditions and a weather forecast. The accurate forecasting of ice conditions relies on a knowledge of the interaction of the atmosphere—particularly wind—and ice. The use of satellites has revolutionized ice forecasting because of the ability to obtain a reasonably regular, comprehensive, single time period picture of the existing ice conditions, the base from which a forecast is made. Then, with the next available set of satellite images, the forecast is immediately checkable and correctable. The maps on this page are made from satellite images; in one case (MAP 290, *left*) the map is a satellite photograph enhanced and outlined for clarity. The Canadian government, among others, issues regular ice condition maps and atlases, many of which are available on the Internet. MAPS 290, 291, and 292 are from published ice atlases.

A much denser network of data-collecting stations has also allowed the development of bathymetric maps of the Arctic Ocean floor—and finally the myth of more land has been proven false.

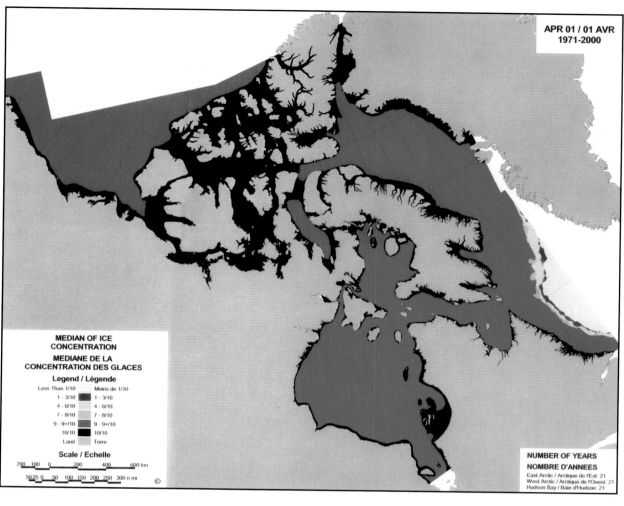

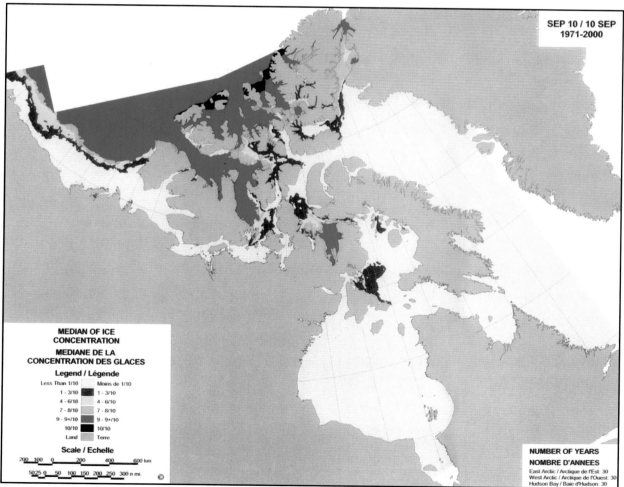

MAP 291 (*left*) and MAP 292 (*left, below*). Two maps from a sea ice atlas for northern Canada, showing the concentration of sea ice at the approximate date of maximum ice (*top*) and at the approximate date of minimum ice (*bottom*). Both maps are long-term medians. The 1 April map (*top*) is based on twenty-one years of observations, while the 10 September map (*bottom*) is based on thirty. White areas are not surveyed.

The Canadian government began publishing atlases showing ice conditions in 1980. The initial efforts were not computerized, and updating them was a laborious task. Digitization was completed in the late 1990s and the first computerized ice atlas, based on the thirty-year period 1971–2000, was published in 2001, for the east coast of Canada. The maps shown here are from the second computer-based atlas to be produced, for Hudson Bay and Arctic waters. The maps are produced by the Canadian Ice Service, a department of Environment Canada, from RADARSAT satellite images and aerial survey.

Although it should be stressed that these maps strictly apply only to the period covered, it is interesting to view them in light of the historical attempts to sail through the Northwest Passage. In this respect, the most striking feature is the long arm of 90 percent ice concentration that leads, even at the date of extreme minimum ice, from the main Beaufort Sea ice pack into M'Clure Strait, Viscount Melville Sound, and M'Clintock Channel. It was this multi-year ice that made life difficult for all those nineteenth-century explorers from Edward Parry onward who attempted to sail through the apparent "main highway" Northwest Passage route first found by Parry. It was this ice that foiled Robert M'Clure's attempt to complete his west-to-east transit of the Northwest Passage in 1850–53 without losing his ship. And it was this ice, surging south down M'Clintock Channel, that halted John Franklin's expedition in 1846 off the northwest tip of King William Island, with such disastrous results.

U.S. Coast and Geodetic Survey scientist Rollin Harris had at one time hypothesized the existence of land in the Arctic (MAP 250, page 164), and in 1911, using information about tides, had questioned whether the Arctic Ocean could be one large deepwater basin, a doubt supported later by others. An American oceanographer, L. V. Worthington, came to the same conclusion in 1951–52 based on temperature differences in water samples.

The Russian scientists were by this time able to prove there was a submarine ridge that traversed the Arctic Ocean from Ellesmere Island to the Novosibirskiye Ostrova. The ridge was not far off that of the land hypothesized by German geographer Augustus Petermann in 1868, which, he thought, extended from northern Greenland to Ostrov Vrangelya (see MAP 225, page 146).

The ridge was named the Lomonosov Ridge after an eighteenth-century scientist and poet. It is the major undersea feature of the Arctic Ocean, separating the generally deeper waters of the Europe–western Asia side of the ocean from the somewhat shallower waters of the North America–eastern Asia side. A modern bathymetric map such as MAP 294, *right*, shows this well. (It is the more-or-less solid green line crossing the ocean diagonally.)

The first contour map of the Arctic was made in 1958, after USS *Nautilus* took sonar scans under the ice. In the 1960s, a longe-range side-scanning sonar imaging system was developed. Towed behind ships, it enabled the undersea topography to be determined with greater precision. Modern developments of this system, called side-scan or multibeam sonar, allow wide tracts of seabed to be scanned at a single pass and, with the help of a computer, to be viewed in three-dimensional form. In the Arctic, towing side-scan sonar behind a ship is difficult, and the best results have been achieved using submarines.

Recently a summary bathymetric map that incorporates all the latest information has been digitally compiled. This is the *International Bathymetric Chart of the Arctic Ocean,* or IBCAO, shown as MAP 294. Although the general outlines of undersea topography have changed since the earlier summary effort, the *General Bathymetric Chart of the Oceans,* or GEBCO (MAP 288, page 187), there are very considerable improvements in resolution. This is principally due to the vastly improved coverage of

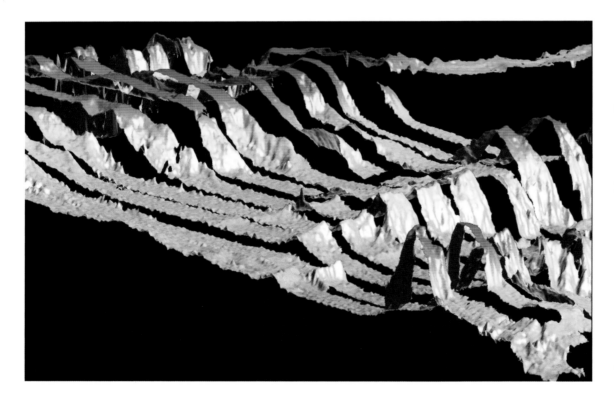

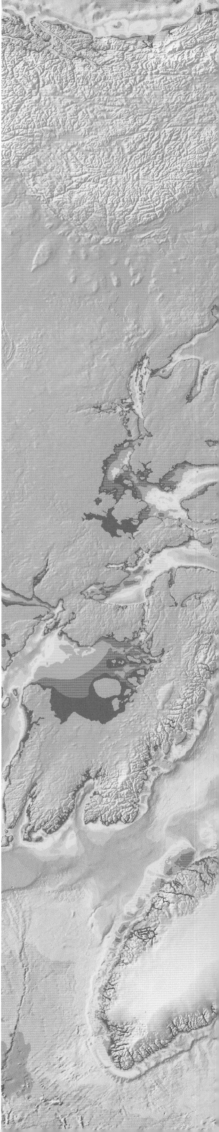

MAP 293 (*above*).
A part of the southern Lomonosov Ridge (khrebet Lomonosova), a major undersea topographic feature in the Arctic Ocean. The submarine ridge divides the ocean into two parts (see MAP 294, *right*). This image was obtained using *Sea*MARC–12 swath mapping sonar and a high-resolution sub-bottom profiler mounted on the hull of a U.S. navy fast-attack submarine in 1999. In order to cover more area than might otherwise be possible in the time allowed, there are "gaps" between the surveyed strips of the sea bottom, giving the image a surreal effect. The colors show different depths and have been added by a computer. This modern work of art nevertheless holds much detailed and valuable information for scientists.

MAP 294 (*right*).
This superb image is the *International Bathymetric Chart of the Arctic Ocean,* shortened in the acronym-speak that oceanographers love to IBCAO. Available on-line, and digitally constructed and updated, it is intended as a summary of the most up-to-date information on the topography of the Arctic seabed. As with cartography in general, the use of digital information has revolutionized the ability to keep maps up-to-date. IBCAO covers the area north of 64° N. This is version 1.0, which appeared in 2001. Eight countries contribute information: Canada, Denmark, Germany, Iceland, Norway, Russia, Sweden, and the United States. The scale (at top) goes from purple (deepest) to red (shallowest). The Lomonosov Ridge is green.

MAP 295 (*left*).
Another multibeam sonar image of part of the Arctic sea floor, this of the Gakkel Ridge (krebhet Gakkel), off the northeastern tip of Greenland. This undersea ridge was first discovered by soundings through the ice in 1948 by Russian scientist Y. Y. Gakkel. Currently under considerable investigation by scientists, the Gakkel Ridge is a "spreading ridge," a fracture line in the Earth's crust where an upwelling of magma causes the movement outwards of the sea floor either side. Lava flows and even an undersea volcano have been discovered on the Gakkel Ridge. Investigation of the ridge increases our understanding of global tectonics.

MAP 296 (*below*).
The North Pole, as seen in a true-color NASA satellite image from 5 May 2000. The image was acquired by a moderate-resolution imaging spectroradiometer (MODIS) on board a TERRA satellite. The sea ice appears white, and areas of open water or recently refrozen sea surface appear black. The many leads in the ice stand out as black. The Pole is shown by a superimposed convergence of meridians. The white areas towards the bottom of the image are clouds; MODIS cannot see through clouds, as it utilizes visible spectrum wavelengths.

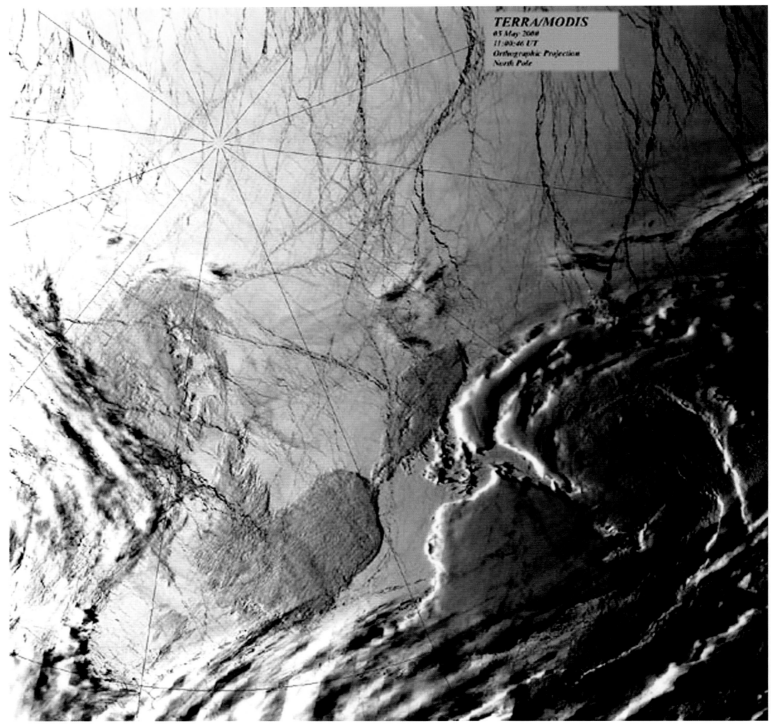

modern multibeam technology. Even a dense network of soundings could never provide coverage of an area in the same detail.

Today's oceanographic and geophysical research continues to add to the knowledge of the sea floor in the Arctic Basin. The maps produced, as some of those shown here illustrate, are often visually spectacular as well as being useful scientifically.

There is today considerable concern over the possibility of global warming. A comprehensive database of information now available from the Arctic is proving to be invaluable in research to confirm a universal warming trend and determine its likely effects.

Recent reports in fact show that the Arctic may be warming faster than the rest of the Earth. The average Arctic surface air temperature over the past century, and during the past few decades in particular, has been exceptionally high. This is thought to be caused by oscillations in atmospheric pressure causing more westerly winds, which warm the Arctic. Permafrost is melting in some areas at a rate of 20 cm per year. In the Siberian cities of Noril'sk and Yakutsk almost three hundred apartment buildings have been damaged by melting permafrost.

Data collected in the Arctic can be used to construct global climate models to help determine how the effects of such warming might be mitigated. Given current trends, ice-free Arctic summers in the not-too-distant future seem a distinct possibility. This would allow, for example, an easy transit through the Northwest and Northeast Passages—raising sovereignty issues—and make more areas available for fishing. In addressing these challenges, scientific information gathered from the Arctic regions will have a vital role to play.

Our knowledge of the Arctic has changed rapidly in the last hundred years or so as technology advanced and resources allowed. Theories of an ice-free open polar sea did not die until the 1870s, yet these were still just as much guesswork as sixteenth-century theories of an ice-free Arctic and a clear passage over the Pole. The Arctic is still an unforgiving environment—witness the fact that NASA has been using Canada's Devon Island to prepare astronauts for a voyage to Mars—but with today's technology the dangers have been minimized and the adventure of attaining the Pole has become merely a matter of being able to afford the cost. Today tourists can relax as their icebreaker crunches its way northwards to the Pole.

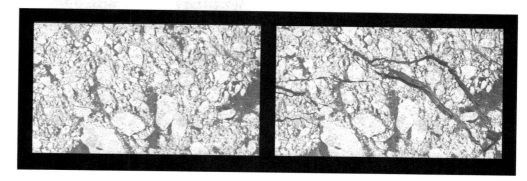

MAP 297 (*above*).
The ever-changing map of the Arctic ice pack. These two RADARSAT images are of the same 96 x 128 km area of the Beaufort Sea, north of Alaska, captured nine days apart. The whiter areas are older, thicker ice, and the darker areas are young, recently formed ice, or water. The later image, that on the right, shows that in just nine days the movement of the ice has created large, extensive cracks or leads. Ice cracks are no ordinary cracks in the Arctic; some have been found that were 2 000 km long. RADARSAT takes a complete image of the Arctic every three days.

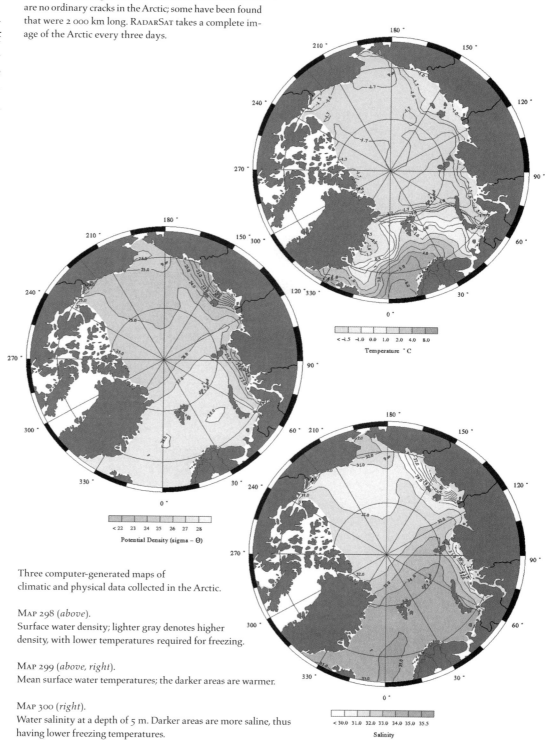

Three computer-generated maps of climatic and physical data collected in the Arctic.

MAP 298 (*above*).
Surface water density; lighter gray denotes higher density, with lower temperatures required for freezing.

MAP 299 (*above, right*).
Mean surface water temperatures; the darker areas are warmer.

MAP 300 (*right*).
Water salinity at a depth of 5 m. Darker areas are more saline, thus having lower freezing temperatures.

Map Catalog

Key to frequently cited sources:

BL	British Library
BNF	Bibliothèque nationale de France
CT	Cameron Treleaven Collection, Calgary
HBCA	Hudson's Bay Company Archives, Winnipeg
JFBL	James Ford Bell Library, University of Minnesota
LC	Library of Congress
NARA	National Archives and Records Administration (U.S. National Archives)
NLC	National Library of Canada*
NMC	National Map Collection, National Archives of Canada*
PGM	*Petermann's Geographische Mittheilungen*
PRO	Public Record Office, U.K.
UKHO	United Kingdom Hydrographic Office, Taunton, U.K.

*Now combined as the Library and Archives of Canada.

Unreferenced maps are from other private collections.

MAP 1 (HALF-TITLE PAGE)
Septentrio Nalium Terrarum descriptio
Gerard Mercator, 1633 or 1635
CT

MAP 2 (TITLE PAGES)
The Arctic Regions showing the North-West Passage of Cap. R. McClure and other Arctic Voyagers
A. Fullarton & Co., 1856

MAP 3 (PAGE 4)
[Untitled map of the polar regions]
Petrus Bertius, 1618
Beach Maps/Alec Parley

MAP 4 (PAGE 5)
Poli Arctici, et Circumiacentium Terrarum Descriptio Novissima. Per F. de Wit
Frederick de Wit, 1715
NMC 21059

MAP 5 (PAGE 6)
Discovery of the North Pole and the Polar Gulf Surrounding It
John B. Shelden, 1869, published 1870
Glenbow-Alberta Institute, Library: An.65.42 Anon

MAP 6 (PAGE 7)
[Norse conception of Arctic geography]
From: Nansen, 1911

MAP 7 (PAGE 7)
Siurdi Stephanii terrarum hyperborearu delineatio Ano 1570
Sigurdur Stefánsson, 1670, copy of a 1590 map
Det Kongelige Bibliotek, Copenhagen:
gl. kgl. saml. 2881 4° (10V)

MAP 8 (PAGE 7)
Gronlandia Iona Gudmundi Islandi
Jón Gudmonson, c.1640
Det Kongelige Bibliotek, Copenhagen:
gl. kgl. saml. 2881 4° (11R)

MAP 9 (PAGE 8)
[World map of Macrobius]
c.1485
Facsimile-Atlas, Nordenskiöld, 1889

MAP 10 (PAGE 8)
Universalior Cogniti Orbis Tabula Ex Recentibus Confecta Observationibus
Johann Ruysch, c.1507
NMC 19268

MAP 11 (PAGE 8)
[World map of Gerard Mercator]
1538
Facsimile-Atlas, Nordenskiöld, 1889

MAP 12 (PAGE 9)
[Globe gores from the "Ambassadors' Globe"]
c.1530
Facsimile-Atlas, Nordenskiöld, 1889

MAP 13 (PAGE 9)
[World map by Martin Waldseemüller]
1513
Facsimile-Atlas, Nordenskiöld, 1889

MAP 14 (PAGE 10)
[MS map of northern Scandinavia]
William Borough, 1560
BL: Royal 18.D.3, f. 123 recto and 124 verso

MAP 15 (PAGE 10)
[MS map of the voyage of Arthur Pet and Charles Jackman to the Kara Sea]
Journal of Hugh Smyth, 1580
BL: Cotton MS Otho E.VIII, f. 78

MAP 16 (PAGE 11)
Russiae, Moscoviae et Tartariae Descriptio
Anthony Jenkinson, 1562
From: Abraham Ortelius,
Theatrum Orbis Terrarum, 1570
LC: G1006.T5 1570b vault

MAP 17 (PAGE 11)
[Inset map of the Arctic from map of the world]
Gerard Mercator, 1569 (facsimile)

MAP 18 (PAGE 12)
[Part of map of North Atlantic]
William Borough, 1576, with additions by Martin Frobisher of his discoveries, 1576
Hatfield House, Courtesy the Marquess of Salisbury

MAP 19 (PAGES 12–13)
Typus Orbis Terrarum
Abraham Ortelius, 1570
From: *Theatrum Orbis Terrarum*
LC: G1006.T5 1570b vault

MAP 20 (PAGE 13)
Carta da Navigar di Nicolo et Antonio Zeni Furono In Tramontana Lano MCCCLXXX
From: Niccolò Zeno, the younger (compiler), *Dello scoprimento dell' Isole Frislanda, Eslanda, Engrouelanda, Estotilanda & Icaria*, 1558
Facsimile-Atlas, Nordenskiöld, 1889

MAP 21 (PAGES 14–15)
Septentrionalium Regionum Descrip.
Abraham Ortelius, 1570
From: *Theatrum Orbis Terrarum*
LC: G1006.T5 1570b vault

MAP 22 (PAGE 16)
Humfray Gylbert knight his charte
John Dee, 1583
Free Library of Philadelphia

MAP 23 (PAGE 17)
Asiae Novissima Tabuia
Gerard de Jode, 1578

MAP 24 (PAGE 17)
Europam ab Asia et Africa
Peter Plancius (engraved by Baptiste Doetecum), 1594

MAP 25 (PAGE 18)
[Untitled map of Yugorskiy Shar (Nassau Strait) and ships of Barentsz's second voyage]
Gerrit de Veer, 1598
From: *Vraye description de trois voyages*, 1600 (French edition of De Veer's book)
NLC: G690 1594 V414 1609 fol., p. 10 recto

MAP 26 (PAGE 18)
Caert van Nova Zembla, de Weygats, de custe van Tartarien en Ruslandt tot Kilduyn toc
Gerrit de Veer, 1598
From: *Vraye description de trois voyages*, 1600 (French edition of De Veer's book)
NLC: G690 1594 V414 1609 fol., p. 35 recto

MAP 27 (PAGES 18–19)
Deliniatio cartæ trium navigationum per Batavos, ad Septentrionalem plagant, Norvegiæ, Moscoviæ et novæ Semblæ
Willem Bernardo (engraved by Baptista Doetecum), 1598
NMC 21063

MAP 28 (PAGE 20)
[Untitled map of the Arctic]
H. Megiser, 1613
Elmer E. Rasmuson Library, University of Alaska, Fairbanks: G 3270 (1613)

MAP 29 (PAGE 20)
Descriptio Novæ Zemblæ
Inset in: *Europam ab Asia et Africa*
Peter Plancius (engraved by Baptiste Doetecum), 1594

MAP 30 (PAGE 21)
The coast of groineland with the latitudes of the havens and harbors as I founde them
James Hall, 1605
BL: Royal 17.A.48, f. 10 verso

MAP 31 (PAGE 22)
Isola di Mayen Scoperta l'Anno 1614
Vincenzo Coronelli, 1694

MAP 32 (PAGE 22)
[Part of] *Carte naurique de bords de Mer du Nort, et Noruest, mis en longitude latitude et en leur route selon les rins de vent par Hessel Gerritsz*
Hessel Gerritz, 1628
BNF

MAP 33 (PAGES 22–23)
[Map of "Greneland" (Spitsbergen) with whaling illustrations]
From: Edward Pellham, *God's Power and Providence: Shewed, In the Miraculous Preservation and Deliverance of eight Englishmen, left by mischance in Green-land Anno 1630, nine moneths and twelve dayes*, London, 1631

MAP 34 (PAGE 24)
[Oceanus Septentrionalus]
Joris Carolus Flandro, 1614
BNF

MAP 35 (PAGE 24)
[Part of] *Carte naurique de bords de Mer du Nort, et Noruest, mis en longitude latitude et en leur route selon les rins de vent par Hessel Gerritsz*
Hessel Gerritz, 1628
BNF

MAP 36 (PAGE 25)
Poli Arctici et Circumiacentium Terrarum Descriptio Novissima
Henricus Hondius, 1642
BL: Maps.C.4.TAB.1, Tome 1, Map 2

MAP 37 (PAGE 26)
Tabula Nautica, qua repræsentātur oræmaritimæ meatus ac freta, noviter a H Hudsono Anglo ad Caurum supra Novam Franciam indagata Anno 1612
Hessel Gerritz, 1612
From: Hessel Gerritz, *Descriptio ac delineatio Geographica Detectionis Freti* (map by Henry Hudson?), 1612

MAP 38 (PAGE 27)
The North Part of America
Henry Briggs, 1625
NMC 6582

194

MAP 39 (PAGE 27)
[Chart of Hudson Strait]
William Baffin, 1615
BL: Add MS 12206, f. 6

MAP 40 (PAGE 28)
[Map of Hudson Bay and Strait,
called the "Stockholm Chart"]
Jens Munk, 1624
From: *Danish Arctic Voyages*,
Hakluyt Society, London, 1896

MAP 41 (PAGE 29)
[Part of] *Carte nauriqve de bords de Mer du Nort,
et Noruest, mis en longitude latitude et en leur route
selon les rins de vent par Hessel Gerritsz*
Hessel Gerritz, 1628
BNF

MAP 42 (PAGE 30)
A Map of Groenland
Anon., 1732
From: Awnsham and John Churchill (compilers),
A Collection of Voyages and Travels, Vol. 2, facing p. 399,
1732

MAP 43 (PAGE 30)
*The Platt of Sailing for the diſcoverye of a Passage
into the South Sea 1631 1632*
Thomas James, 1631–32
BL: Add MS 5415, G.1

MAP 44 (PAGE 31)
[Map of the north polar region]
Luke Foxe, 1635
From: *North-west Fox or, Fox from the North-west Passage*,
B. Alsop and Tho. Fawcet, London, 1635

MAP 45 (PAGE 31)
Le Canada faict par le Sʳ de Champlain
Pierre du Val, 1677
NMC 14080

MAP 46 (PAGE 32)
Groenlande
Alain Mallet, 1685

MAP 47 (PAGE 32)
*Nieuwe wassende graadt kaert Van d'noorder
Zee Custen van America*
Jacob Robyn, 1683
From: *Zee Atlas*, 1683
NMC 7293

MAP 48 (PAGE 33)
[Terre Artiche] polar map
Vincenzo Coronelli, 1692
NMC 8244

MAP 49 (PAGE 34)
*Caerte van't Noorderste Russen, Samojeden, ende
Tingoesen landt: alsoo dat vande Russen afghetekent,
en door Isaac Massa vertaelt is*
From: Hessel Gerritz, *Beschrijvinghe
Van de Samoyeden Landt*, 1612

MAP 50 (PAGES 34–35)
Asiae Recentissima Delineatio
Johann Baptist Homann, 1707
JFBL

MAP 51 (PAGE 35)
Nova et Accuratissima Totius Terrarum
Joan Blaeu, 1664
LC: G3200 1664.B5 TIL vault

MAP 52 (PAGE 35)
*Niewe Lantkaerte van het Noorder en
Ooster deel van Asia* (northeast sheet)
Nicholaas Witsen, 1687
JFBL

MAP 53 (PAGE 35)
*Verbeelding hoe het ys van Nova Zemla na Spitsbergen,
in den jaare 1676 vast geset is geweest*
Nicholaas Witsen, 1676

MAP 54 (PAGES 36–37)
Generalis Totius Imperii Russorum
Johann Baptist Homann, 1723
JFBL

MAP 55 (PAGE 36)
[Map of eastern Siberia]
Gerhard Müller, based on the reports
of Semen Dezhnev, 1736
From: Efimov, 1964, #48

MAP 56 (PAGE 36)
*A Map of the Country which Captⁿ· Beerings past
through in his Journey from Tobolsk to Kamtschatka*
Joseph-Nicolas De L'Isle, based on a map by Jean-Baptiste
Bourguignon d'Anville
From: Jean Baptiste du Halde, *The General History of
China*, 1736

MAP 57 (PAGE 37)
[Russian map of the discoveries of Vitus Bering]
Anon., c.1729
Kungl. Biblioteket (Royal Library), Stockholm

MAP 58 (PAGE 38)
Carte Des Pais Habités par les Samojedes et Ostiacs
Jacques-Nicolas Bellin, 1754

MAP 59 (PAGE 39)
[Russian coast south of Noyava Zemlya]
Stepan Malygin, 1736
From: Efimov, 1964, #81

MAP 60 (PAGE 39)
[Taymyr Peninsula]
Khariton Laptev and Semen Chelyuskin, 1743
From: Efimov, 1964, #89

MAP 61 (PAGE 40)
[Russian coast from the Lena to Kolyma Rivers]
Dmitri Laptev, c.1742
From: Efimov, 1964, #86

MAP 62 (PAGE 40)
[Summary map, Great Northern Expedition]
Anon., c.1742
From: Efimov, 1964, #91

MAP 63 (PAGE 41)
Carte des Nouvelles Découvertes Au Nord de la Mer du Sud
Philippe Buache, published by Joseph-Nicolas De L'Isle,
1752
NMC 21056

MAP 64 (PAGE 41)
*Chart containing part of the Icy Sea with the
adjacent Coast of Asia and America*
John Green (Bradock Mead), 1753
Thomas Jefferys, London, 1753

MAP 65 (PAGE 41)
[Eastern Siberia]
Nikolay Daurkin, 1765
From: Efimov, 1964, #128

MAP 66 (PAGE 42)
[Russian map of the discoveries of Nikolay Daurkin]
Fedor Plenisner, 1765
From: Efimov, 1964, #132

MAP 67 (PAGE 42)
*Carte Generale Des Decouvertes de l'Amiral de Fonte
et autres navigateurs Espagnols Anglois et Russes pour
la recherche du passage a la Mer du Sud*
Didier Robert de Vaugondy, 1775

MAP 68 (PAGE 42)
Carte Réduite de l'Ocean Septentrional
Jacques-Nicolas Bellin, 1766
LC: G1059.B43 1772

MAP 69 (PAGE 43)
*Chart containing part of the Icy Sea with the
adjacent Coast of Asia and America*
John Green (Bradock Mead), updated
by Thomas Jefferys, 1775
From: *American Atlas*, Thomas Jefferys, London, 1775

MAP 70 (PAGE 43)
[Novosibirskiye Ostrova and mainland coast]
Pyotr Pshenitsyn, 1811
From: Efimov, 1964, #187

MAP 71 (PAGE 44)
Nova et Accuratissima Totius Terrarum
Joan Blaeu, 1664
LC: G3200 1664.B5 TIL vault

MAP 72 (PAGE 44)
[Map of the coast and rivers flowing into the western
side of Hudson Bay and the Arctic Ocean]
James Knight, 1719 [?], with later additions
HBCA: G1/19

MAP 73 (PAGE 45)
[Hudson Bay and Strait] *Made by Saml: Thornton at the
Signe of the Platte in the Minories London Anno: 1709*
Samuel Thornton, 1709
HBCA: G2/2

MAP 74 (PAGE 45)
*Chart of Hudson's Bay &Straits, Baffin's Bay, Strait Davis
& Labrador Coast*
Christopher Middleton, 1743
NMC 27782

MAP 75 (PAGE 46)
*Chart of the Seas, Straits, &c. thro' which his Majesty's
Sloop* Furnace *pass'd for discovering a Passage from
Hudsons Bay, to the South Sea*
John Wigate, 1746
BL: Maps 70095.(7)

MAP 76 (PAGE 46)
*A New Map of Part of North America From The Latitude of 40
to 68 Degrees. Including the late discoveries made on Board the
Furnace Bomb Ketch in 1742. And the Western Rivers & Lakes
falling into Nelson River in Hudson's Bay as described by
Joseph La France a French Canadese Indian who Travaled into
those Countries and Lakes for 3 years from 1739 to 1742*
From: Arthur Dobbs, *An Account of the Countries
Adjoining to Hudsons Bay*, London, 1744

MAP 77 (PAGE 47)
Wager Straits Discovered in 1742
[Inset in MAP 75]
John Wigate, 1746
BL: Maps 70095.(7)

MAP 78 (PAGE 47)
*Chart of the Coast where a North West Passage
was attempted*
Henry Ellis, 1748
NMC 21058

MAP 79 (PAGE 48)
*North America with Hudson's Bay and Straights
Anno 1748*
Richard Seale (engraver), 1748
HBCA: G4/20b

MAP 80 (PAGE 48)
*A General Map of North America/
Cartes Generales de L'Amerique Septentrionale*
John Roque, 1762
NMC 116793

MAP 81 (PAGE 49)
*A Map of the Icy Sea in which the several Communications
with the Land Waters and other new Discoveries are exhibited*
Philippe Buache, 1752
From: David Henry, *An Historical Account of All the
Voyages Round the World Performed by English Navigators*,
Vol. 4, London, 1773–74
Yale University Library

MAP 82 (PAGE 50)
Tartariae Sive Magni Chami Regni tÿpus
Abraham Ortelius, 1570
From: *Theatrum Orbis Terrarum*
LC: G7270 1570.O7 vault

MAP 83 (PAGE 50)
*A Map of the New Northern Archipelago discover'd by
the Russians in the seas of Kamtschatka and Anadir*
Jacob von Stählin, 1774
From: *An Account of the New Northern Archipelago*, 1774

MAP 84 (PAGE 50)
[Bering Strait]
William Bayley, 1778
Vancouver Maritime Museum

MAP 85 (PAGE 51)
Carte de L'Ocean Pacifique au Nord de l'Equateur, et des Cotes qui . . . Espagnols, les Russes, et les Anglois, jusqu'en 1780
Tobie Conrad Lotter, 1781
NMC 8607

MAP 86 (PAGE 51)
Chart of the NW Coast of America and the NE Coast of Asia explored in the years 1778 and 1779
From: James Cook, Voyage to the Pacific Ocean, 1784

MAP 87 (PAGE 52)
A Map Exhibiting all the New Discoveries in the Interior Parts of North America
Aaron Arrowsmith, 1795, with additional patch c.1797
NMC 97818

MAP 88 (PAGE 52)
A Plan of the Coppermine River by Sam^{el} Hearne, July 1771
Samuel Hearne, 1771
JFBL

MAP 89 (PAGE 53)
A Map of Part of the Inland Country to the N^h W^t of Prince of Wales Fort H^s, B^y, Humbly Inscribed to the Gov^{nr} Dep^y, Gov^{nr} and Committee of the Honble Hud^{ns} B^y Comp^y By their Hon^{rs}, moste obedient humble servant. Saml Hearne; 1772
Samuel Hearne, 1772
HBCA: G2/10

MAP 90 (PAGE 53)
Map of the Lands around The North Pole by Dalrymple 1789
Alexander Dalrymple, 1772

MAP 91 (PAGE 53)
[Copy of a map of western Canada and the North Pacific Ocean thought to have been prepared by Peter Pond for presentation by Alexander Mackenzie to the Empress of Russia] Copied from the original signed P. Pond Araubaska 6th December 1787 (marginal notation)
[Peter Pond], 1787
PRO: CO 700 America North and South 49

MAP 92 (PAGE 54)
A Map of America between Latitudes 40° & 70° North and Longitudes 40° & 180° West Exhibiting Mackenzie's Track From Montreal to Fort Chipewyan & from thence to the North Sea in 1789, & to the North Pacific Ocean in 1793
Anon. (Aaron Arrowsmith[?]/David Thompson/Alexander Mackenzie), c.1800
PRO: CO 700 Canada 59A

MAP 93 (PAGE 54)
"Chart called Mackenzie's Map, illustrative of his tract from Athabasca Lake down Mackenzie River to the North Sea"
Anon. (Alexander Mackenzie), c.1789
PRO: CO 700 America North and South 54

MAP 94 (PAGE 55)
[Map of the polar regions]
Mikhail Lomonosov, 1763
From: Efimov, 1964, #143

MAP 95 (PAGE 55)
Chart shewing the Track of His Majesty's Sloops Racehorse and Carcass during the Expedition towards the North Pole 1773
From: Phipps, 1774

MAP 96 (PAGE 56)
Chart shewing the different Courses Steered by His Majesty's Sloop Racehorse From July 3^d. to August 22^d.
From: Phipps, 1774

MAP 97 (PAGE 56)
Chart Shewing the Track of the Expedition under the Command of Captn. Buchan, R.N. and the Position of the Packed Ice in June and September, 1818
From: Beechey, 1843

MAP 98 (PAGE 57)
Plan of the Bay in which the Racehorse and Carcass Were Inclosed by the Ice From July 31st to August 10th 1773
Philip D'Auvergne, 1773
BL: Maps.K.Mar.II.4

MAP 99 (PAGE 58)
A Map of East and West Greenland Comprising the Extent of the Greenland Seas and an Outline of the Polar Ice
William Scoresby Jr., 1815
From: William Scoresby, "On the Greenland or Polar Ice," Memoirs of the Wernerian Society, Vol. 2, pp. 328–36, 1815

MAP 100 (PAGE 58)
British Possessions in America
Samuel Lewis and Aaron Arrowsmith, 1804

MAP 101 (PAGE 58)
Nord America
Martin Hartl, after a sketch by Joseph Marx Freiherrn von Leichenstern, 1805
LC: G3300 1805.N6 TIL

MAP 102 (PAGE 59)
[Polar map, with "Probable direction of Ice-bergs"]
[John Barrow], 1817
From: Quarterly Review, October 1817

MAP 103 (PAGE 59)
Map of the Countries Around the North Pole According to the Latest Discoveries
From: Daines Barrington, The Possibility of Approaching the North Pole Asserted, T. and J. Allman, London, 1818

MAP 104 (PAGE 60)
Map of the Arctic Regions
John Barrow, 1818
From: Barrow, 1818

MAP 105 (PAGE 61)
A General Chart Shewing the Track and Discoveries of H.M. Ships Isabella & Alexander to Davis's Straits & Baffins Bay in an attempt to Discover a Passage into the Pacific Ocean
John Ross, 1819
From: Ross, 1819

MAP 106 (PAGE 61)
Track of H.M. Ships Isabella & Alexander from 29th August to the 1st September 1818
From: Ross, 1819

MAP 107 (PAGE 61)
A Chart of the Route of His Majesty's Hired Armed Vessel The Alexander On a Voyage of Discovery to the Arctic Regions Performed in the Year 1818
From: Anon. [Alexander Fisher], no date [1819]

MAP 108 (PAGES 62–63)
Lancaster Sound by Capt. W. H. Parry (sic)
No. 43 C [title on back of map]
[Parry's map of his discoveries, Baffin Bay to Melville Island]
William Edward Parry, 1819–20
BL: Maps.188.o.1 (3)

MAP 109 (PAGE 64)
General Chart shewing the track of H.M. Ships Hecla & Griper, from the Orkneys to Melville Island, North Georgia
From: Parry, 1821

MAP 110 (PAGE 64)
Survey of Winter Harbour Melville Island June 1820
William Edward Parry, 1821
From: Parry, 1821

MAP 111 (PAGE 65)
Îles George
Phillippe Marie Guillaume van de Meulen, 1825
CT

MAP 112 (PAGE 65)
Route From Lake Athapescow, to the Coppermine River, North America Surveyed by the Northern Land Expedition under the command of Lieu^t Franklin, R.N.
John Franklin, 1820–21
PRO: CO 700 Canada 79, Map 1

MAP 113 (PAGES 66–67)
Carte d'une partie de l'Ocean Arctique et de l'Amerique Septentrionale
Pierre Lapie, 1821
From: Nouvelles annales des voyages, de la géographie et de l'histoire, ou Recueil des relations originales inédites, Vol. 11, 1821. Gide fils, Paris, 1819–26
NLC

MAP 114 (PAGE 66)
Carte Generale Des Decouvertes de l'Amiral de Fonte et autres navigateurs Espagnols Anglois et Russes pour la recherche du passage a la Mer du Sud
Didier Robert de Vaugondy, 1775

MAP 115 (PAGE 68)
An Outline to Shew the Connected Discoveries of Captains Ross, Parry, & Franklin In the Years 1818, 19, 20, and 21
From: Franklin, 1823

MAP 116 (PAGE 69)
Chart of Part of the North Eastern Coast of America and its adjacent Islands shewing the Track and Discoveries of His Majesty's Ships Fury and Hecla in search of a North West Passage under the Command of Captn W. E. Parry in the Years 1822–23
J. Bushnan and William Edward Parry, 1822–23
From: Parry, 1824

MAP 117 (PAGE 69)
Chart of Part of the North Eastern Coast of America and its adjacent Islands shewing the Track and Discoveries of His Majesty's Ships Fury and Hecla in search of a North West Passage under the Command of Captn W. E. Parry in the Years 1822. 1823. 1824
J. Bushnan and William Edward Parry, 1822–23
From: Parry, 1824

MAP 118 (PAGE 70)
Eskimaux Chart No. 1 Drawn by Iligliuk at Winter Island, 1822. The Original in the possession of Cap.^{n} Lyon
Iligliuk and William Edward Parry, 1822
From: Parry, 1824

MAP 119 (PAGE 70)
Eskimaux Chart No. 3 The shaded parts drawn by Ewerat, at Winter Island. 1822. The Original in the possession of Cap.^{n} Parry
Ewerat and William Edward Parry, 1822
From: Parry, 1824

MAP 120 (PAGE 70)
Eskimaux Chart No. 2 The shaded parts drawn by Iligliuk at Winter Island 1822. The Original in the possession of Cap.^{n} Parry
Iligliuk and William Edward Parry, 1822
From: Parry, 1824

MAP 121 (PAGE 71)
A Chart of the Arctic Regions
John Harris, April 1827
NMC 136209

MAP 122 (PAGE 71)
General Chart shewing the track of H.M. Ships Fury and Hecla on a Voyage for the Discovery of a North West Passage, A.D. 1821-22-23
From: Parry, 1824

MAP 123 (PAGE 72)
A Chart of the Discoveries of Captains Ross, Parry & Franklin, in the Arctic Regions In the Years 1818, 1819, 1820, 1821 & 1822
With inset: Capt Franklin's Journey from Coppermine River to the Head of Bathurst Inlet, & Return by Hood's River
John Thomson, 1827
NMC 24922

MAP 124 (PAGE 73)
Chart of Hudson's Strait & Sir Thos Rowe's Welcome Shewing the Track & Discoveries of HMS Griper in an attempt to reach Repulse Bay by the Welcome A.D. 1824 Under the command of Capt G.F. Lyon
Edward Nicholas Kendall, 1824
From: Lyon, 1825

MAP 125 (PAGE 73)
Prince Regent's Inlet Re-Surveyed AD 1824–25
From: Parry, 1826

MAP 126 (PAGES 74–75)
North America [northern coast]
Aaron Arrowsmith, 1824, with MS additions and patches by the Hudson's Bay Company
HBCA: G4/31

MAP 127 (PAGE 75)
Chart of Part of the North West Coast of America From Point Rodney to Point Barrow By Captain F. W. Beechey R.N. F.R.S. in His Majesty's Ship Blossom
From: Frederick William Beechey, *Narrative of a voyage to the Pacific and Beering's Strait, to co-operate with the Polar expeditions : performed in His Majesty's ship Blossom, under the command of Captain F.W. Beechey . . . in the years 1825, 26, 27, 28*, H. Colburn & R. Bentley, London, 1831

MAP 128 (PAGE 76)
Route of the Land Arctic Expedition under the command of Lieutenant [?] J. Franklin R.N. from Great Bear Lake River to the Polar Sea Surveyed and drawn by Mr. E. A. Kendall R.N., Assistant Surveyor
Edward Kendall, John Richardson, and John Franklin, 1825
PRO: MPG 386

MAP 129 (PAGE 77)
Chart Drawn by the Natives
Anon. [Ikmalick and Apelaglu?], 1829–30
From: Ross, 1835

MAP 130 (PAGE 77)
[Plan of Somerset House]
From: Ross, 1835

MAP 131 (PAGE 78)
To His Most Excellent Majesty William IVth King of Great Britain Ireland &c This Chart of the Discoveries made in The Arctic Regions, in 1829, 30, 31, 32, & 33 is Dedicated with his Majesty's gracious permission by his Majesty's Loyal and devoted Subjects John Ross, Captain Royal Navy. James Clark Ross, Commander Royal Navy.
John Ross, 1834
From: Ross, 1835

MAP 132 (PAGE 79)
Sketch of North America shewing the Proposed Route of Cap.t Back
George Back, 1833
From: *Journal of the Royal Geographical Society*, Vol. 3, 1833

MAP 133 (PAGE 79)
Sketch Shewing the Route of the recent Arctic Land Expedition
George Back, 1835
From: *Journal of the Royal Geographical Society*, Vol. 5, 1835

MAP 134 (PAGE 80)
Hudson's Strait, Shewing the Track of HMS Terror in 1836-7
George Back, 1837
From: *Journal of the Royal Geographical Society*, Vol. 7, 1837

MAP 135 (PAGE 80)
[Map of the Arctic coast showing the discoveries of Thomas Simpson and Peter Warren Dease, 1838–39]
Anon., 1839
HBCA: G1/5

MAP 136 (PAGE 81)
[Untitled map of the Arctic regions]
From: Barrow, 1846

MAP 137 (PAGE 81)
[Map of Committee Bay]
John Rae, 1847
HBCA: G1/177

MAP 138 (PAGE 82)
Discoveries of the Hon.ble Hudson's Bay Company
John Arrowsmith, 1848
CT

MAP 139 (PAGES 82–83)
Chart Illustrative of the Voyages and Travels of Captains Ross, Parry, Franklin and Back
John Arrowsmith, 1848
NMC 27802

MAP 140 (PAGE 84)
A Chart of Baffin's Bay with Davis & Barrow Straits [with MS additions from James Clark Ross's expedition of 1848–49]
PRO: MFQ 1/199 (3)

MAP 141 (PAGE 85)
Chart of Part of the Arctic Regions as known in 1845, being a copy of the chart supplied to the Franklin Expedition
From: J. Browne, *The Northwest Passage and the Search for Sir John Franklin*, London, 1860, reproduced in Charles F. Hall, 1879

MAP 142 (PAGE 85)
Sketch Map of the Arctic Regions, at the Time of Franklin's Last Expedition & of His Supposed Track
From: Francis Leopold M'Clintock, 1859, reproduced in Hall, 1879

MAP 143 (PAGE 86)
Arctic America Sheet II [with MS additions showing position of James Clark Ross in 1848–49 and proposed positions for 1849–50]
PRO: MFQ 1/199 (5)

MAP 301.
A polar map published by French mapmaker Pierre du Val in 1676. It reflects the seventeenth-century English voyages to Hudson Bay. Off the northern coast of Russia, a misshapen Novaya Zemlya is based on Willem Barentsz's voyages of the late sixteenth century. Here a Northeast Passage seems possible, but not a Northwest Passage.

MAP 144 (PAGE 86)
Arctic America Sheet II [with MS addition of survey of west coast of Somerset Island by James Clark Ross in 1849]
PRO: MPI 1/313 (3)

MAP 145 (PAGE 86)
Arctic America Sheet II [with printed west coast of Somerset Island]
PRO: MPI 1/313 (4)

MAP 146 (PAGE 87)
Chart illustrating Lieut. Hooper's Narrative showing the Country of the Tuski and the Progress of the Boat Expedition
From: Hooper, 1853

MAP 147 (PAGE 87)
Map of the World Illustrating B. Seeman's Narrative of the Voyage of HMS Herald under the command of Captain H. Kellett
From: Seeman, 1853

MAP 148 (PAGE 87)
Part of the Arctic Coast of America between the River McKenzie and Cape Bathurst shewing the track of the boats, under the command of W. J. Pullen, Commander, R.N. from the 22nd July to the 31st of August 1850
William Pullen, 1850
UKHO: L8991 on shelf Ai2

MAP 149 (PAGE 88)
Arctic America Sheet II [with MS addition:] *Track of the Prince Albert Discovery Ship. 1850*
Charles Forsyth, 1850 (MS additions only)
PRO: MPI 1/318 (4)

MAP 150 (PAGES 88–89)
Discoveries in the Arctic Sea, between Baffin Bay & Melville Island Shewing the Coasts Explored on the Ice By Capt.n Ommanney & the Officers of the Expeditions under the Command of Captain H. T. Austin R.N. C.B. & Captain W. Penny. Also by the Hon.ble Hudson's Bay Co.s Expedition under the Command of Rear Admiral Sir John Ross C.B. and Dr. John Rae, in Wollaston & Victoria Land, in Search of Sir John Franklin. 1850–51
John Arrowsmith, April 1852
NMC 8455

MAP 151 (PAGE 90)
A Track Chart of H.M. Sledges Perseverance & Resolute
Francis Leopold M'Clintock, 1852
From: *Parliamentary Papers, Additional Papers*, p. 142, 1852

MAP 152 (PAGE 90)
Chart of the Arctic Coast examined by Dr. J. Rae in Spring & Summer 1851
John Rae, 1851
From: *Journal of the Royal Geographical Society*, Vol. 22, 1852

MAP 153 (PAGE 91)
[Map of the Arctic showing probable position of Franklin and an open polar sea]
Augustus Petermann, 1852
From: *Parliamentary Papers, Further Correspondence and Proceedings*, 1852
CT

MAP 154 (PAGE 91)
Chart Exhibiting the recent discoveries in the Arctic Regions
Edwin Jesse De Haven, 1853
From: Elisha Kent Kane, 1853 (see Kane, 1854; this map was from the rare 1853 edition but is the same as the 1854 map)
CT

MAP 155 (PAGE 92)
Chart Shewing the Discoveries & Explorations of the Travelling parties from the Prince Albert under the Command of W. Kennedy Esq.r in Search of Sir John Franklin. 1851-2
From: Kennedy, 1853
CT

MAP 156 (PAGE 92)
Carte de la Mer Arctique
From: Bellot, 1854

MAP 157 (PAGE 93)
A Chart shewing the West Shores of The Queen's Channel Searched and Explored by Commander Sherard Osborn In June and July 1853
Sherard Osborn, 1853
PRO: MPII 1/19 (26)

MAP 158 (PAGE 93)
Part of the Discoveries of Capt.n Sir Edw.d Belcher C.B. Commanding the Squadron in the Arctic Seas in Search of Sir John Franklin 1852-3
From: Belcher, 1855

MAP 159 (PAGE 94)
[Position of Resolute at Dealy Island from] Track of Sledge Discovery From Sept.r 22nd To Oct.r 14th 1852 Under the Charge of Lieut. G. F. Mecham, HMS Resolute
George Frederick Mecham, 1852

MAP 160 (PAGES 94–95)
Chart shewing the track of H.M. Sledge "Discovery," Between April 4, and July 6 1853
George Frederick Mecham, 1853
PRO: MPII 1/19 (37)

MAP 161 (PAGE 96)
Chart shewing the Coasts Discovered by Commander F. L. M.c Clintock with his Track in Search of Sir John Franklin's Expedition 1853
Francis Leopold M'Clintock, 1853

MAP 162 (PAGE 97)
Chart Shewing the track of the Sledge James Fitzjames from March 10 to April 15 1853 also of the Sledge John Dyer Between May 6 and June 9 1853 forming a part of the Searching operations from HMS Resolute – Capt* H. Kellett C.B. both Sledges under the charge of William T. Domville Surgeon*
William Domville, 1853
PRO: MPII 1/19 (22)

MAP 163 (PAGE 98)
Chart Shewing the land discovered by H.M. Ship Investigator Between September 1850 and October 1851
Robert M'Clure, 1851
UKHO: L 9302

MAP 164 (PAGES 100–101)
Discoveries in the Arctic Sea between Baffin Bay, Melville Island, & Cape Bathurst: Shewing the Coasts Explored by the Officers of the Various Expeditions
John Arrowsmith, November 1853
NMC 21054

MAP 165 (PAGE 102)
Chart to Illustrate the Narratives of Captn*. Collinson, C.B. and D*^{r}*. Rae*
From: *Parliamentary Papers, Further Papers Relative to the Recent Arctic Expeditions in Search of Sir John Franklin* p. 942, 1855

MAP 166 (PAGE 102)
Erebus Bai nach der Aufnahme des Comm. W. J. S. Pullen 1854
PGM 1855

MAP 167 (PAGE 103)
Chart of part of the Coast of North America shewing the journeys of Comr Maguire HMS Plover during the months of March and April 1853 with the latest native Intelligence of HMSS Investigator & Enterprise
Rochfort Maguire, 1853
PRO: MPII 1/24 (8)

MAP 168 (PAGE 103)
Map of the Arctic Exploration from which resulted the first information of Sir John Franklin's missing Party: by Dr. John Rae. 1854
From: *Journal of the Royal Geographical Society,* Vol. 25, 1855

MAP 169 (PAGE 104)
Map of a Portion of the Arctic Shores of America to Accompany Captn* M.cClintock's Narrative*
John Arrowsmith, 1859
From: M'Clintock, 1859 (from dummy proof copy)
CT

MAP 170 (PAGES 104–5)
Arctic Seas Shewing the North-West Passage, The Coasts explored by the several Searching Expeditions, and the spot where the remains of Sir John Franklin have been discovered
W. and A. K. Johnston, 1854, with MS additions
NMC 6296

MAP 171 (PAGES 106–7)
Wellington Channel, Melville Island &c from Admiralty Chart
A. Fullarton & Co., 1856

MAP 172 (PAGE 108)
[Map of Frobisher Bay]
From: Hall, 1864

MAP 173 (PAGE 108)
Sketch of Coast Lines From Ft. Churchill to Lancaster Sound By Ar-mou in 1866
Copied by Charles Hall, 1866
From: Hall, 1879

MAP 174 (PAGE 109)
Journey to King Williams Land 1869
Charles Hall, 1869
From: Hall, 1879

MAP 175 (PAGE 110)
Sketch of King Williams Land By the Innuit In-nook-poo-zhu-jook in 1869
Copied by Charles Hall, 1869
From: Hall, 1879

MAP 176 (PAGE 110)
[Map showing the routes of Charles Hall in 1868–69 and Frederick Schwatka in 1879–80]
PGM 1885

MAP 177 (PAGE 111)
[Map of the explorations of Frederick Schwatka]
PGM 1880

MAP 178 (PAGE 111)
King William Land Showing the Line of Retreat of the Franklin Expedition
W. and A. K. Johnston
From: *Proceedings of the Royal Geographical Society and Monthly Record of Geography,* Vol. 2, 1880
NLC

MAP 179 (PAGE 112)
Map of the North-Eastern Part of Siberia to Illustrate the Narrative of the Expedition to the Polar Sea in 1820, 21, 22, and 23
From: Wrangel (Vrangel), 1840

MAP 180 (PAGES 112–13)
[Russian map of the Arctic Ocean]
From: Efimov, 1964, #191

MAP 181 (PAGE 114)
[Map showing the tracks of Petr Anzhu, 1821–23]
PGM Tafel 9, 1879

MAP 182 (PAGES 114–15)
[Special map of North Siberia]
PGM Tafel 10, 1879

MAP 183 (PAGE 116)
[Bering Sea, Bering Strait, and the adjacent part of the Arctic Ocean]
United States North Pacific Exploring Expedition, 1855
NARA: RG 37, 283.22, #1

MAP 184 (PAGE 116)
The Asiatic Coast of Behring's Straits. Surveyed in the U.S. Ship Vincennes July and August 1855
NARA: RG 37, 181.36, #67

MAP 185 (PAGE 117)
Herald Island by the U.S. Ship Vincennes August 1855
NARA: RG 37, 191.33, #85

MAP 186 (PAGE 117)
[Track of the American merchant bark *Nile*]
Thomas Long, 1867
NARA: RG 37, 142.21, #N-3

MAP 187 (PAGE 118)
[Bering Sea, Bering Strait, and part of the Arctic Ocean]
Revised edition of North Pacific Exploring Expedition map, 1868
NARA: RG 37, 191.33, #83

MAP 188 (PAGE 119)
[Track of *Corwin* and *Rodgers*, 1880–81, with inset of Ostrov Vrangelya (Wrangel Land)]
PGM Tafel 2, 1882

MAP 189 (PAGE 120)
[Track of Weyprecht and Payer from Novaya Zemlya to Zemlya Frantsa Iosifa, 1872]
PGM Tafel 5, 1877

MAP 190 (PAGE 120)
Chart of the North Polar Sea
British Admiralty, 1874
CT

MAP 191 (PAGE 121)
[Discoveries of Julius Payer in Zemlya Frantsa-Iosifa, 1872–74]
PGM Tafel 11, 1876

MAP 192 (PAGE 121)
[Petermann Land]
From: *Century Atlas,* 1897

MAP 193 (PAGE 121)
[Map to illustrate Petermann's theory of connected lands near the Pole]
PGM Tafel 20, 1874

MAP 194 (PAGES 122–23)
Map of the Arctic Ocean in the limits of the Russian Empire
Russian
Anon., 1874
LC: G7062.C6S12 1874.R8 MLC

MAP 195 (PAGE 124)
A Survey of the Principal Points on the Northern Coast of Spitzbergen Chiefly Constructed from the Observations of Captn*. W. E. Parry & Lieut*^{t}*. H*^{y}* Foster and Shewing the Track of His Maj*^{ys}* Ship Hecla and her Boats AD 1827*
R. H. Foote,
From: Parry, 1828

MAP 196 (PAGE 125)
Chart of the Northern Portion of Baffin Bay: to illustrate Comm^r* Inglefields Report 1853*
From: *Journal of the Royal Geographical Society,* Vol. 23, 1853

MAP 197 (PAGE 125)
[Map of the open polar sea]
From: Kane, 1856

MAP 198 (PAGE 126)
[Page from the journal of Elisha Kent Kane for 13 August 1853]
CT

MAP 199 (PAGE 127)
Discoveries of the American Arctic Expedition (to the Northward of Sir Thomas Smith Sound) in Search of Sir John Franklin 1853-4-5 under the Command, & to illustrate the Paper of Dr. E. K. Kane – U.S. Navy
From: *Journal of the Royal Geographical Society,* 1856

MAP 200 (PAGE 128)
Chart of Smith Sound shewing Dr. Haye's [Hayes's] Track & Discoveries 1860–61
From: Hayes, 1867

MAP 201 (PAGES 128–29)
[Comparative maps of Smith Sound and region to the north]
PGM 1867

MAP 202 (PAGE 129)
Map of the World (Mercator's projection) Showing the principal Surface Currents of the Oceans & Thermometric Gateways to the North Pole
Silas Bent, 1872
From: *An Address Delivered before the St. Louis Mercantile Library Association January 6th 1872 Upon the Thermal Paths to the Pole*
CT

MAP 203 (PAGE 130)
[Map of the track of Karl Koldeway in the Greenland Sea, 1868]
PGM 1871

MAP 204 (PAGE 131)
[Pack ice conditions found by David Gray in the Greenland Sea, 1874]
PGM 1875

MAP 205 (PAGE 131)
[Discoveries of the Hall expedition, 1871–73]
PGM 1874

MAP 206 (PAGE 132)
[Discoveries of the Nares expedition, 1875–76]
PGM 1876

MAP 207 (PAGE 133)
[Illustrated map of Eastern Ellesmere Island and northwestern Greenland after the Nares expedition]
Anon., no date

MAP 208 (PAGE 133)
Smith Sound, Kennedy and Robeson Channels
From: Nares, 1878

MAP 209 (PAGE 134)
Chart Showing the Discoveries of the British Polar Expedition 1875–76
From: Anon., 1877

Map 210 (page 134)
The North Polar Regions
From: Anon., 1877

Map 211 (page 135)
Track chart of the USS Jeannette, Lieut.-Comm. George W. De Long, from San Francisco, up to the sinking of the ship; together with the route followed by the officers and crew in their escape over the ice to the Siberian Coast
From: De Long, 1884 (fold-out map at end of Vol. 1)

Map 212 (page 136)
[The drift of the *Jeannette* and track of survivors to the Siberian Coast]
PGM 1882

Map 213 (page 137)
Arctic Ocean North of Behrings Strait. Track of the USS Rodgers, Lieut. R. M. Berry Comdg. North of Wrangel Island, September 1881
NARA: RG 37, 191.1, #4

Map 214 (page 137)
Map showing the explorations by Lieut. J. B. Lockwood, U.S. Army, 1882
From: Greely, 1886

Map 215 (page 138)
[Map of stations participating in the International Polar Year, 1882–83]
From: Greely, 1886

Map 216 (page 138)
Chart of Discoveries made in North Greenland by J. B. Lockwood, Lieut., 23rd Inf. U.S.A.
From: Greely, 1886

Map 217 (page 139)
Map of the Arctic Regions. Circumpolar Map No. II (Geographical Discoveries made since 1818 in Red)
From: Hall, 1879

Map 218 (page 140)
[Map of Adolf Erik Nordenskiöld's expedition to Spitsbergen, 1868]
PGM Tafel 8, 1870

Map 219 (pages 140–41)
[Professor Nordenskiöld's route around the north point of Asia in the steamer *Vega* to 27 August 1878]
PGM Tafel 2, 1879

Map 220 (pages 142–43)
[Special map of North Siberia between the Lena and Bering Strait with Professor Nordenskiöld's course in the steamer *Vega*]
PGM Tafel 17, 1879

Map 221 (page 144)
[Right-hand map, untitled, of two-part] *Map of the North Coast of the Old World From Norway to Behrings Straits with the Track of the* Vega *Expedition*
Adolf Erik Nordenskiöld, 1882
From: Nordenskiöld, 1882

Map 222 (page 144)
Track of the Vega *through Behrings Straits*
Adolf Erik Nordenskiöld, 1882
From: Nordenskiöld, 1882

Map 223 (page 145)
[Southwest coast of Greenland, with inset Greenland]
PGM Tafel 5, 1880

Map 224 (page 145)
[East coast of Greenland, with expedition tracks 1822–68]
PGM Tafel 5, 1880

Map 225 (page 146)
Karte der Arktischen & Antarktischen Regionen zur Übersicht der Enteckungsceschichte von A. Petermann [Map of Petermann's hypothesis: Greenland connected to Wrangel Land]
PGM Tafel 12, 1868

Map 226 (page 147)
[East Greenland, showing Nordenskiöld's 1883 expedition]
PGM Tafel 3, 1886

Map 227 (page 147)
Map of Southern Greenland Shewing the Route of the Norwegian Expedition in 1888
From: Nansen, 1890, Vol. 1

Map 228 (page 148)
Map of Franz Josef Land Showing Journeys and Discoveries of Frederick G. Jackson F.R.G.S. Leader of the Jackson-Harmsworth Polar Expedition 1894–7
CT

Map 229 (pages 148–49)
Map Showing Route of Fram, and Dr. Nansen's and Lieut. Johansen's Sleigh Journey
From: Fridtjof Nansen, *Farthest North*, Macmillan Colonial Library edition [1897?]

Map 230 (page 150)
Bathymetrical Chart of the North Polar Seas By Dr Fridtjof Nansen
From: *Geographical Journal*, November 1907

Map 231 (page 150)
Map Showing Nansen's Proposed Route to the Pole
From: Nansen, 1898

Map 232 (page 151)
[Summary map of the state of polar exploration]
PGM 1897

Map 233 (page 152)
[Peary Land and Melville Land]
Century Atlas, 1897

Map 234 (page 152)
[Peary's proposed routes for 1893]
From: *Geographical Journal*, 1893

Map 235 (page 153)
[Untitled map of northern Greenland with signatures of Peary Arctic Club members]
From: *National Geographic Magazine*, Vol. 14, p. 330, 1902

Map 236 (page 153)
Map Shewing Route of the North Greenland Expedition of 1891–92
Robert E. Peary, 1893
From: Cyrus C. Adams, "Lieutenant Peary's Arctic Work," *Geographical C. Journal*, Vol. 1, 1893

Map 237 (pages 154–55)
The Polar Regions Showing the Routes and Explorations of Robert E. Peary, U.S.N. From 1892 to 1906
From: Peary, 1907

Map 238 (page 156)
[Map of the southern coast of Ellesmere Island showing Sverdrup's wintering place in 1899–1900; found by Bernier, 1906]
Otto Sverdrup and/or Gunnerius Isachsen, 1900
NMC 193769

Map 239 (page 156)
The Arctic Regions with the Tracks of Search Parties and the Progress of Discovery
[with later MS additions probably by Bernier: sector proposals]
U.S. Department of the Navy, 1896
NMC 11644

Map 240 (page 157)
Map Showing the Field of Work of the Second Norwegian Polar Expedition in the "Fram" Captain Sverdrup 1898–1902
by Captain G. J. Isachsen
From: Sverdrup, 1904

Map 241 (page 158)
[Track of Amundsen from Norway to Nome, Alaska]
Anon., c.1906
NMC 14664

Map 242 (page 158)
Gjøahavn Med Nærmest Omgivelser
From: Roald Amundsen, *Nordvest-Passagen: Beretning om Gjøa-Ekspeditionen 1903–1907*, 1908

Map 243 (page 159)
Map of King Haakon VII Coast and Queen Maud's Sea and of Lieut. Hansen and Sergt. Ristveldt's Sledge Expedition 1905
From: Amundsen, 1908

Map 244 (page 160)
Polar Advance of the National Standards
From: Cook, 1911

Map 245 (page 160)
[Cook's route to the North Pole]
From: Cook, 1911

Map 246 (page 161)
Peary Map Showing Where He Says Dr. Cook Went
From: *Chicago Daily Tribune*, 13 October 1909
LC

Map 247 (page 162)
[Peary's route to the North Pole]
From: Peary, 1910

Map 248 (pages 162–63)
Specimen Map [of the north polar regions, with Peary's route to the Pole added]
Matthews-Northrup Works, 1907, with Peary route overprinted, 1909

Map 249 (page 164)
[Bradley Land and Crocker Land]
From: Balch, 1913

Map 250 (page 164)
[Harris's hypthesis of land in the Arctic Basin]
Rollin Harris, 1904

Map 251 (page 164)
[Crocker Land, from] *The Polar Region Showing the Routes and Explorations of Robert E. Peary, U.S.N. From 1892 to 1906*
From: Peary, 1907

Map 252 (page 165)
The Arctic Regions with the Tracks of Search Parties and the Progress of Discovery
[with later MS additions probably by Bernier: sector proposals]
U.S. Department of the Navy, 1896
NMC 11644

Map 253 (page 165)
[Sketch map of Bylot Island, Pond Inlet, and adjacent coasts]
Joseph-Elzéar Bernier, 1911
Archives nationale de Québec: les papiers Bernier

Map 254 (page 166)
Carte indiquant les frontières internationales dans L'Ocean arctique proposées par le capitaine Bernier
Joseph-Elzéar Bernier, 1909
From: *Le Temps*, Ottawa, 18 October 1909

Map 255 (page 166)
Surveys and Discoveries in the Arctic Regions with additions and changes to 1911 on Coast of Baffin Island by J. T. E. Lavoie, C.E. Capt. J. Bernier's Arctic Expedition, 1910–11
J. T. E. Lavoie, 1911
From: Bernier, no date [1911–12]

Map 256 (page 166)
[Drift of the *Karluk*, from] *Department of the Naval Service. Key Map of the Canadian Arctic Expedition Discoveries in the Arctic Sea 1914–18*
Separate fold-out map in Stefansson, 1921, first edition.

Map 257 (page 167)
Discoveries in the Arctic Sea
Vilhjalmur Stefansson, 1916
From: Stefansson, 1921

Map 258 (page 167)
Department of the Naval Service. Key Map of the Canadian Arctic Expedition Discoveries in the Arctic Sea 1914–18
Separate fold-out map in Stefansson, 1921, first edition.

MAP 259 (PAGE 168)
The Arctic Regions
From: *National Geographic Magazine*, November 1925

MAP 260 (PAGE 168)
Physical Map of the Arctic
American Geographical Society, 1929

MAP 261 (PAGE 169)
[Severnaya Zemlya and the eastern coast of the Taymyr Peninsula, with discoveries in 1913]
PGM 1914

MAP 262 (PAGE 169)
[Central Russian north coast about 1916]
From: Gvozdetsky, 1974

MAP 263 (PAGE 169)
[Central Russian north coast about 1974]
From: Gvozdetsky, 1974

MAP 264 (PAGE 170)
Map Showing the Routes and Drifts of the Various Polar Expeditions North of Spitsbergen and Franz Josef Land
From: Andrée, 1930

MAP 265 (PAGE 171)
[Aerial mapping technique used by Walter Mittelholzer]
From: Mittelholzer, 1925

MAP 266 (PAGE 172)
[Lommes Bay, Spitsbergen]
From: Mittelholzer, 1925

MAP 267 (PAGE 172)
[Amundsen's attempt on the North Pole, May 1925]
From: *New York World*, June 1925

MAP 268 (PAGE 173)
Amundsen's Route North and Back
From: "Amundsen Back Safe in Spitzbergen"
New York Times, 19 June 1925

MAP 269 (PAGE 174)
Ellesmere Island Region
Inset from: *The Arctic Regions* (MAP 259)
From: *National Geographic Magazine*, November 1925

MAP 270 (PAGE 174)
[Byrd's flight to the North Pole, 1926]
(endpaper)
From: Byrd, 1928

MAP 271 (PAGE 175)
Chart of Route Flown by Lieutenant Commander Richard Evelyn Byrd from Kings Bay, Spitsbergen, to the North Pole on May 9, 1926
From: Byrd, 1928

MAP 272 (PAGE 176)
[Track of the *Norge* from Rome to Teller, Alaska]
From: Nobile, 1926

MAP 273 (PAGE 176)
[Track of *Norge* and unexplored land]
Atlas map (original source unknown), 1930

MAP 274 (PAGE 177)
[Track of *Norge* to the North Pole and projected route to Alaska]
From: *New York Times*, 12 May 1930

MAP 275 (PAGE 177)
[Untitled map showing the track of the *Norge* from Spitsbergen to Alaska in May 1926]
From: Amundsen, 1928

MAP 276 (PAGE 178)
[Track of Hubert Wilkins in April 1928]
From: Wilkins, 1928

MAP 277 (PAGE 179)
Carta Dimostrativa del Volo del Dirigible "Italia" alla Terra di Nicola II
From: Nobile, 1930

MAP 278 (PAGE 180)
Photostat map *Carta delle Regioni Artiche settore della Groenlandia*, with MS additions by Umberto Nobile showing the track of his airship *Italia* to the North Pole in 1928. Map is marked "Gift from Gen. Nobile, Jan 29–1929."
American Geographical Society Collection,
University of Wisconsin-Milwaukee Library

MAP 279 (PAGE 180)
Carta Dimostrativa del Volo del Dirigible "Italia" al Polo Nord
From: Nobile, 1930

MAP 280 (PAGE 181)
[Proposed track of Wilkins' submarine *Nautilus*]
From: Wilkins, 1931

MAP 281 (PAGE 182)
Map of the Drifting Soviet Expedition at the North Pole, Drawn under the supervision of Y. M. Shokalsky
From: Brontman, 1938

MAP 282 (PAGE 182)
Route of the Flight
From: Brontman, 1938

MAP 283 (PAGE 183)
Composite map of Foxe Basin, 1946, from two aeronautical edition maps:
Foxe Basin North
Sheets 37 SW and 37 SE
G/3401/.P6/S506/.C36, 1943 [1946]
Foxe Basin South
Sheets 36 NW and 36 NE
G/3401/.P6/S506/.C36, 1946
National Archives of Canada

MAP 284 (PAGE 183)
Composite map of Foxe Basin, 1949, from two aeronautical edition maps:
Foxe Basin North
Sheets 37 SW and 37 SE
G/3401/.P6/S506/.C36, 1949
Foxe Basin South
Sheets 36 NW and 36 NE
G/3401/.P6/S506/.C36, 1949
National Archives of Canada

MAP 285 (PAGE 184)
The Arctic Regions, by the US Naval Oceanographic Office, 6th edition, June 1948, revised 26 October 1970 [with MS annotation showing route of USS *Nautilus* through the North Pole, and signed *William R. Anderson, C.O. Nautilus, first transpolar voyage, 1915 EDT, 3 August 1958*]
Newberry Library, Chicago: Fitzgerald Polar Map 6F 81

MAP 286 (PAGE 186)
[Drift of the *Georgii Sedov*, 1937–40]
From: Armstrong, 1958

MAP 287 (PAGE 186)
[Tracks of Russian drifting stations and locations of aircraft landing, 1948–57]
From: Armstrong, 1958

MAP 288 (PAGES 186–87)
General Bathymetric Chart of the Oceans (GEBCO) [Arctic sheet]
Canadian Hydrographic Service, Ottawa
5th edition, August 1979, reprinted September 1983

MAP 289 (PAGE 188)
[Thickness of sea ice]
RadarSat satellite image
Canadian Space Agency/NASA

MAP 290 (PAGE 188)
Western Arctic/Arctique de l'ouest
From: *Annual Arctic Ice Atlas*, Winter 2001
RadarSat image
Environment Canada, 30 July 2002
NLC

MAP 291 (PAGE 189)
[Median Sea Ice Concentration in the Canadian Arctic for 1 April, 1971–2000]
From: *Sea Ice Climatic Atlas, Northern Canadian Waters 1971–2000*, Environment Canada, Canadian Ice Service, Ottawa, 2002

MAP 292 (PAGE 189)
[Median Sea Ice Concentration in the Canadian Arctic for 10 September, 1971–2000]
From: *Sea Ice Climatic Atlas, Northern Canadian Waters 1971–2000*, Environment Canada, Canadian Ice Service, Ottawa, 2002

MAP 293 (PAGE 190)
[Multibeam bathymetric map of part of the southern Lomonosov Ridge]
SCICEX, 1999

MAP 294 (PAGE 190–91)
International Bathymetric Chart of the Arctic Ocean (IBCAO)
U.S. National Geophysical Data Center (NGDC)/National Oceans and Atmospheric Administration (NOAA), 2002

MAP 295 (PAGE 192)
[Multibeam bathymetric map of part of the Gakkel Ridge]
SCICEX, 1999

MAP 296 (PAGE 192)
[True-color satellite image of the North Pole]
MODIS image, 5 May 2000
NASA

MAP 297 (PAGE 193)
[Two RadarSat images of an area of the Beaufort Sea, nine days apart]
Canadian Space Agency/NASA

MAP 298 (PAGE 193)
[Surface water density]

MAP 299 (PAGE 193)
[Mean surface water temperature]

MAP 300 (PAGE 193)
[Water salinity at 5 m]

MAP 301 (PAGE 197)
Terra Arctica
Pierre du Val, 1676
Beach Maps/Alec Parley

MAP 302 (PAGE 203)
Map Showing Route of Fram, and Dr. Nansen's and Lieut. Johansen's Sleigh Journey
From: Nansen, *Farthest North*, Macmillan Colonial Library edition [1897?]

MAP 303 (PAGE 208)
["Oud Greenland"]
Hidde Dirks Katt, 1777

Adolf Erik Nordenskiöld's ship *Vega* rounds Mys Dezhneva—East Cape—on 20 July 1879, thus completing the first transit of the Northeast Passage, a shortcut to Cathay sought for centuries.

Bibliography

Abramson, Howard S.
Hero in Disgrace: The Life of Arctic Explorer Frederick A. Cook
Paragon House, New York, 1991

Amundsen, Roald E. G.
Roald Amundsen's "The North West Passage"; Being the Record of a Voyage of Exploration of the Ship "Gjoa" 1903–1907
A. Constable, London, 1908

Amundsen, Roald E. G.
First Flight Across the Polar Sea
Hutchinson, London, 1928

Anderson, William R.
Nautilus 90 North
World, Cleveland, 1959

Andrée, Salomon August, et al.
The Andrée Diaries: Being the Diaries and Records of S. A. Andrée, Nils Strindberg and Knut Fraenkel Written during their Balloon Expedition to the North Pole in 1897 and Discovered on White Island in 1930, Together with a Complete Record of the Expedition and Discovery
Bodley Head, London, 1930

Anon.
Recent Polar Voyages
Thomas Nelson, London, 1877

Armstrong, Terence
The Russians in the Arctic: Aspects of Soviet Exploration and Exploitation of the Far North, 1937–57
Methuen, London, 1958

Bacon, Edgar Mayhew
Henry Hudson: His Times and His Voyages
G. P. Putnam's Sons, New York, 1907

Balch, Edwin Swift
The North Pole and Bradley Land
Campion, Philadelphia, 1913

Barrow, John
A Chronological History of Voyages into the Arctic Regions Undertaken Chiefly for the Purpose of Discovering a North-East, North-West, or Polar Passage Between the Atlantic and Pacific
John Murray, London, 1818

Barrow, John
Voyages of Discovery and Research within the Arctic Regions
John Murray, London, 1846

Beechey, Frederick
A Voyage of Discovery Towards the North Pole
Richard Bentley, London, 1843

Belcher, Edward
The Last of the Arctic Voyages: Being a Narrative of the Expedition in H.M.S. Assistance
Lovell Reeve, 2 vols., 1855

Bellot, Joseph-René
Journal d'un voyage aux mers polaires
Perrotin, Paris, 1854

Bernier, Joseph-Elzéar
Report on the Dominion Government Expedition to the Northern Waters and Arctic Archipelago of the D.G.S. "Arctic" in 1910
No publisher stated [Government publication], [1911]

Berton, Pierre
The Arctic Grail: The Quest for the Northwest Passage and the North Pole, 1818–1909
McClelland & Stewart, Toronto, 1988

Best, George
The Three Voyages of Martin Frobisher, in Search of a Passage to Cathaia and India by the North-West, A.D. 1576–8
Reprint of edition edited by Richard Collinson, n.d., Burt Franklin, New York, no date [1963?]

Binney, George
With Seaplane and Sledge in the Arctic
Hutchinson, London, 1925

Brögger, W. C., and Nordahl Rolfsen
Fridtiof Nansen 1861–1893
Longmans Green, London, 1896

Brontman, L.
On the Top of the World: The Soviet Expedition to the North Pole 1937
Victor Gollancz, London, 1938

Bryce, Robert M.
Cook and Peary: The Polar Controversy, Resolved
Stackpole, Mechanicsburg, PA, 1997

Byrd, Richard Evelyn
Skyward
G. P. Putnam's Sons, New York, 1928

Cameron, Ian
To the Farthest Ends of the Earth: The History of the Royal Geographical Society 1830–1980
Macdonald & Jane's, London, 1980

Conway, William Martin
The First Crossing of Spitzbergen
J. M. Dent, London, 1897

Cook, Frederick A.
My Attainment of the Pole
Mitchell Kennerley, New York, 1911

Cook, Frederick A.
Return from the Pole
Pellegrini & Cudahy, New York, 1951

Delgado, James P.
Across the Top of the World: The Quest for the Northwest Passage
Douglas & McIntyre, Vancouver, 1999

De Long, George W.
The Voyage of the Jeannette. The Ship and Ice Journals of George W. De Long. Edited by Emma De Long
Houghton Mifflin, 2 vols., Boston, 1884

De Veer, Gerrit
A True Description of Three Voyages by the North-East Towards Cathay and China, Undertaken by the Dutch in the Years 1594, 1595, and 1596
Hakluyt Society, London, 1853

Diubaldo, Richard J.
Stefansson and the Canadian Arctic
McGill-Queen's University Press, Montreal, 1978

Dorion-Robitaille, Yolande
Captain J. E. Bernier's Contribution to Canadian Sovereignty in the Arctic
Indian and Northern Affairs, Ottawa, 1978

Dunbar, Moira, and Keith Greenaway
Arctic Canada from the Air
Canada Defence Research Board, Ottawa, 1956

Eames, Hugh
Winner Lose All: Dr. Cook and the Theft of the North Pole
Little, Brown, Boston, 1973

Eckener, Hugo
My Zeppelins
Putnam, New York, 1931

Efimov, A. V. (ed.)
Atlas geograficheskikh v Sibiri i severo-zapadnoy Amerike XVII–XVIII vv. [Atlas of geographical discoveries in Siberia and northwestern America in the 17th–18th centuries]
Nauka, Moscow, 1964

Ellsworth, Lincoln
Air Pioneering in the Arctic: The Two Polar Flights of Roald Amundsen and Lincoln Ellsworth
National Americana Society, New York, 1929

Euller, John
Ice, Ships, and Men
Abelard-Schuman, New York, 1964

Fairley, T. C.
Sverdrup's Arctic Adventures
Longmans, London, 1959

[Fisher, A.]
Journal of a Voyage of Discovery to the Arctic Regions
Richard Phillips, London, n.d. [1819]

Franklin, John
Narrative of a Journey to the Shores of the Polar Sea, in the Years 1819, 20, 21, and 22
John Murray, London, 1823

Franklin, John
Narrative of a Second Expedition to the Shores of the Polar Sea, in the Years 1825, 1826, and 1827
John Murray, London, 1828

Goerler, Raimund E. (ed.)
To the Pole: The Diary and Notebook of Richard E. Byrd, 1925–1927
Ohio State University Press, Columbus, 1998

Greely, Adolphus W.
Three Years of Arctic Service: An Account of the Lady Franklin Bay Expedition of 1881–84 and the Attainment of the Farthest North
Charles Scribner's Sons, 2 vols., New York, 1886

Grierson, John
Challenge to the Poles: Highlights of Arctic and Antarctic Aviation
G. T. Foulis, London, 1964

Gvozdetsky, N. A.
Soviet Geographical Explorations and Discoveries
Progress, Moscow, 1974

Hall, Charles F.
Life with the Esquimaux: the narrative of Captain Charles Francis Hall, of the whaling barque "George Henry," from the 29th May, 1860, to the 13th September, 1862
Samson Low, Son & Marston, 2 vols., London, 1864

Hall, Charles F.
Narrative of the Second Arctic Expedition Made by Charles F. Hall . . . and his Residence Among the Eskimos During the Years 1864–'69
Government Printing Office, Washington, 1879

Hayes, Derek
First Crossing: Alexander Mackenzie, His Expedition Across North America, and the Opening of the Continent
Douglas & McIntyre, Vancouver, 2001

Hayes, Derek
Historical Atlas of the North Pacific Ocean
Douglas & McIntyre, Vancouver, 2001

Hayes, Derek
Historical Atlas of Canada
Douglas & McIntyre, Vancouver, 2002

Hayes, Isaac Israel
The Open Polar Sea: A Narrative of a Voyage of Discovery Towards the North Pole in the Schooner "United States"
Sampson Low, Son, and Marston, London, 1867

Herbert, Wally
The Noose of Laurels: Robert E. Peary and the Race to the North Pole
Atheneum, New York, 1989

Holland, Clive
Arctic Exploration and Development c.500 B.C. to 1915: An Encyclopedia
Garland, New York, 1994

Holland, Clive (ed.)
Farthest North: The Quest for the North Pole
Robinson, London, 1994

Hooper, William H.
Ten Months Among the Tents of the Tuski
John Murray, London, 1853

James, Thomas
The Strange and Dangerous Voyage of Capt. Thomas James (1633)
Edited by W. A. Kenyon
Royal Ontario Museum, Toronto, 1975

Joerg, W. L. G.
Brief History of Polar Exploration Since the Introduction of Flying
Special Publication No. 11,
American Geographical Society, New York, 1930

Kane, Elisha Kent
The U.S. Grinnell Expedition in Search of Sir John Franklin: A Personal Narrative
Sampson Low, London, 1854

Kane, Elisha Kent
Arctic Explorations: the Second Grinnell Expedition in Search of Sir John Franklin, 1853, '54, '55
Childs & Peterson, Philadelphia, 1856

Kennedy, William
A Short Narrative of the Second Voyage of the Prince Albert
W. H. Dalton, London, 1853

Lyon, George F.
A Brief Narrative of an Unsuccessful Attempt to Reach Repulse Bay
John Murray, London, 1825

M'Clintock, Leopold
Voyage of the 'Fox' in Arctic Seas: A Narrative of the Discovery of the Fate of Sir John Franklin and his Companions
John Murray, London, 1859

M'Clure, Robert
The Discovery of the North-West Passage
Edited by Sherard Osborn
Longman, Brown, Green, Longmans & Roberts, London, 1856

McDermott, James
Martin Frobisher: Elizabethan Privateer
Yale University Press, New Haven, 2001

McGhee, Robert
The Arctic Voyages of Martin Frobisher: An Elizabethan Adventure
Canadian Museum of Civilization/McGill-Queen's University Press, Montreal and Kingston, 2001

McGoogan, Ken
Fatal Passage: The Untold Story of John Rae, the Arctic Adventurer Who Discovered the Fate of Franklin
HarperCollins, Toronto, 2001

Mittelholzer, Walter
By Airplane Towards the North Pole: An Account of an Expedition to Spitzbergen in the Summer of 1923
George Allen & Unwin, London, 1925

Mountfield, David
A History of Polar Exploration
Dial Press, New York, 1974

Mowat, Farley
Ordeal by Ice: The Search for the Northwest Passage
McClelland & Stewart, Toronto, 1973

Mowat, Farley
The Polar Passion: The Quest for the North Pole
McClelland & Stewart, Toronto, 1973

Nansen, Fridtjof
The First Crossing of Greenland
Longmans, Green, 2 vols., London, 1890

Nansen, Fridtjof
Farthest North: Being a Record of a Voyage of Exploration of the Ship Fram 1893–96
Archibald Constable, 2 vols., London, 1897
Also: Macmillan Colonial Library edition [1897?]

Nansen, Fridtjof
"Future North Pole Exploration"
McClure's Magazine, Vol. 10, No. 4, pp. 295–305, February 1898

Nansen, Fridtjof
In Northern Mists: Arctic Exploration in Early Times
Heinemann, London, 1911

Nares, George Strong
Narrative of a Voyage to the Polar Seas during 1875–6 in H.M. Ships "Alert" and "Discovery"
Sampson Low, 2 vols., London, 1878

Neatby, Leslie H.
In Quest of the North West Passage
Longmans, Green, Toronto, 1958

Neatby, Leslie H.
Conquest of the Last Frontier
Longmans, Toronto, 1966

Nobile, Umberto
Il "Norge" al Polo Nord: Da Roma allo Stretto di Behring in Dirigible
L. Cappelli, Bologna, 1926

Nobile, Umberto
L'"Italia" al Polo Nord
A. Mondadori, Milano, 1930

Nobile, Umberto
With the "Italia" to the North Pole
Dodd Mead, New York, 1931

Nobile, Umberto
My Polar Flights: An Account of the Voyages of the Airships Italia *and* Norge
Frederick Muller, London, 1961

Nordenskiöld, Adolf Erik
The Voyage of the Vega *Round Asia and Europe*
Macmillan, New York, 1882

Officer, Charles, and Jake Page
A Fabulous Kingdom: The Exploration of the Arctic
Oxford University Press, New York, 2001

Pasetsky, Vasily
The Land that Never Was
Progress, Moscow, 1986

Parry, Ann
Parry of the Arctic: The Life Story of Admiral Sir Edward Parry 1790–1855
Chatto & Windus, London, 1963

Parry, William Edward
Journal of a Voyage for the Discovery of a North-West Passage from the Atlantic to the Pacific Performed in the Years 1819–20 in His Majesty's Ships Hecla *and* Griper
John Murray, London, 1821

Parry, William Edward
Journal of a Second Voyage for the Discovery of a North-West Passage from the Atlantic to the Pacific Performed in the Years 1821–22–23 in His Majesty's Ships Fury *and* Hecla
John Murray, London, 1824

Parry, William Edward
Journal of a Third Voyage for the Discovery of a North-West Passage from the Atlantic to the Pacific Performed in the Years 1824–25 in His Majesty's Ships Fury *and* Hecla
John Murray, London, 1826

Parry, William Edward
Narrative of an Attempt to Reach the North Pole in boats fitted for the purpose, and attached to H.M. *Ship "Hecla" in 1827*
John Murray, London, 1828

Peary, Robert E.
Nearest the Pole: A Narrative of the Polar Expedition of the Peary Arctic Club in the S.S. Roosevelt, *1905–1906*
Hutchinson, London, 1907

Peary, Robert E.
The North Pole: Its Discovery in 1909 Under the Auspices of the Peary Arctic Club
Frederick A. Stokes, New York, 1910

Phipps, Constantine John
A Voyage Towards the North Pole Undertaken by His Majesty's Command, 1773
J. Nourse, London, 1774

Pullen, H. F.
The Pullen Expedition in Search of Sir John Franklin: The Original Diaries, Log, and Letters of Commander W. J. S. Pullen
Arctic History Press, Toronto, 1979

Rae, John
Narrative of an Expedition to the Shores of the Arctic Sea in 1847 and 1848
Boone, London, 1850

Raurala, Nils-Erik (ed.)
The Northeast Passage: From the Vikings to Nordenskiöld
Helsinki University Library, John Nurminen Foundation, Helsinki, 1992

Rawlins, Dennis
Peary at the North Pole: Fact or Fiction?
Robert B. Luce, Washington, 1973

Rawlins, Dennis
"Byrd's Heroic 1926 North Pole Failure"
Polar Record, Vol. 36, No. 196, pp. 25–50, 2000

Rey, Louis (ed.)
Unveiling the Arctic
Arctic Institute of North America/University of Alaska Press, Calgary and Fairbanks, 1984

Ross, John
Voyage of Discovery, Made Under the Orders of the Admiralty, in His Majesty's Ships Isabella *and* Alexander, *for the Purpose of Exploring Baffin's Bay, and Inquiring into the Probability of a North-West Passage*
John Murray, London, 1819

Ross, John
Narrative of a Second Voyage in Search of a North-West Passage, and of a Residence in the Arctic Regions During the Years 1829, 1830, 1831, 1832, 1833
A. W. Webster, London, 1835

Sale, Richard
Polar Reaches: The History of Arctic and Antarctic Exploration
The Mountaineers Books, Seattle, 2002

Savours, Ann
The Search for the North West Passage
St. Martin's Press, London, 1999

Seeman, Berthold
Narrative of the Voyage of H.M.S. Herald *During the Years 1845–51*
Reeve, 2 vols., London, 1853

Stefansson, Vilhjalmur
The Friendly Arctic
Macmillan, New York, 1921

Stefansson, Vilhjalmur
Unsolved Mysteries of the Arctic
Macmillan, New York, 1938

Sverdrup, Otto
New Land: Four Years in the Arctic Regions
Longmans Green, London, 1904

Symons, Thomas H. B. (ed.)
Meta Incognita: A Discourse of Discovery: Martin Frobisher's Arctic Expeditions, 1576–1578
Canadian Museum of Civilization, 2 vols., Ottawa, 1999

Unwin, Rayner
A Winter Away from Home: William Barents and the North-east Passage
Seafarer/Sheridan, London and Dobbs Ferry, NY, 1995

Vaughan, Richard
The Arctic: A History
Sutton, Stroud, U.K., 1999

Verner, Coolie
Explorers' Maps of the Canadian Arctic 1818–1860
Cartographica Monograph No. 6
B. V. Gutsell, Toronto, 1972

Wrangel (Vrangel), Ferdinand P.
Narrative of an Expedition to the Polar Sea, in the Years 1820, 1821, 1822 & 1823
James Madden, London, 1840

Wilkins, George Hubert
Flying the Arctic
G. P. Putnam's Sons, New York, 1928

Wilkins, George Hubert
Under the North Pole: The Wilkins-Ellsworth Submarine Expedition
Brewer, Warren & Putnam, New York, 1931

Williams, Glyn
Voyages of Delusion: The Search for the Northwest Passage in the Age of Reason
HarperCollins, London, 2002

Wright, Theon
The Big Nail: The Story of the Cook-Peary Feud
John Day, New York, 1970

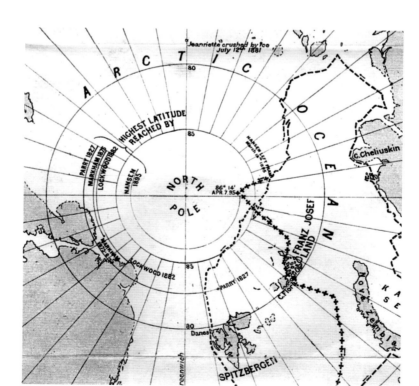

MAP 302.
This map from Fridtjof Nansen's book about the drift of the *Fram* and his attempt on the Pole in 1893–96 also shows the comparative latitudes reached by Edward Parry in 1827, Albert Markham in 1876 (not 1875 as on the map), and James Lockwood in 1882. The dashed line is the drift of the *Fram*, while the line of crosses is the route of Nansen and Hjalmur Johansen after they left the *Fram* on 14 March 1895.

Index

A

Abruzzi, duke of 154
Academy Gletscher (Glacier), Greenland 152
Act of Parliament, £20,000 for finding of Northwest Passage 47
Adelaide Peninsula, Canada 85, 111
Admiralty Inlet, Baffin Island 71
Advance (Edwin Jesse de Haven, 1850–51) 88, 91, 125
Advance (Elisha Kent Kane, 1853–55) 127
Ahngmalokto (Inuk with Robert Peary) 153
Ah-wehlah (Inuk with Frederick Cook) 160, 161
Air Force Island, Foxe Basin 183
Airships 171, 176, 181
Alaska 41, 51
Albert, Prince, of Monaco 187
Alderman Jones Sound. *See* Jones Sound
Aldrich, Pelham (Nares expedition) 134, 154
Aleksandr Sibiryakov (with Otto Shmidt, 1932) 169
Alert (George Nares, 1875–76) 133
Alert Point, Ellesmere Island 134, 154
Aleuts 7
Alexander (Edward Parry, 1818) 59
Allman Bay, Ellesmere Island 152
Ambassadors' globe 8
America, airship (Walter Wellman, 1907–09) 171
American Geographical Society 109
American Traveller... By an old and experienced trader, The (1769) 66
Amund Ringnes Island, Canada 156, 161
Amundsen, Roald 74, 158, 159, 164, 168, 171, 172, 176, 181
Amundsen Gulf, Canada 158
Anadyr River, Siberia 35, 36, 40
Anderson, William (commander, *Nautilus*, 1958) 185
Andrée, Saloman August (balloonist, 1896) 171
Andreyev, Stepan 41
"Andreyev Land" 42, 43
Anzhu, Petr Fedorovich 112, 114
Apelaglu (Inuk, drew maps for John Ross, 1830) 77
Archangel'sk, Russia. *See* Arkhangel'sk
"Arctic Highlanders" 7, 60
Arctic Meteorological and Climate Atlas 188
Arctic Ocean Hydrographic Expedition (1913) 168
Arkhangel'sk, Russia 9, 34, 181
Arktika (first ship to reach North Pole, 1977) 185
Ar-mou ("the Wolf"; Inuk, drew map for Charles Hall, 1866) 108
Arrowsmith, Aaron, mapmaker 58, 74
Arrowsmith, John, mapmaker 82, 104
Assistance (Edward Belcher, 1852–54) 92
Assistance (Erasmus Ommanney, 1850–51) 88
Astrup, Eivind (with Robert Peary, 1891) 152
Athabasca River, Canada 54
Austin, Horatio 88
Aviation in the Arctic 170–82
Axel Heiberg Island, Canada 154, 156, 160, 164, 174
Ayde (Martin Frobisher, 1577) 15

B

Bache Peninsula, Ellesmere Island 156
Back, George 65, 74, 79
Back River, Canada 79, 85
Baffin, William 27, 28, 32, 33, 58
Baffin Bay 28, 33, 58
Baffin Island 13, 32, 69, 71, 166
Bakhov, Ivan 41
Balloons, use in Arctic 170
Bancroft, Ann (first woman to reach Pole, 1986) 185
Banks, Joseph 56, 59, 62
Banks Island, Canada 93, 97, 99, 102, 166
Bank's Land (Banks Island) 64, 66, 84
Barents Sea 26
Barentsz, Willem 2, 4, 16, 33
Barnarda, Pedro de (apocryphal voyage, 1640) 66
Barrington, Daines 55, 59
Barrow, John (Jr.) 99

Barrow, John, Second Secretary of the Admiralty 56, 58, 60, 81, 84, 124
Barrow's polar map of 1818 60, 81 (1846 update), 139 (1879 update)
Barrow Strait 62, 86
Bartlett, Robert 154, 161, 166
Basque whalemen 22
Bathurst Island, Canada 63, 166
Bathymetric maps of the Arctic 149, 150, 188
Batty Bay, Somerset Island 92
Bay Fiord, Ellesmere Island 160
Bayley, William 50
Baymaud 168
Bay of Mercy (Mercy Bay; Banks Island) 99
Bear Island (Bjørnøya), Norwegian Sea 17, 21
Beaufort Sea 193
Beaumont, Lewis (Nares expedition) 135
Beechey, Frederick 56, 72, 75
Beechey Island (south coast of Devon Island) 85, 92, 158
Beechey Point, Alaska 178
Belcher, Edward 92, 125, 133, 156
 decision to abandon four ships 99
Bellin, Jacques-Nicolas, mapmaker 38, 42
Bellot, Joseph-René 92
Bellot Strait (Somerset Island/Boothia Peninsula) 78, 92, 104
Belyy Ostrov (north of Yamal Peninsula) 34
Bennett, Floyd (Amundsen's pilot, 1925–26) 174
Bennett, Gordon, owner of the *New York Herald* 135
Bennett Island (Ostrov Bennetta), Siberia 137
Bering, Vitus 36, 37, 38, 50
Bering Strait 36, 37, 50, 135, 142, 144
Bernier, Joseph-Elzéar 156, 165, 166
Berry, Robert Mallory 118, 137
Bertius, Petrus, mapmaker 4
Bessels, Emil (doctor, Hall expedition, 1871) 132
Biarmia 11
"Big lead" 155
Billings, Joseph 42
Binney, George 172
Bird, Edward 86
Blaeu, Joan, mapmaker 35, 44
Blossom (Frederick Beechey, 1825–28) 75
Bona Confidentia (Cornelius Durfoorth, 1553) 9
Bona Esperanza (Hugh Willoughby, 1553) 9
Booth, Felix 76, 88
Boothia Felix (Boothia Peninsula) 77, 78, 92, 102 103
Borden Island, Canada 167
Borough, Stephen 10
Borough, William 12
Bougainville, Louis-Antoine de 55
Bowden's Inlet (Rankin Inlet, Hudson Bay) 47
Bowdoin Fjord, Greenland 152
Bradford, Abraham (surgeon, *Resolute*, 1850–51) 90
Bradley, John R. 160
"Bradley Land" 161, 164
Brainard, David Legge (Greely expedition, 1881–84) 137, 138
Brendan, Saint 7
Briggs, Henry, mathematician and mapmaker 27, 28
Brock Island, Canada 167
Brodeur Peninsula, Baffin Island 73
Brooke, John (U.S. North Pacific Exploring Expedition, 1855) 116
Brunel, Oliver 16
Buache, Philippe 41, 49
Buchan, David 56
Buckland, John 10
Button, Thomas 27
Byam Martin Island, Canada 62, 63, 166
Bylot, Robert 27, 28, 58
Bylot Island, Canada 165
Byrd, Richard 158, 174

C

Cabot, Sebastian 8, 9
Cagni, Umberto 154, 172
California (Francis Smith, 1746–47) 47
Calvert, James 185
Cambridge Bay, Victoria Island 102, 168
Camden Bay, Alaska 102
Cape Aldrich, Ellesmere Island 134
Cape Alexander 80, 126

Cape Bathurst, reached by William Pullen 87
Cape Bounty, Melville Island, named by Parry 63
Cape Britannia, Greenland 137
Cape Columbia, Ellesmere Island 134, 161, 178, 185
Cape Fanshaw Martin, Ellesmere Island 134
Cape Hecla, Ellesmere Island, 153
Cape Henrietta Maria, Hudson Bay, named by Thomas James 30
Cape Independence (Kap Constitution), Greenland 126
Cape Inglefield (Kap Inglefield), Greenland 127
Cape Isabella 126
Cape Joseph Henry, Ellesmere Island 133, 134, 138
Cape Kane, Greenland 138
Cape Kellett, Banks Island 99
Cape Maundy Thursday, Axel Heiberg Island 156
Cape Morris Jesup, Greenland 175
Cape Osborn, Devon Island 92
Cape Sabine, Ellesmere Island 125, 138
Cape Sheridan, Ellesmere Island 133, 161
Cape Sparbo (Cape Hardy), Devon Island 161
Cape Stallworthy, Axel Heiberg Island 154, 160
Cape Svartevoeg (Cape Stallworthy) 160
Cape Tabin (Mys Chelyuskin, Taymyr Peninsula) 35
Cape Walker, Russell Island 84, 92
Cape Washington, Greenland 138
Carcass (C. John Phipps, 1773) 55, 56
Carolus Flandro, Joris 22, 24
Castor and Pollux River, named by John Rae 82
Cator, John Bertie 88
Chake, Martin 50
Challenger (George Nares, 1873–75) 133
Challenger, sledge (Pelham Aldrich, 1875–76) 134
Champlain, Samuel de 31
Chancellor, Richard 9
Chantrey Inlet, Canada 79
Charles (Luke Fox, 1631) 28, 31
Charles Hanson, whaling ship 158
Charlton Island, James Bay 30
Chelyuskin (1933) 169
Chelyuskin, Semen 39
Chesterfield Inlet, Hudson Bay 44, 47
Chicago Daily Tribune 161
Chicago Record Herald 171
Chichagov, Vasiliy 55
Chirikov, Alexei 38
Chkalov, Valery (pilot) 182
Chukchi 7
Chukchi Peninsula (Chukotskiy Poluostrov), Siberia 36, 42, 116
Chukchi Sea 178
Churchill, John and Awnsham, writers 30
Churchill River, Canada 27, 28, 44, 52
"Clarence Wyckoff Island," Greenland 153
Clark, Charles 154
Clavering, Douglas 73, 145
Clerke, Charles 51
Cluny, Alexander 66
Cnoyen, Jacobus 11
Cockburn Island (part of Baffin Island) 69
Coleman, Patrick 109
Colin Archer Peninsula, Devon Island 156
Collinson, Richard 88, 93–94, 102, 158
Columbia Meridian 162
Committee Bay 81, 82
Company of Adventurers (Luke Foxe) 28
Company of Merchants Discoverers of the North-West Passage (Northwest Company) 27
Cook, Frederick 152, 158, 160, 167, 171, 182
 "discovery" of Bradley Land 164
Cook, James 50–51
Cook and Peary: The Polar Controversy, Resolved 160
Coppermine River 52, 53, 64, 65, 74
"Cornow" (Inuk; drew map for Joseph-Elzéar Bernier) 165
Cornwallis Island, Canada 62, 63, 93, 165
Coronation Gulf, Canada 52, 65, 102
Coronelli, Vincenzo, mapmaker 22, 33
Cowles, Thomas 50
Cresswell, Samuel Gurney 99, 105
Cresswell Bay, Somerset Island 77
"Crocker Land" 154, 155, 160, 164
Croker, John, First Lord of the Admiralty 61
"Croker Mountains" (John Ross, 1818) 60
Crozier, Francis Rawdon Moira 84, 104

204

Crozier Channel (Eglinton/Prince Patrick Islands) 93
Cumberland House (on Saskatchewan River) 65, 74
Cunningham, John 21

D

Dalrymple, Alexander, hydrographer and mapmaker 52, 53
Danish sovereignty 21
Danskøya, Danes Island 170, 171
Daurkin, Nikolay 41, 42
D'Auvergne, Philip (midshipman with C. John Phipps, 1773) 57
Davis, John 13, 15, 17, 21
Davis Strait 28, 58
Dealy Island (south coast of Melville Island) 94
Dease, Peter Warren 80
Dease Strait 102
Dee, John, polymath, mapmaker 16
De Haven, Edwin Jesse 88, 91, 125
De Jode, Gerard, mapmaker 17
De L'Isle, Joseph-Nicolas, mapmaker 41, 49
De Long, George Washington 43, 118, 135, 158
De Veer, Gerrit 18
Devon Island, Canada 62, 91, 92, 193
Dezhnev, Semen 35, 36, 42, 50
Dietrichson, Lief 172, 181
Diomede Islands, Bering Strait 37
Discourse of a Discoverie for a New Passage to Cataia 11
Discovery (William Baffin, 1615) 27
Discovery (Charles Clerke, 1778) 27, 44, 51
Discovery (William Moor, 1741–42) 44
Discovery (Henry Stephenson, 1875–76) 133
Discovery, sledge (Frederick Mecham) 93, 94
Discovery Bay, Ellesmere Island 135, 138
Discovery Harbour, Ellesmere Island 133
Dobbs, Arthur 44
Dobbs Galley (William Moor, 1745–46) 47
Dolphin, boat (John Richardson and Edward Kendall, 1826) 74
Dolphin and Union Strait 74, 102
Domville, William (surgeon, *Resolute*, 1852–54) 97, 99
Dorothea (David Buchan, 1818) 56
Drift of Russian scientific station SP-1 182
Drifting research stations 186, 188
Dutch East India Company 26
Du Val, Pierre, mapmaker 31
Dvina River (Russia) 9, 11, 34

E

"Eagle" (*Ørnen*), balloon 170
Eagle City, Alaska 158
East India Company 53
Ebierbing (Joe) (Inuk with Charles Hall) 108, 111
Eckener, Hugo 181
Edge, Thomas 22
Edgeøya (Spitsbergen) 22, 24
Edward Bonaventure (Richard Chancellor, 1553) 9
Egingwah (Inuk with Peary, 1909) 161
Eglinton Island, Canada 93, 94
Eielson, Ben 178
Eiríksson, Leifr 7
Electric telegraph, between ships 94
Ellef Ringnes Island, Canada 156
Ellesmere Island, Canada 133, 134, 138, 152, 153, 156, 160, 161, 173, 174, 178
Ellis, Henry 47
Ellis, John (names "Wiches Land," 1617) 22
Ellsworth, Lincoln 172, 176, 181
Elson, Thomas (master, *Blossom*, 1825–28) 76
Elson Lagoon (Elson's Bay), Alaska 103
Endeavour, sledge (Parry, Spitsbergen, 1827) 124
Engel, Samuel, polymath, mapmaker 55, 56
Enterprise (Richard Collinson, 1850–55) 86, 88, 93, 94, 102, 158
Enterprise (James Clark Ross, 1848) 86
Enterprise, sledge (James Clark Ross, Spitsbergen, 1827) 124
Erebus (John Franklin, 1845–) 111
Erebus (James Clark Ross, Antarctica, 1840) 84
Ericsson, Leif 7
Erik the Red 7
Eskimos 7. *See also* Inuit
Estotiland 20
Etah (now Taseq) 152, 160, 174
Etienne, Jean-Louis 185
E-tuk-i-shuk (Inuk with Frederick Cook) 161
Eureka Sound (Ellesmere/Axel Heiberg Islands) 160
Evenki 7

F

Farthest North (1897) 149
"Farthest norths" 124, 151, 153, 154
Fearless, sledge (George Nares, 1852) 93
Felix Booth 88
Felix Harbour 77
Feucht, Ludwig 172
Fisher, Alexander 61
Fitzjames, James 104
Flagler Bay, Ellesmere Island 156, 160, 174
Fletcher, Joseph 188
Fletcher's Ice Island T-3 188
Floeberg Beach 133
Flying the Arctic (1928) 178
Foley Island, Foxe Basin 183
Fonte, Bartholomew de 49, 66
Ford, Edsel 174
Forecasting of ice conditions 188
Forsyth, Charles Codrington 88
Fort Conger 137, 138, 152, 153
Fort Enterprise 65, 68
Fort Franklin (Déline) 74
Fort Hope 82
Fort Providence 68, 80
Fort Simpson 87
Foundation for the Promotion of the Art of Navigation 162
Fox (Leopold M'Clintock, 1857–59) 102
Foxe, Luke 25, 27, 28, 31, 32
Foxe Basin 32, 69, 183
Foxe Channel 32, 80
Foxe Peninsula 32
Foynøya (Foyn Island), Spitsbergen 181
Fram (Otto Sverdrup) 156
Fram (Otto Sverdrup and Fridtjof Nansen) 138, 148, 149, 150, 151, 164, 168, 170
Francke, Rudolph 160
Franklin, District of 165
Franklin, Jane 84, 88, 92, 125
Franklin, John 2, 56, 64, 72, 74, 84, 189
Franklin Strait 85
Franz Josef Land (Zemlya Frantsa-Iosifa) 118, 121, 131, 138, 146, 148, 154, 170, 179, 182
Friendly Arctic, The (1921) 167
Friesland (Frisland) 12, 13, 20
Frobisher, Martin 11, 12, 16, 108
Frobisher Bay, Baffin Island 12, 108
"Frobisher's Streytes" 13
Frozen Strait (north of Southampton Island) 28, 32, 45, 70, 80
Fullarton & Co., mapmakers 2, 105
Furnace (Christopher Middleton, 1741) 44
Fury (Henry Hoppner, 1823–24) 73
Fury and Hecla Strait (Melville Peninsula/Baffin Island) 68, 69, 71, 72, 109
Fury Beach, Somerset Island 73, 78

G

Gabriel (Martin Frobisher, 1576) 12
Gakkel, Y. Y. 192
Gakkel Ridge (krebhet Gakkel) 192, 193
Gastaldi, Giacomo 50
Gateshead Island, named by Richard Collinson 102
Gedenshtrom, Matvey 42, 114
General Bathymetric Chart of the Oceans (GEBCO) 187, 190
George Henry finds drifting *Resolute* 102
Georgii Sedov 144, 151, 169, 186
Gerritz, Hessel 22, 24, 26, 28, 34
Gilbert, Humfray 11, 16
Giles, Cornelius 121, 170
"Gilles Land" 121, 131, 170
Gjøa (Roald Amundsen, 1903–06) 158
GjøaHavn (Gjoa Haven) 158
Godhavn (Qeqertarsuaq), Greenland 127

God's Power and Providence: Shewed . . ., 22
Godthåb (Nuuk), Greenland 147
Good Friday Bay, Axel Heiberg Island 156
Goodlad, William 22
Goose Fiord, Ellesmere Island 156
Gordiyenko, Pavel 184
Gore, John 51
Graf Zeppelin (airship) 181, 182
Gray, David 131
Great Bear Lake 74
Great Fish (Back) River 79, 85, 102
"Great Land" 38, 40, 41, 42, 49
Great Northern Expedition 37, 38, 40
Greely, Adolphus Washington 137, 138
Greely Fiord 138
Green, John (pen name of Bradock Mead) 41, 43
Greenland (Kalaallit Nunaat) 6, 7, 21, 22, 33, 132, 133, 137, 145, 147, 152, 165, 174, 190
 connected to Wrangel Island (Ostrov Vrangelya) 146
Greenland Company 25
Greenland Sea 131, 182
Grenville, Richard 11
Grey Point, Spitsbergen 175
Griffin, Jane (Lady Jane Franklin) 84, 88, 92, 125
Griffin, Samuel 88
Griffith Island, Barrow Strait 166
Grinnell, Henry 91
Grinnell expedition 125
Grinnell Peninsula, Devon Island 91, 93
Griper (Douglas Clavering, 1823) 73
Griper (Henry Hoppner, 1819) 62, 70
Griper (George Lyon, 1824) 72
Gromov, M. 182
Grovenor, Gilbert 162
Gudmonson, Jón 7

H

Hall, Charles Francis 105, 110, 131, 132, 135
Hall, Christopher 13, 108
Hall, James 21, 28
Hall Basin 127, 132
Hammer, H. H. 171
Hannah (Tookoolio) (Inuk woman with Charles Hall) 108, 109
Harbour Fiord, Ellesmere Island 156
Harris, John 71
Harris, Rollin 164, 190
Haswell, William 99
Hayes, Isaac Israel 126, 127
Healy, Michael 118
Hearne, Samuel 52, 58, 65
Hearst, William Randolph 181
Hecla (Matthew Liddon, 1819–20) 62
Hecla (George Lyon, 1821–23) 70
Hecla (Edward Parry, 1824–25) 73
Hecla (Edward Parry, 1827, Spitsbergen) 124
Hecla and Griper Bay 63, 93
Heemskerck, Jacob van 17
Heiberg, Axel 156
Hendrik, Hans 126, 132
Henrietta Island (Ostrov Genriyetta) 135
Henrietta Maria (Thomas James, 1630–31) 30
Henson, Matthew 153, 161, 162
Herald (Henry Kellett, 1848–50) 86
Herald Island (Ostrov Geral'd) 87, 114, 117, 135, 166
Herbert, Wally 185
Herjolfsson, Bjarni 7
Herschel Island 74
Hobson, William (with Leopold M'Clintock, 1857–59) 104
Homann, Johann Baptist 35, 36
Hood, Robert (midshipman, Franklin, 1819–21) 65, 68
Hooker Island (Ostrov Hooker) 181
Hooper, Leighton 118
Hooper, William 86, 87
Hooper Island (Mackenzie Delta) 87
Hope Sanderson 16
Hoppner, Henry 70, 73
Hudson, Henry 16, 22, 26, 55
Hudson Bay 26, 27, 28, 32, 33, 45, 108
Hudson's Bay Company 44, 52, 74, 82, 86, 88, 168
Hudson Strait 15, 28

205

Humboldt Glacier, Greenland 126
Humfray Gylbert knight his charte 16
Hunter, John (surgeon) 56

I

Ice all the way to the Pole (contradicting Petermann's theories) 133
Ice Cape 35
Ice Haven (Ledyanaya Gavan'), Novaya Zemlya 19
Icy Cape, Alaska 51, 75
Igloolik Island, Foxe Basin 69, 72
Ikmalick (Inuk who drew maps for John Ross) 77
Independence Fjord, Greenland 152
Inglefield, Edward 92, 105, 125, 126
In-nook-poo-zhee-jook (Inuk who drew maps for Charles Hall) 110
International Bathymetric Chart of the Arctic Ocean (IBCAO) 150, 190–91
International Polar Year (1882–83) 138
Intrepid (John Bertie Cator, 1850–51) 88
Intrepid (Leopold M'Clintock, 1852–54) 92, 94, 99
Inuit 7, 60, 153, 160, 161
    and Charles Hall 108, 109, 111, 132
    maps 71, 77, 108, 165
    sense of geography 70
Inventio Fortunata 2, 11
Investigator (Edward Bird, 1848–49) 86
Investigator (Robert M'Clure, 1850–53) 88, 93, 94, 97, 99, 105
Isabella (John Ross, 1818) 59
Isabella, whaler (rescued John Ross, 1833) 79
Isachsen, Gunerius (Sverdrup expedition) 156
Isachsen Peninsula 156
Isobel (steam yacht; Edward Inglefield, 1852) 125
Italia (airship) 170, 179, 182

J

Jackman, Charles 10, 15
Jackson, Frederick 148, 150
Jackson Island (Ostrov Dzheksona) 148, 150
James, Thomas 25, 28, 32
James Bay 30
Jan Mayen Island 22, 24
Jansson, Jan 25
Jeannette 43, 118, 135, 138, 137, 158, 164
Jeannette Island (Ostrov Zhannetty), Siberia 135
Jenkinson, Anthony 10, 11
Jensen, Jens 145
Joe (Ebierbing; Inuk with Charles Hall) 109, 132
Johannesen, Edvard Holm 120, 169
Johansen, Frederick Hjalmar (with Nansen 1893–96) 148
John Barrow, sledge (Sherard Osborn, 1853) 93
Johnston, W. and A. K., mapmakers 105
Jones Sound 28, 60, 92, 156
Josephine Ford (Amundsen's airplane) 174, 175

K

Kalaallit Nunaat, Greenland 6
    See also Greenland
Kane, Elisha Kent 91, 105, 125, 126
Kane Basin 126
Kap Constitution, Greenland 126
Kap (Cape) Morris Jessup, Greenland 153
Kap Neumayer, Greenland 154
Kap (Cape) Washington, Greenland 152
Kap Wyckoff, Greenland 153
Kara Sea 10, 17, 19, 168, 169
Karluk (Robert Bartlett, 1913–14) 166, 167, 178
Kellett, Henry 86, 87, 92, 93, 114
"Kellett Land" 114
Kellett Strait 93
Kendall, Edward 74, 76
Kennedy, William 92
Kennedy Channel (Ellesmere Island/Greenland) 127
Kent Peninsula (north coast of Canada) 68
Khatanga River, Russia 39
Kholmogory, Russia 34
King Christian's Island (King Christian Island) 156, 167
King Point (on coast near Mackenzie Delta) 158
King's Bay (KongsFjorden), Spitsbergen 26, 176, 178
King's Bay (town; now Ny Ålesund) 172, 174

King William Island (King William Land) 77, 78, 79, 81, 85, 102, 104, 109, 110, 111, 158, 189
Knight, James 44
Kodlunarn Island, Frobisher Bay 108
Kola Peninsula (Kol'skiy Poluostrov), Russia 9, 34
Koldeway, Karl 118, 131, 145, 146
Kolyma River, Siberia 35, 36, 37, 40, 112
Kolyuchinskaya Guba (Nordenskiöld wintering place) 142, 144
Kong Karls Land (King Carl's Land) 22
König Oscar Land (King Oscar Land) 121
Kosmin, Prokopy 114
Kotzebue, Otto von 59
Kotzebue Sound, Alaska 59
Krasny Oktobr (*Red October*), armed icebreaker 118
Krassin, Russian icebreaker 181
Kuro Siwo current, "pathway to the Pole" 135
Kvitøya (White Island), Spitsbergen 170

L

Lady Franklin (William Penny, 1850) 88
Lady Franklin Bay, Ellesmere Island 133
La France, Joseph 45, 46
Lake Hazen, Ellesmere Island 138, 153
Lancaster Sound, Canada 28, 58, 60, 62
"Land" in the Arctic Ocean 164, 188, 190
Lands Lokk (Kleybolte Peninsula) 154, 156
Lapie, Chevalier de 66
Lapps 7
Laptev, Dmitri 40
Laptev, Khariton 39
Laptev Sea 40, 151, 168
Larsen Sound, Canada (north of King William Island) 85
Larson, Henry 167
Lasinius, Peter 40
Lavoie, J. T. E. 166
Leads 193
Lena Delta 40, 137, 143
Lena River 36, 37, 39, 40, 143
Lewis, Samuel, mapmaker 58
Liddon, Matthew 62
Liddon Gulf, Melville Island 63, 93
Lincoln Sea 152
Linschoten, Jan van 17
Lion, boat (John Franklin and George Back, 1826) 74
Litke, Fedor 121
Lockwood, James Booth 137, 138
Lok, Michael (Frobisher voyage promoter) 15
Lommes Bay, Spitsbergen 172
Lomonosov, Mikhail 55
Lomonosov Ridge (khrebet Lomonosova) 55, 146, 164, 190, 193
Long, Thomas 114, 116, 117, 125, 146
Lord Lumley's Inlet (Frobisher Bay) 13
Lord Mayor Bay, Boothia Peninsula 77, 81, 82
Lotter, Conrad, mapmaker 51
Lougheed Island, Canada 156, 167
Low, Albert Peter 156
Lundborg, Einar 181
Lyakhov, Ivan 42
Lyon, George 69, 70, 72, 73
Lyon Inlet, Melville Peninsula 69, 70

M

M'Clintock, Francis Leopold 86, 90, 92, 93, 102, 104, 156
M'Clintock Channel (Victoria/Prince of Wales Islands) 85, 104, 158, 188
M'Clure, Robert Mesurier 2, 88, 93, 94
M'Clure Strait 188
MacCormick Fjord, Greenland 152
Mackenzie, Alexander 53, 58
Mackenzie Delta 53, 72, 74, 76, 86, 158
Mackenzie King Island, Canada 167
MacMillan, Donald, expedition, 1914 155, 173, 174
Macrobius, mapmaker 8
Magellan, Ferdinand 11
Maguire, Rochfort 103
Makarov, Stepan 168
Malakhov, Mikhail 185
Maldonado, Lorenzo Ferrer (apocryphal voyage) 66
Mallet, Alain, mapmaker 32

Malmgren, Finn 180, 181
Malygin, icebreaker 181
Malygin, Stepan 38
Mansurov, Ivan 35
Marble Island, Hudson Bay 44, 111
Marco Polo, sledge (Markham/Parr, 1876) 133
Marconi radio direction-finder 177
Markham, Albert 133, 134
Mary (John Ross's yacht, with *Felix*, 1850) 88
Massa, Isaac 34
Matonabbee (Chipewyan chief) 52
Matyushkin, Fedor 114
Maud (Roald Amundsen, 1918–21) 168, 171
Maury, Matthew Fontaine 91, 126
May, Jan Jacobsz 22
Mead, Bradock (John Green) 41, 43
Mecham, George Frederick 93, 94, 99
Medvezh'i Islands, Siberia 41
Meighen Island, Canada 161, 167
Melville Bay (Melville Bugt), Greenland 60, 133, 152
Melville Island, Canada 62, 63, 64, 92, 94, 166
Melville Peninsula (Foxe Basin/Gulf of Boothia) 69, 71, 82
Mercator, Gerard, mapmaker 2, 4, 8, 11
Merchants Adventurers of England (1553) 9
Mercy Bay, Banks Island 93, 97, 99
Meteorit Ø (Meteorite Island), Greenland 152
Michael (Martin Frobisher, 1576) 12
Michel (Teroahauté; voyageur with Franklin, 1821) 68
Middendorf, Aleksandr 40
Middleton, Christopher 44, 45, 49
Miertsching, Johann (interpreter with M'Clure) 99
Mikkelsen, Ejnar 152
Minin, Fedor 39
Mistaken Straightes (Hudson Strait) 15
Mittelholzer, Walter 171
Moderate-resolution imaging spectroradiometer (MODIS) 192
Montreal Island, Chantrey Inlet, Canada 81, 104, 111
Moor, William 44, 46, 47, 49
Moore, Thomas 86, 103
Morton, William 126
"Mr Joris Eylandt" (Jan Mayen Island) 22, 24
Müller, Gerhard 36, 37
Multibeam sonar 190
Munk, Jens Ericksen 28, 30
Murav'yev, Stepan 38
Murmansk, Russia 185
Muscovy Company 9, 10, 12, 21, 24, 26
My Attainment of the Pole (1910) 160
Mylius-Erichsen, Ludvig 152, 174
Mys Arkticheskiy (Arctic Cape), Russia 168
Mys Chelyuskin (Cape Chelyuskin), Russia 40, 143, 168, 169
Mys Shelagskiy (Cape Shelagskiy), Russia 144
Mys Svyatoy Nos, Russia 40, 41
Mys Yakan, Russia 112

N

Nagurskiy, Yan Iosifovich 171
Nancy Dawson (Robert Sheddon, 1849) 86
Nansen, Fridtjof 7, 147, 148, 186
Nansen Sound 160
Nares, George Strong 93, 133
Nares expedition 132–34
NASA 192, 193
National Geographic Society 162, 164, 174, 175
Nautilus (Hubert Wilkins, 1931) 181
Nautilus, USS (William Anderson, 1958) 185, 190
Navy Cliff, Independence Fjord, Greenland 152
Nearest the Pole (1907) 155
Nelson, Horatio, exploit with bear 56
Nelson Head, Banks Island 158
Nelson River, Canada 27, 47
Nenets 7
Neptune (Albert Peter Low, 1903) 156, 165
New land (Canada) 183
New Land (Otto Sverdrup, 1904) 157
New land (Russia) 168
New York Times 173
New York World 172
Nicholas of Lynn 11
Nicholas II Land (Severnaya Zemlya, Russia) 168

Nile (Thomas Long, 1867) 114, 117
Nobel, Alfred 170
Nobile, Umberto 154, 176, 179, 182
Noordsche Compagnie 24, 25
Noose of Laurels, The (1989) 160
Nordaustlandet (North East Land; Svalbard) 57, 121, 179
Nordenskiöld, Adolf Erik 140, 145, 147
Norge (airship) 154, 176, 177
Noril'sk, Russia 193
Norse settlements in Greeenland 7, 21
Northeast Passage 26, 40, 140, 144, 168, 193
"Northern Land" (Severnaya Zemlya), Russia 168
Northern Sea Route 169
North Georgia Gazette 63
North Magnetic Pole 77, 78, 158
North Pole 124, 171, 174, 182, 184, 186, 192
  first achievement (?) by Amundsen, Ellsworth, and Nobile (1926) 176
North Star (William Pullen, 1852–54) 92, 94
North Star (James Saunders; supply ship for James Clark Ross) 85
North West Company 53, 54
Northwest Company 27, 28
Northwest Passage 2, 8, 13, 15, 16, 27, 32, 44, 53, 59, 74, 84, 99, 158, 166, 167, 193
Norton, Moses 52
Novaya Zemlya, Russia 9, 10, 16, 17, 18, 20, 24, 26, 33, 34, 120
Novgorod, Russia 34, 35
Novopashennyy, Per 168, 169
Novosibirskiye Ostrova (New Siberian Islands) 37, 38, 40, 41, 42, 114, 148
Nunataks 147

O

Ob' 35, 38
Obskaya Guba 38, 140
Oimekon, Siberia (the "Cold Pole") 6
Omdhal, Oskar 172
Ommanney, Erasmus 88, 91
Ommanney Bay, Prince of Wales Island 92
Ooqueah (Inuk with Peary) 161
Ootah (Inuk with Peary) 161
Open polar sea 6, 55, 84, 91, 124, 127, 128, 129, 132, 133, 135, 193
Open Polar Sea, The (1867) 128
Ornen ("Eagle"; Andreé's balloon) 170
Ortelius, Abraham, mapmaker 12, 13, 20
Osborn, Sherard 88, 92, 99, 133
Ostrova Belaya Zemlya, Kara Sea 148, 149
Ostrova Dzheksona, Kara Sea 148
Ostrov Bennetta (Bennett Island), Siberia 137
Ostrov Bol'shoy Lyakhovskiy, Siberia 40, 41, 42
Ostrov Dzheksona, Russia 150
Ostrov Faddeyevskiy, Novosibirskiye Ostrova 43
Ostrov Gallya, Zemlya Frantsa-Iosifa 120
Ostrov Genriyetta (Henrietta Island), Siberia 135
Ostrov Hooker, Zemlya Frantsa-Iosifa 181
Ostrov Mak-Klintoka, Zemlya Frantsa-Iosifa 120
Ostrov Malyy Lyakhovskiy, Novosibirskiye Ostrova 40, 42
Ostrov Malyy Taymyr, Severnaya Zemlya 168, 169
Ostrov Oktabr'skoy Revolyutsii, Severnaya Zemlya 168
Ostrov Rudol'f, Zemlya Frantsa-Iosifa 182
Ostrov Shmidta (west of Severnaya Zemlya) 168
"Ostrov Tsesarevicha Aleksaya," Ostrov Malyy Taymyr 168
Ostrov Ushakova (west of Severnaya Zemlya) 169
Ostrov Uyedinaniya (Lonely Island), Kara Sea 169
Ostrov Vaygach, Novaya Zemlya 11, 143
Ostrov Vize (west of Severnaya Zemlya) 169
Ostrov Vrangelya (Wrangel Island) 87, 91, 112, 114, 116, 117, 125, 135, 166, 186, 190
Ostrov Zhannetty (Bennett Island), Siberia 135
Ottar (Norse king) 9
Ova Medvezh'i, Siberia 41
Overwintering by British naval ships 63
Ovtsyn, Dimitri 38
Oxford University surveying and mapping expedition (1924) 172

P

Palander, Adolf Arnold Louis 140, 143
Pandora (Allen Young) 158
Papanin, Ivan 182
Parliamentary reward, Northwest Passage 58
Parry, William Edward 56, 59, 62, 72, 73, 124
Parry Islands, Canada 63
Pavlov, Mikhail 38
Payer Harbour (Ellesmere Island) 153
Payer, Julius 120, 121, 146, 150
Peary, Robert Edwin 147, 152, 153, 160, 161, 182
  claim to the North Pole 164
Peary Arctic Club 152, 153
Peary Channel (northern Greenland) 153
"Peary system" 161
Pechora River, Russia 34
Peel Sound, Canada 85, 92
Peel Strait, Canada 158
Pellham, Edward 22
Penny, William 88
Penny Strait, Canada 161
Perseverance, sledge 90
Pet, Arthur 10, 15
Petermann, Augustus 91, 110, 111, 118, 121, 128, 129, 132, 146, 190
Petermann Land 120, 121
Petermann's Geographische Mittheilungen 111, 114, 119, 120, 140, 142, 151, 156, 169
Phipps, Constantine John 51, 55
Phippsøya (Phipps Island), Spitsbergen 56
Phoenix (Edward Inglefield, 1853) 92, 99, 105
Pillage Point, Mackenzie Delta 74
Pim, Bedford 97, 99
Pioneer (Sherard Osborn, 1850–51 and 1852–54) 88, 92, 99
Plaisted, Ralph 184
Plancius, Peter, polymath and mapmaker 17, 20, 55
Plover (Rochfort Maguire, 1852–54) 102, 103
Plover (Thomas Moore, 1848–52) 86
"Plover Island," Siberia 87, 114, 116
Point Anxiety, Alaska 74
Point Barrow, Alaska 75, 76, 86, 97, 102, 174, 178, 185
Point Separation, Mackenzie Delta 74
Point Turnagain (north coast of Canada) 68, 72, 76, 77
Poirier, Pascal, Senator (sector principle) 165, 166
Polaris (Charles Hall, 1871–72) 132, 133
Polynyas 126, 185
Pond, Peter 53, 54
Pond Inlet, Baffin Island 71, 165
Poole, Jonas 22
Port Bowen (east side of Prince Regent Inlet) 73
Port Leopold, Somerset Island 92
Prince Albert (Charles Codrington Forsyth, 1850) 88
Prince Albert (William Kennedy, 1851–52) 92
Prince Albert Sound, Victoria Island 99, 102
Prince Charles Island, Foxe Basin 183
Prince of Wales Fort, Hudson Bay 44, 52
Prince of Wales Island, Canada 92, 104
Prince of Wales Strait (Banks/Victoria Islands) 99, 102
Prince Patrick Island, Parry Islands 93, 94, 97
Prince Regent Inlet 62, 72, 73, 86
Prince William Sound, Alaska 53
Princess Royal Islands, Prince of Wales Strait 94, 97
Proliv Karskiye Vorota (Novaya Zemlya/Ostrov Vaygach), Russia 10
Proliv Longa (Long Strait), Siberia 116
Proliv Yuorskiy Shar (Ostrov Vaygach/mainland), Russia 10
Pronchischev, Vasiliy 39
Proteus (Aldolphus Greely, 1881) 138
Provideniya, Siberia 185
Prudhoe Land (Greenland) 126
Pshenitsyn, Pyotr 43, 112, 114
Pullen, William 86, 87, 92
Pullen Island, Mackenzie Delta 87
Purchas, Samuel 28
Pyasina River, Russia 39
Pytheas of Massalia 7

Q

Queen Maud Sea (Queen Maud Gulf) 158
Queens Channel (Devon/Bathurst Islands) 93

R

Racehorse (John Phipps, 1773) 55, 56
RADARSAT 188, 193
Rae, John 82, 86, 90, 99, 102
Rae Isthmus, Melville Peninsula 82
Rankin, John 45
Rankin Inlet, Hudson Bay 45
Reliance, boat (John Franklin and George Back, 1826) 74
Relics of Franklin expedition 109, 111
Remarks upon Capt. Middleton's Defence . . . 45
Rensselaer Bugt (Bay), Greenland 126
Repulse Bay 32, 44, 45, 66, 69, 70, 72, 73, 79, 109
Rescue (Samuel Griffin, 1850–51) 88, 91
Resolute (Horatio Austin, 1850–51) 88, 90
Resolute (Henry Kellett, 1852–54) 92, 93, 94, 99
  drift of 102
Resolute, sledge 90
Resolution (James Cook, 1776–79) 51
Resolution Island, Baffin Island 12
Return Reef, Alaska 75, 80, 103
Richardson, John 65, 68, 74, 86
Riiser-Larsen, Hjalmar 172
Ring of ice supposedly around the polar regions 131
Ringnes brothers 156
Robeson Channel (Ellesmere Island/Greenland) 131, 132
Robyn, Jacob, mapmaker 32
Rodgers (Robert Mallory Berry, 1881) 118, 137
Rodgers, John 114, 118
Roes Welcome Sound (Southampton Island/mainland) 32, 45, 73
Roosevelt (Robert Peary, 1908) 161
Ross, James Clark 70, 76, 84, 85, 86, 92, 124
Ross, John 7, 56, 59, 73, 76, 88
Royal Danish Geographical Society 161
Royal Geographical Society 125, 127, 156, 158
Russell Island, Barrow Strait 84
Russell Point, Banks Island 97
Russian Geographic Society 118
Ruysch, Johann, mapmaker 8

S

Sabine, Edward 62, 63, 73
Sackheuse, John (Inuk with John Ross, 1818) 60
St. Lawrence Island, Bering Strait 37, 41
St. Roch (Henry Larson, 1942–44) 167
Sanderson, William 15, 17
Sannikov, Yakov 43, 112
"Sannikov Land" 43, 112
Sarychev, Gavriil 42
Satellite, sledge 93
Satellite Bay, Prince Patrick Island 97
Satellite images 188–93
Savigsivik, Greenland 152
Schley, Winfield Scott 139
Schwatka, Frederick 109, 110
Scoresby, William 58, 59, 145
Sea ice atlas 189
Sea of Okhotsk (Okhotskoye More), Russia 35
Seale, Richard, mapmaker 49
SeaMARC-12 swath mapping sonar 190
Second Kamchatka Expedition 38
Sector principle 165, 166
Sedov, Georgiy 171
Seegloo (Inuk with Peary, 1909) 161
Selifontov, Vasiliy 39
Semenov, Grigoriy 36
Severnaya Polyus (North Pole), SP designation 186
Severnaya Zemlya, Russia 120, 168, 169, 179
Shalaurov, Nikita 41
Shedden, Robert 86
Shelden, John B. 6
Sherard Osborn Fjord, Greenland 135
Sheriff Harbour, Boothia Peninsula 77
Sherman Basin (north coast of Canada) 111
Sherman Inlet (north coast of Canada) 111
Shmidt, Otto 168, 169, 182, 184
Side-scan sonar 190

Simpson, George, Governor 80
Simpson, Thomas 80
Sindt, Ivan 51
Sir James Lancaster's Sound. *See* Lancaster Sound
Sir Thomas Smith's Sound. *See* Smith Sound
Skate, uss (James Calvert, 1958–59) 185
Skeletons, Franklin expedition 104
Skuratov, Aleksey 38
Skyward (1926) 174
Sledge journey, longest ever 94
Sledge journeys 88, 93, 94
"Sledge maps" 93
Smith, Francis 46, 47
Smith Sound 28, 60, 125, 133, 160
Smyth, Hugh 10, 11
Snowmobiles, use of 184
Somerset House (John Ross, Fury Beach, 1832–33) 77
Somerset Island, Canada 86, 92, 165
Sonar, multibeam 190
Sophia (Alexander Stewart, 1850) 88
Sorgfjorden, Spitsbergen 124
South Pole 158
Southampton Island, Hudson Bay 80, 108
Sovereignty issues 156, 165, 167, 174, 193
Sovetskiy Soyuz (first ship to cross Arctic Ocean, 1991) 185
sp-1, first ice island drifting station 182, 186
Spitsbergen (Svalbard) 7, 22, 24, 25, 56, 124, 131, 140, 168, 170, 172, 174
Stadukhin, Mikhail 41
Stählin, Jacob von 50, 51
Star of the North, sledge 93
Starvation Cove, Adelaide Peninsula 85, 111
Stefánsson, Sigurdar, map 7
Stefansson, Vilhjalmur 118, 161, 166, 167, 178, 185
Steger, Will 185
Stephenson, Henry 135
Sterlegov, Dmitri 39
Stewart, Alexander 88
Strait of Anian 44, 50, 66
Submarines 190
Svalbard (Spitsbergen) 6, 18, 179
Svenskøya (Kong Karls Land) 22, 24
Sverdrup, Otto 147, 148, 150, 156, 165, 167
Sverdrup Islands, Canada 156
Swedish Academy of Sciences 170

T

Tabula Nautica, map 26
Talbot (supply ship, 1854) 99
Tamyr Peninsula (Poluostrov Taymyr), Russia 35, 39, 143, 168, 169
Tegetthoff (Karl Weyprecht, 1871–74) 120
Teller, Alaska 176
TERRA satellite 192
Territorial claims, Canada 165
Terror (George Back, 1836–37) 80
Terror (Francis Crozier, 1845–47) 84, 111
Thank God Harbor (Greenland) 132
Theatrum Orbis Terrarum, Abraham Ortelius 12, 15
Thomas Corwin (Calvin Leighton Hooper/Michael Healy, 1879–84) 118
Thorne, Robert 55
Thornton, Samuel 45
Thule 7
Times (newspaper) 158
Timofeyevich, Yermak 35
Todd Island (south of King William Island) 110
Tookoolio (Hannah; Inuk with Charles Hall) 108
Trent (John Franklin, 1818) 56, 75
Tyson, George 132

U

Uemura, Naomi 162, 185
Union, boat (John Richardson and Edward Kendall, 1826) 74
United States (Isaac Hayes, 1860–61) 128
United States North Pacific Exploring Expedition 114
Upernavik, Greenland 16
Ushakov, Georgiy Alekseyevich 168

V

Vagin, Mercuriy 41
Vancouver, George 51
Vaygach (Per Novopashennyy, 1913) 168
Vaygach Island (Ostrov Vaygach) 10, 11, 18
Veer, Gerrit de. *See* De Veer, Gerrit
Vega (Louis Palander, 1878–79) 140, 143, 144
Victoria Island (Victoria Land), Canada 74, 80, 84, 86, 90, 102, 158, 166
Victory (paddle steamer, John Ross, 1829–33) 76, 77
Vil'kitskiy, Boris Andreyevich 168, 169
Vincennes (John Rodgers, 1855) 114, 118
Viscount Melville Sound, Canada 97, 99, 102
Vize, Vladimir Yul'yevich 169
Vodopyanov, Mikhail 182, 184
Vrangel', Ferdinand Petrovich 112
Vyatka, Yakov 41

W

Wager, Sir Charles 44
Wager Bay, Canada 32
Wager River (Wager Bay) 45
Wager "Straits" 45, 47
Wainwright Inlet, Alaska 86
Walker Bay, Victoria Island 102
Ward Hunt Island (north coast of Ellesmere Island) 185
Weber, Richard 185
Wellington Channel (Devon/Cornwallis Islands) 85, 91, 92, 125
Wellman, Walter 171
Weyprecht, Karl 118, 138, 146, 150
Whale Island (Mackenzie's name for Garry Island) 76, 87
Whaling 33
Whaling expedition to Spitsbergen 22
White Sea (Beloye More) 9, 34, 39
"Wiches Land" (likely Svenskøya) 24
Wigate, John 45, 46
Wiggins, Joseph 140
Wilczek, Graf Johann von 118, 131
Wilkins, George Hubert 164, 166, 167, 178, 181, 185

Willoughby, Hugh 9
Willoughby's Land (Novaya Zemlya) 26
Windward (Frederick Jackson, 1895–96) 149, 150
Windward (Robert Peary, 1898) 152
Winter Harbour, Melville Island 63, 64, 93, 94, 99, 166
Winter Island, Melville Peninsula 69, 70
Winter Lake (Franklin, 1820–21) 65
Witsen, Nicholaas 35
Wollaston Land (Wollaston Peninsula), Victoria Island 74, 84, 90, 102
Worthington, L. V. 190
Wrangel Island (Ostrov Vrangelya) 87, 91, 112, 114, 116, 117, 125, 135, 166, 186, 190
Wyniatt, Robert 99
Wyniatt Bay, Victoria Island 99

Y

Yakuts 7
Yakutsk 193
Yamal (at North Pole) 185
Yamal Peninsula (Poluostrov Yamal) 38, 39
Yenisey, River 34, 38, 39
Yermak (first ocean-going icebreaker, 1898) 168
Young, Allen 104, 158
Young, Walter 60
Yugorskiy Shar (Ostrov Vaygach/mainland), Russia 16, 143

Z

Zemlya Frantsa-Iosifa (Franz Josef Land) 118, 121, 131, 138, 146, 148
"Zemlya Imperatora Nikolaya II" (Franz Josef Land) 154, 168, 170, 179, 182
Zemlya Vil'cheka (Wilczek Land; Franz Josef Land) 120
Zeno brothers, Antonio and Nicolo 13
Zeno map 12, 13

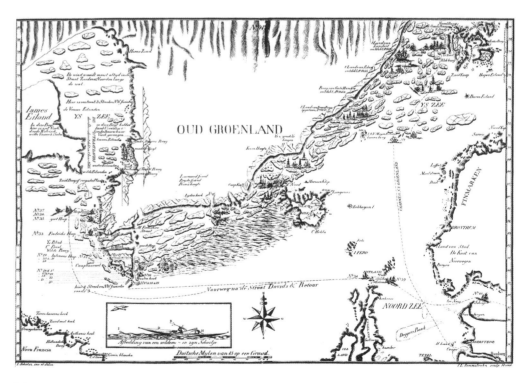

MAP 303.
A map of "Oud Greenland" or "Old Greenland," so named to distinguish it from the new Greenland—Spitsbergen—shown at top right. The map was drawn by a whaling captain named Hidde Dirks Katt in 1777, and is a detailed and interesting depiction of the ice conditions on the east coast of Greenland. Katt, like most Greenland whalers, thought of Spitsbergen as an island off the coast of "Old Greenland."

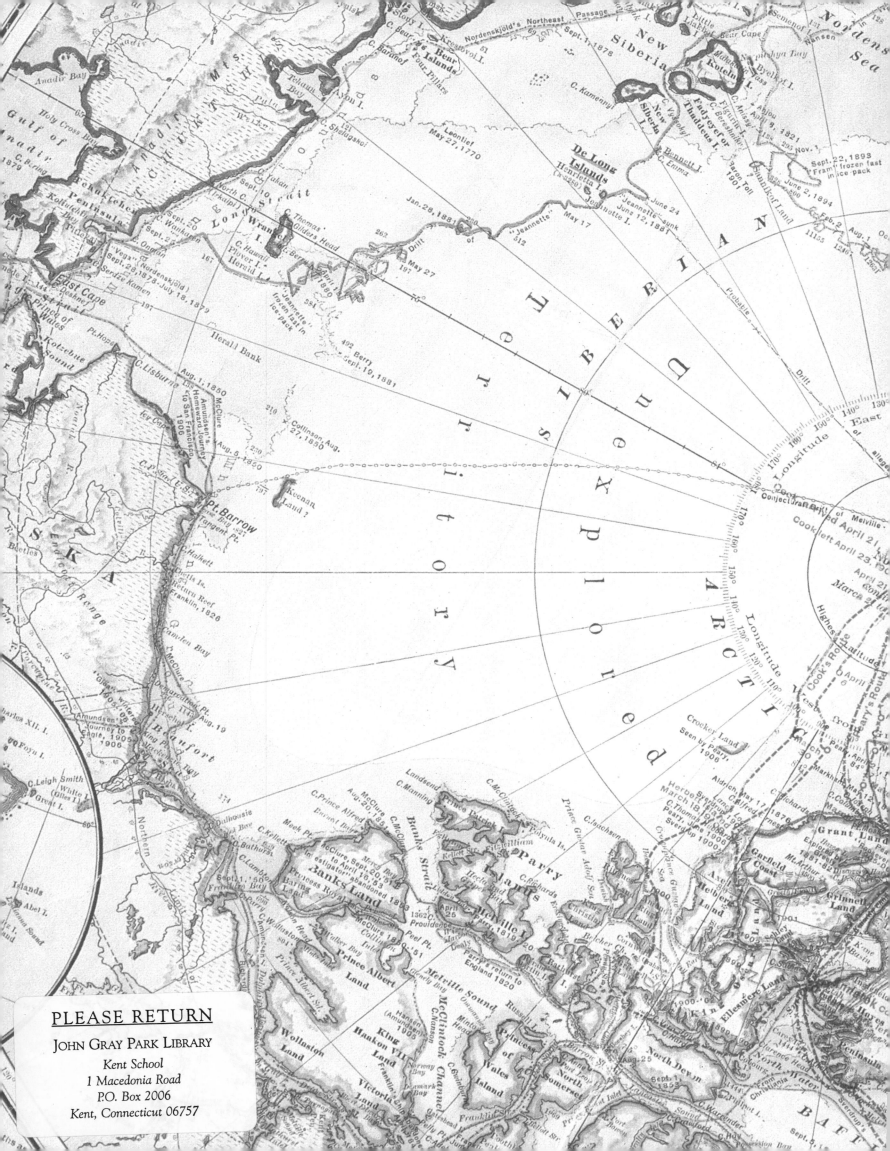